Compendium of Plant Genomes

Series editor

Chittaranjan Kole
Nadia, West Bengal
India

More information about this series at http://www.springer.com/series/11805

Mathilde Causse · Jim Giovannoni
Mondher Bouzayen · Mohamed Zouine
Editors

The Tomato Genome

Springer

Editors
Mathilde Causse
GAFL
INRA
Montfavet Cedex
France

Jim Giovannoni
Boyce Thompson Institute for Plant Research
Cornell University
Ithaca, NY
USA

Mondher Bouzayen
INRA-INP Toulouse
Castanet Tolosan
France

Mohamed Zouine
INRA-INP Toulouse
Castanet Tolosan
France

ISSN 2199-4781 ISSN 2199-479X (electronic)
Compendium of Plant Genomes
ISBN 978-3-662-53387-1 ISBN 978-3-662-53389-5 (eBook)
DOI 10.1007/978-3-662-53389-5

Library of Congress Control Number: 2016950912

Printed on acid-free paper

This Springer imprint is published by Springer Nature
The registered company is Springer-Verlag GmbH Germany
The registered company address is: Heidelberger Platz 3, 14197 Berlin, Germany

We dedicate this volume to Prof. Steven Tanksley without whom the tomato system and genome would never have been developed to the exceptional utility and quality they serve today.

Preface

The Tomato Genome Sequence: How Did It Happen and Why Does It Matter?

The tomato genome sequencing project was initiated as part of the International Solanaceae Project (SOL) by a large international consortium of 10 countries (Korea, China, UK, India, The Netherlands, France, Japan, Spain, Italy and the United States). The tomato was chosen as reference species for the Solanaceae due to the high level of macro and micro-synteny within this plant family which comprises more than 3000 species among which some are important crops such as the fruit-bearing vegetables tomato, eggplant, and pepper, and the tuber-bearing potato, in addition to a number of medicinal and ornamental plants. The goal of the tomato genome sequencing project was to generate new information and resources allowing to shed light on how a common set of genes can give rise to a wide range of morphologically and ecologically distinct organisms, and how a better understanding of the genetic basis of plant diversity can be harnessed to meet the needs of the fast growing world population for a sustainable food crop production. It is important to mention that the launching of the tomato genome sequencing project would have not been possible without the use of the rich resources previously generated using this plant species. Undoubtedly, the project took advantage of the large collection of EST sequences, the high number of genetic markers, the dense and saturated genetic maps, and the well-characterized genomic libraries already available (http://sgn.cornell.edu/).

In many ways, the project represented a unique scientific and human adventure where the participants shared the scientific effort and the financial outlay and worked in close collaboration. Starting with conventional sequencing technologies the project shifted to the new high-throughput sequencing technologies, just emerging at the time. In this regard, the tomato genome sequencing project accompanied the transition from the old to the new sequencing era. Indeed, the Sanger sequencing method was initially used, but the advent of next-generation (NextGen) sequencing technologies has prompted the consortium to adopt these promising techniques. In retrospect, we can now say that the choice of these pioneering technologies was a wise decision, although it posed a risk at the time because there was no prior experience where the NextGen sequencing technologies have been applied de novo to sequence a large and complex eukaryotic genome. The consortium had to overcome the difficulties of high-throughput data processing and assembly of "reads" without any possibility to rely on past experience in this area. An important challenge was the buildup of a pipeline for the genome sequence assembly, and in this respect, one of the most striking aspects of the project's success had been to produce finally a high-quality assembled tomato genome sequence using for the first time the new sequencing technologies.

Due to the estimated elevated cost of producing a high-quality sequence of the complete tomato genome, the initial strategy was the preferred sequencing of the euchromatin region where the majority of genes reside. This approach presents the advantage to target only 25 % of the total tomato genome thus allowing to significantly reduce the sequencing effort. The BAC-by-BAC sequencing strategy built on the existing saturated tomato genetic map, and made use of the genetic markers to select seed BACs within the gene-rich part of the tomato genome. The starting point for sequencing the genome was BACs anchored to the genetic map, and this minimal tiling path then extends from seed BACs to cover the whole genome. Once completed, the BAC-by-BAC tomato genome sequence was anticipated to provide a framework for shotgun sequencing of other Solanaceae species. While this approach enabled a rapid progress at the early phases of the project, it struck quickly with the difficulty of selecting BACs to power the sequencing pipeline. Finally, the slowness of this process became a serious obstacle pushing the consortium to seek other alternatives to reinvigorate the project. The advent of next-generation sequencing technologies offered an attractive option despite the lack of experience in applying these techniques to complex genomes. Switching to high-throughput sequencing launched the project into a new and original adventure where you have to discover simultaneously both the problems and their solutions. In particular, the consortium realized that these approaches require massive use of bioinformatics tools that had to be acquired and implemented in a short period of time.

The switch to a whole genome sequencing approach that combines both next-generation sequencing and Sanger sequencing boosted the project leading to a high-quality assembled tomato genome sequence within a relatively short period of time. The present book tells the tale of the tomato genome sequencing adventure with the various chapters describing in great detail every step of the sequencing project. Chapters 1 and 2 provides a brief review of the birth of the tomatoes in the Andean regions of South America, the history of their botanical classification along with other wild and cultivated Solanaceae as well as information about the main production areas. The following chapters deal with gene and QTL mapping in tomato with a particular emphasis on the new opportunities that the tomato genome sequences are providing for the genetic and molecular dissection of complex traits and how it helps breeders to shape new and better tomato varieties. The chapter on tomato resources for functional genomics describes the main resources, strategies, and tools currently available for linking genes to phenotypes in tomato. The chapters devoted to the generation of the tomato genome sequence per se emphasize the sequencing and assembling strategies used in the project and the genome quality evaluation and the finishing methods. A separate chapter is dedicated to the annotation of the tomato genome with the aim to provide the best gene structures, a high-quality functional description for the protein-coding genes. The sequencing of the chloroplast and mitochondrial genomes, described in a specific chapter, adds to the understanding of the plant evolutionary history of tomato based on the phylogenetic position inferred from the organelles sequences information. The following two chapters review recent research on the timing and formation of ancient genome duplications and their evolutionary effects on the shaping of modern Solanaceae genomes. They also address the synteny among Solanaceae genomes providing insight into the modes and tempo of plant genome evolution and illustrating how a better knowledge of genome synteny and colinearity can facilitate the mobilization of resources from one species to other in this agronomically important family. The last chapter describes the tomato-centric databases and other generic resources freely accessible to Solanaceae community.

While the effort to produce an improved assembly with a larger coverage of the tomato genome is ongoing, the present version of the tomato genome (The Tomato Genome Consortium, Nature 2012) is among, if not the best quality of, all dicot genomes published to date, excluding Arabidopsis. Producing a reference tomato genome sequence represented a major breakthrough and has provided invaluable resource that has opened new avenues for research. Building on this resource enabled the development of a variety of genome-wide approaches like whole genome transcriptomic profiling that is nowadays becoming a routine method for expression studies. Likewise, genotyping-by-sequencing is currently spreading as a method of choice and mapping by sequencing is being increasingly used. The access to a complete genome sequence also fostered epigenetics studies allowing to establish a genome-wide mapping of various epigenetic marks. More recently, genome editing is experiencing a rapid growth to address the functional significance of candidate genes in the tomato model. These are some of the main areas that have been impacted by the acquisition of a high-quality reference genome for tomatoes, but most likely, we are only at the dawn of these dramatic developments and more unexpected ones will break out in the future.

Castanet Tolosan, France

Mondher Bouzayen
Mohamed Zouine

Contents

The Tomato: A Seasoned Traveller

1

Sophie Colvine and François Xavier Branthôme

Abstract

Originating from South America, tomato is now produced all over the world. After a slow propagation in European Mediterranean countries since the sixteenth century, it has started to be largely cultivated in the twentieth century. It has experienced spectacular growth over the last 50 years both for processing tomato and fresh market. The growth of global trade reflects the rise in consumption, with a recent increase in Asia, notably in China, which has become the first producer in the last years.

Keywords

Tomato · Production · Trade · Processing · Fresh market

Although still a matter of debate, the birth of the tomato is generally located in the Andean regions of South America. In this century of rampant and frenetic globalisation, the slower-paced journeys that took it from Peru, then Mexico, to the shores of the Caribbean and South East Asia and from Southern Italy to Northern Europe before reaching North America are pretty mind boggling.

But, the tomato's travels have not just been geographic. It has been consumed and indeed grown since well before the Christian era (500 years BC in Mexico) but first had to convince the cultures and people it encountered that it was safe. Its heart-shaped form and red colours conquered the Moors who discovered it in Spain, but it was subsequently considered to be an aphrodisiac by the Italian Herbalist, Pietro Andrea Mattioli, who gave it the name of 'love apple' in 1544, or as 'highly toxic' by the English Physician and Herbalist, John Gerard in the late sixteenth century. The suspicions it raised relegated it to the status of an ornamental plant hidden away at the bottom of the garden throughout the seventeenth and early eighteenth centuries. The most that can be said is that its colour and taste brightened and spiced up a few soups around 1730 in England

S. Colvine (✉)
World Processing Tomato Council, 1328 Route de Loriol, 84170 Monteux, France
e-mail: colvine@tomate.org

F.X. Branthôme
Tomato News, Brantomate Consulting, 613 Chemin de la Blanchère, Résidence Golf 2, 84270 Vedène, France
e-mail: fxb@tomatonews.com

M. Causse et al. (eds.), *The Tomato Genome*, Compendium of Plant Genomes,
DOI 10.1007/978-3-662-53389-5_1

while at the same time on the other side of the Atlantic, scientists strongly discouraged its consumption due to its links with Mandrake and Deadly Nightshade, both members of the Solanaceae family. For the tomato to be definitively considered as a food in its own right, it needed President Thomas Jefferson's political influence and strength of conviction in 1809, a cultural and industrial revolution and, almost 30 years later, again in the US, a media offensive by the New York Times. It would be a further 30 years until, in 1869, Henry John Heinz founded the company in Pittsburgh whose name and flagship product remain inextricably linked to the tomato.

Now grown in all latitudes, or almost, the tomato has experienced spectacular growth over the last 50 years (Fig. 1.1). As a member of the Solanaceae family, it is often compared to the potato which holds the record for annual worldwide consumption with more than 376 million tonnes of potatoes produced in 2013 according to the FAO. The tomato is more modest by comparison and currently settles for an annual production level of 164 million tonnes.

Nonetheless, the tomato outclasses its cousin in terms of production growth (Table 1.1). Admittedly, potato cultivation already stood at nearly 271 million tonnes in 1961, precisely ten times that of the tomato, but during the last 50 years, the amounts of tomatoes produced worldwide have multiplied by 5.8, jumping from less than 28 million tonnes in 1961 to nearly 164 million tonnes in 2013. This growth is all the more pronounced in Asia and especially China, the world's biggest producer with just over a quarter of total production, where there has been a sevenfold increase in production while in India production has been multiplied by 18. Over the same period, potato production *only* increased by 20 %, weighing-in at just 376 million tonnes in 2013, or barely twice that for tomatoes! These figures however only account for commercial production and exclude family farming and subsistence production which can be fairly significant in certain regions.

The reasons for this growth lie in a dramatic improvement in agricultural productivity which reflects the wide interest in both vegetables making it possible to expand production way beyond what would have been expected based on existing surface area increases alone. Average figures given by the FAO (currently 34 t/ha compared to 16 t/ha in 1961) give only a rough idea of the astonishing progress made by agronomy. Average yields for processing tomato fields in California which are frequently used as an example, have quite simply jumped from 25 t/ha in 1961 to 105 t/ha in 2014 and some farmers even manage to reach spectacular yields of 150 t/ha. In other words, the quality of fruit harvested from the same field has increased fourfold in the space of just two generations. Under glass, average yields are now around 400 t/ha and can even reach 1000 t/ha!

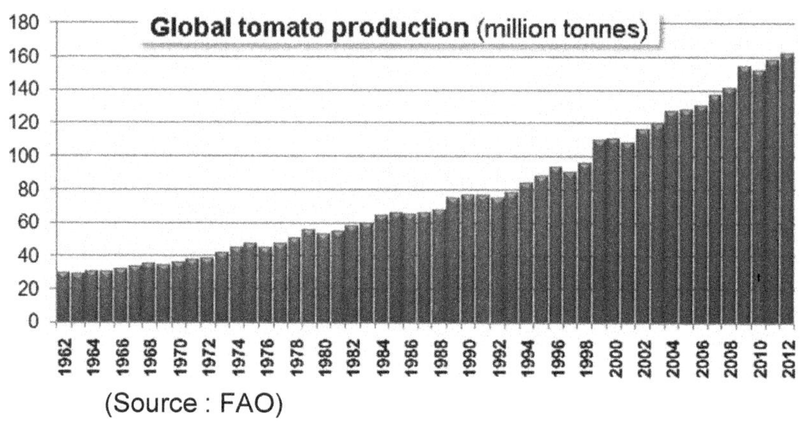

Fig. 1.1 Global tomato production (million tonnes)

Table 1.1 Main tomato-producing countries (2012)

	Production (tonnes)	Area harvested
China	50,000,000	1,000,000
India	17,500,000	870,000
USA	13,206,950	150,140
Turkey	11,350,000	300,000
Egypt	8,625,219	216,395
Iran	6,000,000	160,000
Italy	5,131,977	91,850
Spain	4,007,000	48,800
Brazil	3,873,985	63,859
Mexico	3,433,567	96,651
Uzbekistan	2,650,000	60,000
Russia	2,456,100	117,700
Ukraine	2,274,100	85,700
Nigeria	1,560,000	270,000
Portugal	1,392,700	15,400
Morocco	1,219,071	15,639
Tunisia	1,100,000	28,900
Iraq	1,100,000	62,500
Greece	979,600	16,000
Indonesia	887,556	56,042
Cameroon	880,000	150,000

Source FAO

And examples abound in the main tomato-producing countries of China, India, Turkey, Egypt, Italy, Iran, Spain, Brazil and Mexico and so on. The tomato has continued to travel which has subsequently led to its being selected, improved, made more resistant, more productive, fleshier, redder and eventually taken from the fields and tables to the processing factories. As a standard-bearer for the Mediterranean diet, the tomato has quickly adapted to modern lifestyles. It has even become emblematic for a few leaders in the global food industry, including some that have themselves engaged in the lengthy process of selecting varieties and, just a few decades ago, 'invented' the illustrious ancestors of those jointly used by the processing industry today.

As such, the tomato has long been the leading processed 'vegetable' in the world. The diversity of processed tomato products makes it impossible to list the countless forms in which the tomato is consumed everywhere on a daily basis throughout the world. Indeed, the quantities of tomatoes used for sauces, diced tomatoes, pastes, on pizzas, for passata, in ketchup, peeled, chopped, frozen, or powdered tomatoes, to name just a few of the most common forms, increase regularly each year. Here, also, growth has been astonishing with the global industry increasing its production from 22 million tonnes in the 1990s to nearly 40 million tonnes by the end of 2010. No other vegetable can boast consumption figures in processed form that represent nearly a third of its fresh volumes. This is indeed the case for the tomato which to be consumed the world over is only processed (and cultivated solely for this purpose) in what boils down to a quite a small number of countries. The leader among them is California which accounted for nearly a third of worldwide production over the last

10 years with an average annual volume of over 10 million tonnes (Fig. 1.2). One of the strengths of the American industry is the size of its companies, including 9 which rank among the top 12 biggest tomato businesses in the world.

As a relative newcomer in 2000, the Chinese industry has quickly become one of the global leaders. It owes its heavy-weight status to the strength of its exports of pastes which account for virtually all of its products. The other advantage China has is to have spotted and developed markets that were practically ignored until the late 1990s thanks to a particularly competitive commercial policy.

The historical processor and uncontested leader in the European industry is Italy. It only recently relinquished its place as world leader to China, a position it occupied for a long time in quantitative terms due to robust sales of pastes but also the diversity of exported products and the domination it holds in the canned sector, especially peeled tomatoes.

Italy however remains the world leader in terms of revenue. In 2013, business generated nearly 2.1 billion US Dollars for the Italian industry whereas Chinese and American sales only amounted to 984 and 715 million US Dollars, respectively.

Nevertheless, the processing tomato is also takes in Spanish, Portuguese, Chilean, Iranian, Turkish and Greek industries, to name just those key players in international trade. They all operate on a global level each with their specific characteristics in terms of processing techniques, products, packaging, customers or geographic zones. These nine countries account for 80 % of global processing power for the export market for paste alone which is the main processed tomato product marketed today. The price of this growing concentration of processing hubs is that a significant number of regions are increasingly dependent on supplies of processed tomato products.

The growth of global trade reflects the rise in consumption (Fig. 1.3). 40 million tonnes of the 159 million tonnes of fresh tomatoes identified by the FAO are consumed each year throughout the world in processed form In good years or bad, this amount rises by the equivalent of one million tonnes of fresh tomatoes each year, but the components of global consumption of processed tomato products (the different product categories) evolve at the whims of cultural choices, social and economic constraints, political events and dietary patterns, etc.

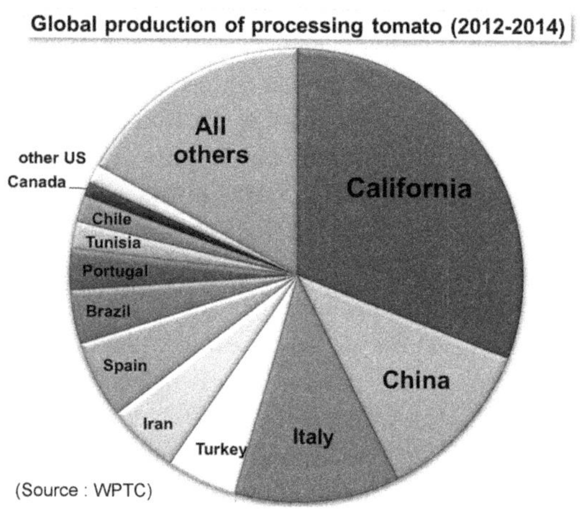

Fig. 1.2 Global production of processing tomato (2012–2014)

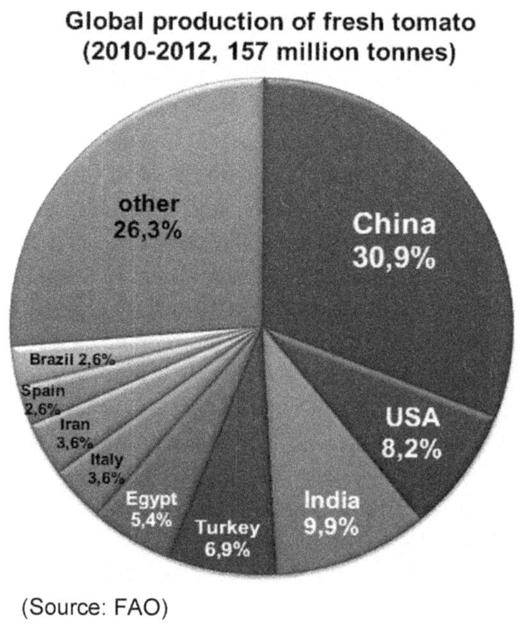

Fig. 1.3 Global production of fresh tomato

According to FAO figures, average global consumption per capita was 20.5 kg in 2009, with variations from 0 to more than 100 kg in North Africa and the Middle East. This compares to around 31 kg in the European Union and 44 kg in the US. In terms of processed tomatoes, it was 6 kg in 2011 according to WPTC figures, which shows a 50 % increase over the last 15 years (4 kg in 1995). This is both a little and a lot since this level of individual consumption is just over one kilogramme of paste, whereas in 2013 just one third of the world's population consumed more than this threshold. Although eating habits and consumption levels can be incredibly varied from one continent to the other, the most surprising example is without doubt China. It is both the world's leading supplier of pastes and the biggest consumer of tomatoes and now accounts for more than 42 million tonnes per year. Out of this impressive total, only one million tonnes (2.5 %) are consumed in processed form, i.e. the equivalent of about 800 grams of fresh tomatoes per year and per person. On the other hand, the impression tomatoes has made on culinary cultures and dietary traditions, however different they may be, in Italy and the USA can be clearly seen in the individual consumption ratios. Although far from holding any records in the discipline, American or Italian consumers each consume more than 30 kilos of tomatoes every year in the form of pastes, sauces, pizzas, etc. For someone living in Parma, Rome or Naples, fresh tomatoes remain a must which accompanied with mozzarella, basil or olive oil, still represent more than 56 % of annual consumption. In Sacramento, Houston or Springfield, fresh tomatoes are rarer and ketchup, sauces and other processed forms of tomato now account for more than three-quarters of annual tomato consumption!

The tomato's forms, tastes and circumstances may differ, but whether fresh or processed, it constitutes a universally recognised foodstuff that is independent of age, religion and culture. With each minute that passes, 300 tonnes of tomatoes disappear. 228 tonnes are taken up by fresh consumption and 72 tonnes are consumed in processed form. Whatever the latitude or longitude, these two markets complement each other, grow together and feed off each other. Nevertheless, everything, or nearly everything, sets these two faces of the same crop apart. First, the varieties are all derived from common ancestors destined for the fresh market. Some varieties occasionally got confused as 'dual-purpose' varieties but now they are totally differentiated between the fresh and processed sectors. Second, there is the period and type of cultivation; annual and under glass in once case and highly seasonal and open-field in the other. Cultivating and harvesting fresh tomatoes is highly dependent on the availability of manpower while it is increasingly mechanised for the processed sector and then there are the regions of production, logistical restrictions, techniques and costs, etc. But in the end, the amounts consumed, whether fresh or processed, are rising in line with each other at just over a 25 and 75 %, respectively of global consumption.

The tomato's journeys via winds and currents, through different cultures, skills, culinary arts, across changing land and seasons as well as for different economic reasons and industrial logistics have sometimes been unexpected and eventful but have built up a long and rich history. They brought the wild cherry tomato all the way from Peru to the individual ketchup portion consumed in the fast food restaurants of Shanghai. Every day it becomes a little more universal, it unveils yet more new qualities while research demonstrates its contribution to health, advances its farming attributes and positions it in a more environmentally friendly global approach. The journey and the story do not stop there. Its colours and forms, its contents, its strengths and its virtues are yet more complex and secret, but that is for genetics to discover.

As a geographical, historical, cultural and artistic link, the tomato already has a great history. It also has a bright future.

The Tomato (*Solanum lycopersicum* L., Solanaceae) and Its Botanical Relatives

2

Sandra Knapp and Iris Edith Peralta

Abstract

The cultivated tomato, *Solanum lycopersicum* L., is a member of the small section *Lycopersicon* along with its 12 wild relatives. An additional four species from sections *Juglandifolia* and *Lycopersicoides* are traditionally considered as tomato wild relatives. These species are all endemic to South America, but the cultivated tomato itself has achieved worldwide distribution with the help of human populations. Tomato and its wild relatives are part of a larger monophyletic group (the Potato clade) that also contains the potatoes and their wild relatives. Here we review the taxonomic and phylogenetic history, relationships and species-level taxonomy of the cultivated tomato and its wild relatives, and highlight important studies of diversity that remain to be undertaken in the group, especially in light of global environmental and climatic change.

Keywords

Taxonomy · Tomato · *Solanum lycopersicum* · Wild relatives · Systematics

Introduction

The cultivated tomato, *Solanum lycopersicum* L., belongs to the diverse family Solanaceae, which includes more than 3000 species, occupying a wide variety of habitats (Knapp 2002). The Solanaceae contain many species of economic use, such as food (tomatoes, potatoes, peppers and eggplants), medicines (deadly nightshade, henbane, datura) and ornamental purposes (petunias). *Solanum lycopersicum* was previously recognized as *Lycopersicon esculentum* Mill., but data from both morphology and molecular sequences support its

S. Knapp (✉)
Department of Life Sciences, Natural History Museum, Cromwell Road, London SW7 5BD, UK
e-mail: s.knapp@nhm.ac.uk

I.E. Peralta
Facultad de Agronomía, Universidad Nacional del Cuyo, Almirante Brown 500, 5505 Chacras de Coria, Argentina

I.E. Peralta
Instituto Argentino de Investigaciones de las Zonas Áridas, (IADIZA-CCT CONICET Mendoza), Calle Adrián Ruiz Leal s/n, 5500 Mendoza, Argentina

M. Causse et al. (eds.), *The Tomato Genome*, Compendium of Plant Genomes,
DOI 10.1007/978-3-662-53389-5_2

inclusion in the large genus *Solanum* L., and a revised new nomenclature has resulted (Peralta and Spooner 2001, 2005; Spooner et al. 2005; Peralta et al. 2006, 2008a). Morphological characters, phylogenetic relationships and geographical distribution have demonstrated that tomatoes (*Solanum* sect. *Lycopersicon* (Mill.) Wettst.) and their immediate outgroups in *Solanum* sect. *Lycopersicoides* (A. Child) Peralta and sect. *Juglandifolia* (Rydb.) A. Child form a sister clade to potatoes (sect. *Petota* Dumort.), with *Solanum* sect. *Etuberosum* (Buk. and Kameraz) Child being sister to potatoes + tomatoes (Spooner et al. 1993; Peralta and Spooner 2001; Spooner et al. 2005; Peralta et al. 2008a; Rodriguez et al. 2010; Särkinen et al. 2013). Analyses of multiple data sets from a variety of genes unambiguously establish tomatoes to be deeply nested in *Solanum* (Bohs and Olmstead 1997, 1999; Olmstead and Palmer 1997; Olmstead et al. 1999; Peralta and Spooner 2001; Bohs 2005; Särkinen et al. 2013). The monophyletic *Solanum* with the inclusion of all traditional segregate genera (*Cyphomandra* Mart. ex Sendtn., Bohs 1995; *Lycopersicon* Mill., Spooner et al. 1993; *Normania* Lowe and *Triguera* Cav., Bohs and Olmstead 2001) is one of the ten most species-rich genera of angiosperms (Frodin 2004, see also Solanaceae Source, http://www.solanaceaesource.org). It contains several crops of economic importance in addition to the tomato, such as the potato (*S. tuberosum* L.) and the aubergine or eggplant (*S. melongena* L.), as well as other minor crops (naranjilla, *S. quitoense* Lam.; tamarillo or tree tomato, *S. betaceum* Cav. and pepino, *S. muricatum* Aiton). The majority of taxonomists as well as most plant breeders and other users have accepted the re-integration of tomatoes to *Solanum* (e.g. Caicedo and Schaal 2004; Fridman et al. 2004; Schauer et al. 2006; Mueller et al. 2005; Tomato Genome Consortium 2012; see also http://tgrc.ucdavis.edu/key.html). The tomato and all of its wild relatives were treated in a taxonomic monograph by Peralta et al. (2008a).

The tomatoes and their close relatives are easily distinguished from any other group of *Solanum* species by their bright yellow flowers and pinnate or pinnatifid, non-spiny leaves; the only other species in the genus with bright yellow flowers is *S. rostratum* Dunal, a spiny member of sect. *Androceras* (Nutt.) Whalen of the Leptostemonum clade (Whalen 1979) and *S. huayavillense* Del Vitto, a member of the Morelloid clade (Barboza et al. 2013). Here we provide a brief review of the history of generic classification of the tomatoes and their wild relatives, species diversity and relationships amongst wild tomatoes, the position of the tomato in the Solanaceae and timing of relevant diversification events in the family and review the history of tomato introduction from its native range to a worldwide distribution as a cultivated plant.

Generic Position of the Tomato and Its Relatives

The system of giving plants a genus and species name began with Linnaeus in the first edition of *Species Plantarum* (1753); before that plant names were long sentences (polynomials) in Latin that described the plant and distinguished it from others. In his first edition of *The Gardener's Dictionary* (Miller 1731) Philip Miller, the English botanist and curator of the Chelsea Physic Garden, used the generic name *Lycopersicon* meaning "wolf peach", a term previously coined by de Tournefort (1694), and included a number of taxa with multi-locular fruits ("roundish, soft, fleshy Fruit, which is divided into several Cells, wherein are contain'd many flat Seeds"), all colour variants of the cultivated tomato (*S. lycopersicum*). In the same work, Miller also recognized *Solanum*, and included within it the eggplant as "*Solanum Americanum, spinosum, foliis Melongenae, fructu mammoro*" and the potato as "*Solanum tuberosum, esculentum*" (Miller 1731). His definition of *Lycopersicon* was confined to plants that we would today recognize as cultivars of *S. lycopersicum*, the cultivated tomato.

In *Species Plantarum*, Linnaeus (1753) classified tomatoes in the genus *Solanum*, and described *S. lycopersicum* and *S. peruvianum*. The French botanist Adrian de Jussieu (1789), in his classification, also included tomatoes in *Solanum*. Miller (1754), however, continued to use both the generic name *Lycopersicon* and polynomial nomenclature in the abridged 4th edition of *The Gardener's Dictionary*. He expanded his definition of *Lycopersicon* by including "*Lycopersicon radice tuberose, esculentum*" (the potato) within it, using the following reasoning (Miller 1754): "This Plant was always ranged in the Genus of *Solanum*, or Nightshade, and is now brought under that Title by *Dr. Linnaeus*; but as *Lycopersicon* has now been establish'd as a distinct Genus, on account of the Fruit being divided into several Cells, by intermediate Partitions, and as the Fruit of this Plant [the potato] exactly agrees with the Characters of the other species of this Genus, I have inserted it here." The editor of the posthumously published edition of *The Gardener's and Botanist's Dictionary* (Miller 1807), Thomas Martyn, merged *Lycopersicon* and *Solanum*, and recognized all Miller's species as members of *Solanum*. Miller (1754) did not recognize the tomatoes by their elongate anther cones, used by later authors (e.g. D'Arcy 1972; Nee 1999; Hunziker 2001) to justify the segregation of the genus *Lycopersicon*, but instead, based his genus on fruit characters.

A number of classical and twentieth century authors have recognized the genus *Lycopersicon* mainly based on the anther morphology (e.g. Dunal 1813, 1852; Bentham and Hooker 1873; Müller 1940; Luckwill 1943; Correll 1958; D'Arcy 1972, 1987, 1991; Hunziker 1979, 2001; Rick 1979, 1988; Child 1990; Rick et al. 1990; Symon 1981, 1985; Hawkes 1990), but others continued to recognize the tomatoes as members of the genus *Solanum* (MacBride 1962; Seithe 1962; Heine 1976; Fosberg 1987). Today, tomatoes are widely accepted as members of the large and diverse genus *Solanum*, based on the results of both morphological and molecular analyses (see Peralta et al. 2008a for details).

Species Diversity and Relationships of Wild Tomato Relatives

Solanum sect. *Lycopersicon* consists of 13 closely related taxa; the cultivated tomato, *Solanum lycopersicum*, exists only as a domesticated or feral plant (Peralta et al. 2008a), and 12 wild species (Table 2.1): *Solanum arcanum*, *S. cheesmaniae*, *S. chilense*, *S. chmielewskii*, *S. corneliomulleri*, *S. galapagense*, *S. habrochaites*, *S. huaylasense*, *S. neorickii*, *S. pennellii*, *S. peruvianum* and *S. pimpinellifolium* (Peralta et al. 2005; Spooner et al. 2005; Peralta et al. 2008a). All of the wild species of section *Lycopersicon* occur on the western slopes of the Andes in dry desert or pre-desert environments (Fig. 2.1; for distributions and environments of all species see Table 2.1). Four species have been segregated from the green-fruited species *S. peruvianum* sensu lato (s.l.); two of them, *S. arcanum* and *S. huaylasense,* were described as new (Peralta et al. 2005) from Peru, while the other two, *S. peruvianum* and *S. corneliomulleri*

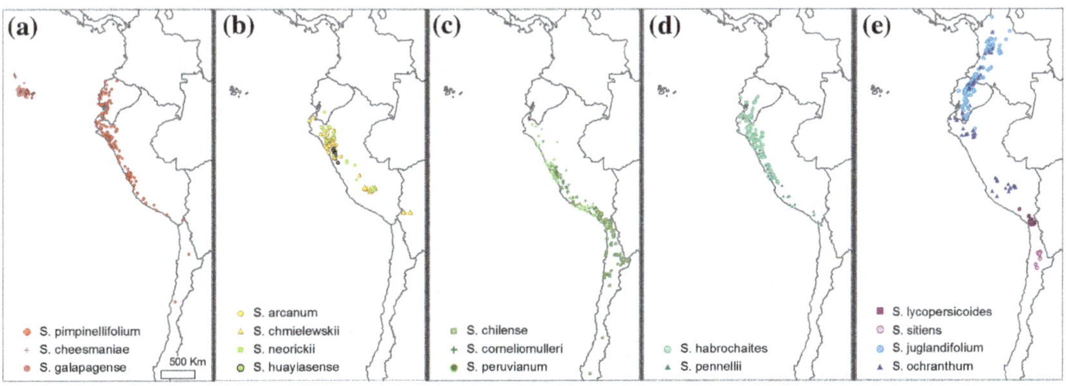

Fig. 2.1 Distribution maps of tomato wild relatives

Table 2.1 Tomatoes and their wild relatives (Peralta et al. 2008a 'Lycopersicon group' corresponds to the red- and orange-fruited species)

Species	Distribution	Habitat (elevational range	Section according to Peralta et al. (2008a, b)
Solanum arcanum Peralta	Northern Peru	Dry inter-Andean valleys and in coastal lomas (seasonal fog-drenched habitats); 100–4000 m	*Lycopersicon* 'Arcanum group'
Solanum cheesmaniae (L. Riley) Fosberg	Galápagos Islands	Dry, open, rocky slopes; sea level—1300 m	*Lycopersicon* 'Lycopersicon group'
Solanum chilense (Dunal)Reiche	Coastal Chile and southern Peru	Dry, open, rocky slopes; sea level—4000 m (B. Igic, pers. comm. has suggested the higher elevation plants represent a new species)	*Lycopersicon* 'Eriopersicon group'
Solanum chmielewskii (C.M. Rick, Kesicki, Fobles & M. Holle) D.M. Spooner, G. J. Anderson & R.K. Jansen	Southern Peru and northern Bolivia	Dry inter-Andean valleys, usually on open, rocky slopes; often on roadcuts; 1200–3000 m	*Lycopersicon* 'Arcanum group'
Solanum corneliomulleri J.F. Macbr.	Southern Peru (Lima southwards)	Dry, rocky slopes; 20–4500 m (low elevation populations associated with landslides in southern Peru)	*Lycopersicon* 'Eriopersicon group'
Solanum galapagense S.C. Darwin & Peralta	Galápagos Islands	Dry, open, rocky slopes; seashores; sea level—1600 m	*Lycopersicon* 'Lycopersicon group'
Solanum habrochaites S. Knapp & D.M. Spooner	Andean Ecuador and Peru	Montane forests, dry slopes and occasionally coastal lomas; 10–4100 m	*Lycopersicon* 'Eriopersicon group'
Solanum huaylasense Peralta	Rio Santa river drainage, north-central Peru	Dry, open, rocky slopes; 950–3300 m	*Lycopersicon* 'Eriopersicon group'
Solanum juglandifolium Dunal	Andean Colombia, Ecuador and Peru	Montane cloud forests; 1000–3200 m	*Juglandifolia*
Solanum lycopersicoides Dunal	Southern Peru and northern Chile	Rocky slopes and ravines; 1250–3600	*Lycopersicoides*
Solanum lycopersicum L.	Globally cultivated domesticate	Cultivated; sea level—4000 m	*Lycopersicon* 'Lycopersicon group'
Solanum neorickii D.M. Spooner, G.J. Anderson & R.K. Jansen	Southern Ecuador to southern Peru	Dry inter-Andean valleys; 500–3500 m	*Lycopersicon* 'Arcanum group'
Solanum ochranthum Dunal	Andean Colombia, Ecuador and Peru	Montane cloud forests; 1850–4100 m	Juglandifolia
Solanum pennellii Correll	Northern Peru to northern Chile	Dry slopes and washes, usually in flat areas; sea level—4100 m	*Lycopersicon* 'Neolycopersicon group'
Solanum peruvianum L.	Central Peru to northern Chile	Dry coastal deserts and lomas; sea level—3000 m	*Lycopersicon* 'Eriopersicon group'
Solanum pimpinellifolium L.	Southwestern Ecuador to northern Chile (many northern populations in Ecuador are admixture with *S. lycopersicum*; Peralta et al. 2008a, b; Blanca et al. 2013)	Dry slopes, plains and around cultivated fields; sea level—3000 m	*Lycopersicon* 'Lycopersicon group'
Solanum sitiens I.M. Johnst	Northern Chile	Dry ravines and slopes (hyperarid areas); 2000–3500 m	*Lycopersicoides*

For further details of crossability and other biological parameters of wild tomatoes see Grandillo et al. (2011)

had already been named by Linnaeus (1753) and MacBride (1962), respectively. In addition, *S. galapagense*, a yellow to orange-fruited plant, was segregated from *S. cheesmaniae*; both species are endemic to the Galápagos Islands (Darwin et al. 2003). Lucatti et al. (2013) have suggested that *S. galapagense* and *S. cheesmaniae* should be considered conspecific but we think the morphological and combined molecular evidence argues against the lumping of these taxa; this will only obscure the useful differences already seen and used by plant breeders from these two taxa at whatever rank they are recognized (Grandillo et al. 2011). Peralta et al. (2008a) put these 12 species into three informal species groups ('Arcanum', 'Eriopersicon' and 'Neolycopersicon', see Table 2.1) based on a combination of morphological and molecular analyses. All members of sect. *Lycopersicon* are diploid ($2n = 24$) (Peralta and Spooner 2001; Nesbitt and Tanksley 2002), characterized by a high degree of genomic synteny (Chetelat and Ji 2007; Stack et al. 2009; Tomato Genome Consortium 2012), and are to some degree intercrossable (Taylor 1986). Non-phylogenetic schemes (Müller 1940; Luckwill 1943; Rick 1979) for the relationships of tomatoes and their wild relatives have been treated in detail by Peralta et al. (2008a), so we will not treat them here.

Two other sets of species complete the tomato wild relatives in the broad sense (Table 2.1). *Solanum* sect. *Juglandifolia* contains the two woody tomato-like nightshades *S. ochranthum* and *S. juglandifolium.* These two species are partially sympatric and they are morphologically similar, both being woody perennials with rampant, liana-like stems up to 30 m in length (Correll 1962; Rick 1988; Peralta and Spooner 2005; Peralta et al. 2008a). Based on evidence from molecular sequence data (Peralta et al. 2008a) sect. *Juglandifolia* is the sister group of the wild tomatoes in the strict sense. Sister to both groups is *Solanum* sect. *Lycopersicoides*, comprising the allopatric sister species *S. lycopersicoides* and *S. sitiens*. These four tomato-like nightshade species have in common several morphological features that make them intermediate between tomato and potato (Rick 1988; Stommel 2001; Smith and Peralta 2002). Tomato-like morphological characters that together differentiate them from most of other *Solanum* species include yellow corollas, pedicels articulated above the base, pinnately segmented non-prickly leaves, and lack of tubers (Correll 1962; Rick 1988). These four allied outgroup species are diploids ($2n = 24$), but strong reproductive barriers isolate them from the core tomato group (Rick 1988; Correll 1962; Child 1990; Stommel 2001; Smith and Peralta 2002; Grandillo et al. 2011). Overall, crosses between the cultivated tomato and all but two (*S. ochranthum* and *S. juglandifolium*) of these wild species are possible, although with varying degrees of difficulty (Rick 1979; Rick and Chetelat 1995; Pertuzé et al. 2002; Grandillo et al. 2011). Although, using special techniques, introgression lines have been developed between *S. lycopersicoides* and *S. lycopersicum* (Chetelat et al. 1998; Canady et al. 2006). These have been useful in the elaboration of genetic maps (Chetelat and Meglic 2000), and for the understanding of cold, pest and pathogen resistances (Davis et al. 2009).

Cladistic and phenetic studies of species boundaries and relationships within the tomatoes and all their wild relatives have used a combination of molecular and morphological data (Palmer and Zamir 1982; Spooner et al. 1993; McClean and Hanson 1986; Miller and Tanksley 1990; Bretó et al. 1993; Marshall et al. 2001; Alvarez et al. 2001; Peralta and Spooner 2001, 2005; Spooner et al. 2005; Rodríguez et al. 2010). These studies used a variety of techniques, data sets and analysis types; the reader is referred to the primary literature and to the summary of the results of these studies in Peralta et al. (2008a) for further details of specific algorithms used and parameters set. The four species with brightly coloured fruits (*S. cheesmaniae*, *S. galapagense*, *S. lycopersicum*, *S. pimpinellifolium*) unambiguously form a closely related monophyletic group in all molecular analyses and this relationship has been suggested by all who have studied tomatoes previously (Müller 1940; Luckwill 1943; Rick 1979).

Rodriguez et al. (2010) used a set of nuclear COSII (conserved orthologous set II, Wu et al.

2006) markers to investigate the test their utility for phylogeny reconstruction in both potato and tomato. They did not intend to provide a definitive phylogenetic reconstruction for these groups, but instead focused on identifying markers that would be useful for future studies. Their analysis of the tomato clade, however, provided robust and well-supported hypotheses of species relationships in which the "red-orange-clade" comprising *S. lycopersicum*, *S. pimpinellifolium*, *S. galapagense* and *S. cheesmaniae* was consistently recovered with bootstrap values of 100 % and posterior probabilities of 1 (Rodriguez et al. 2010). Relationships amongst the green-fruited species revealed several different topologies, suggesting different gene genealogies, and whether section *Juglandifolia* or *Lycopersicoides* is sister to the tomatoes sensu stricto was unresolved, in contrast to previous studies (see above). Their Bayesian analysis (Rodríguez et al. 2010) using 18 COSII markers showed two sister group relationships in the "red-orange clade"—*S. galapagense* + *S. cheesmaniae* and *S. lycopersicum* + *S. pimpinellifolium*. This is in accordance with geography (Darwin et al. 2003; Peralta et al. 2008a) with the two Galápagos endemics most closely related to each other, and *S. lycopersicum* most closely related to its wild progenitor (Tomato Genome Consortium 2012). Koenig et al. (2013) recovered *S. galapagense* as sister to *S. lycopersicum* and *S. pimpinellifolium* sister to them (they did not include *S. cheesmaniae*), but they suggest this result stems from potential incomplete lineage sorting resulting from the extremely close relationship amongst the red- and orange-fruited species. Causse et al. (2013) also showed that repeated introgressions from wild species over the course of modern tomato breeding have resulted in extensive variation at the molecular level, perhaps obscuring the relationships of the cultivated species to one or other of its close wild relatives.

All those studying the cultivated tomato have unambiguously placed its evolutionary origins with the other tomato species with brightly coloured berries. These are all species of dry, desert habitats, suggesting there is much genetic variation yet to mine in the very close relatives of *S. lycopersicum* to help tomatoes deal with environmental change to come.

Tomatoes in the Solanaceae

Tomato is a flagship species in the Solanaceae, and has been extensively used in studies on the evolution and development of fruit characters in particular (Lippman and Tanksley 2001; van der Knaap et al. 2002; Seymour et al. 2013). The Solanaceae themselves are members of the derived Asterid Clade of flowering plants (Angiosperm Phylogeny Group 2009) and molecular dating analyses coupled with fossil evidence suggests they arose just after the Cretaceous/Tertiary boundary, approximately 59 Million years ago (Bell et al. 2010) to ca. 49 Million years ago (Mya; 46.2–53.7 Mya) (Särkinen et al. 2013; see Fig. 2.2). Fossils available for stratigraphic calibration of the phylogenetic tree of the family are few (Särkinen et al. 2013) and all dates presented here must be considered minimum ages; it may be that older fossils are found that change the absolute, but not relative, ages of the clades mentioned here.

Solanum lycopersicum belongs to the large clade Solanoideae (sometimes defined as a subfamily) whose members possess berries as a fruit type (with some modifications, see Knapp 2002). The stem age of the Solanoideae is estimated at ca. 21 Mya (19.0–23.3 Mya), around the same time that many of the major clades within the family began to diversify rapidly (Särkinen et al. 2013). *Solanum* itself has a stem age of ca. 17 Mya (14.5–17.7 Mya) and a crown age of ca. 15.5 Mya (13.3–17.5 Mya, see Fig. 2.2). Stem and crown ages differ due to differential inclusion of putative common ancestors (extinct taxa) in the group to be analyzed (see Baum and Smith 2012). This hyper-diverse genus with its more than 1200 species (see Knapp et al. 2004) is relatively young and the start of its diversification occurred in the mid-Miocene.

The tomato (*S. lycopersicum*) and its relatives belong to Särkinen et al.'s (2013) *Solanum* Clade I, and within that to the Potato clade (see Fig. 2.2), whose stem age was calculated at ca. 14.3 Mya (12.5–16.3 Mya), with the tomato and its relatives diverging from the potatoes (section *Petota*) at ca.

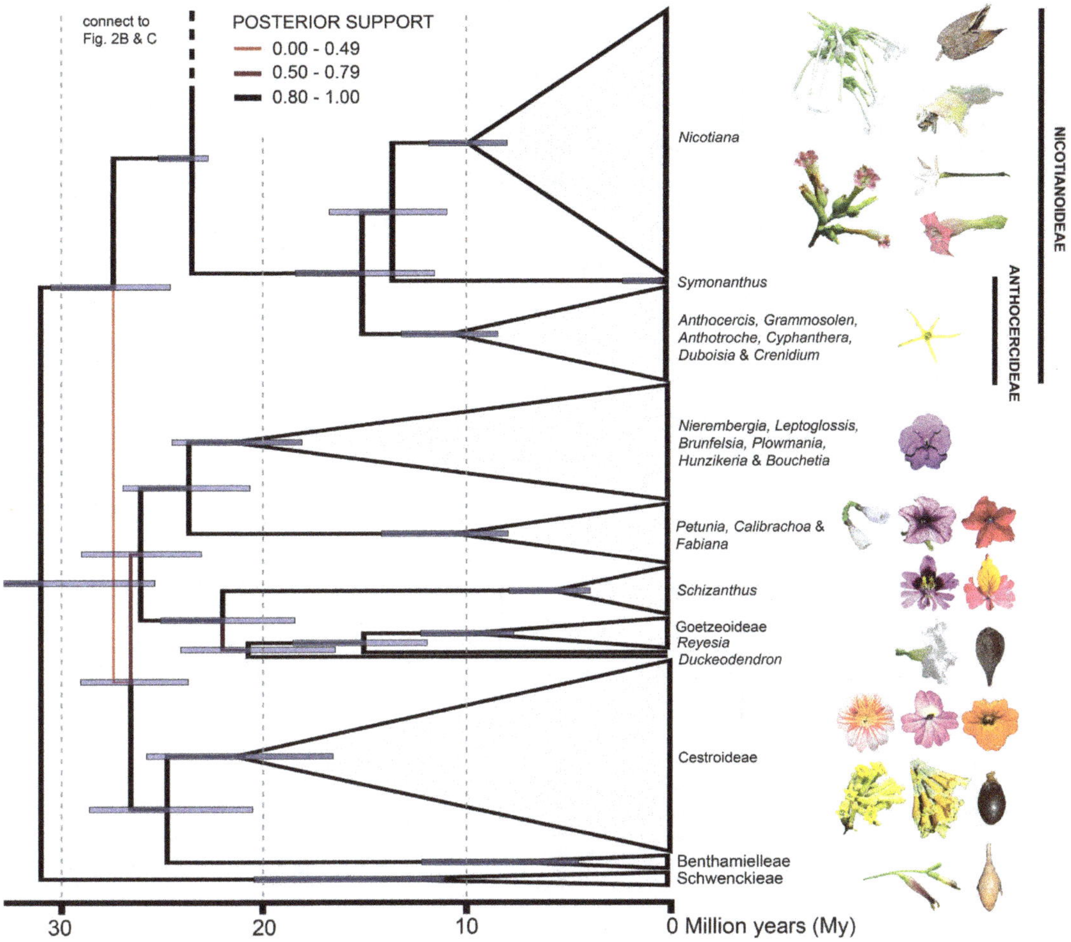

Fig. 2.2 Dated Solanaceae phylogeny; only major clades shown with representative flowers/fruits alongside. *Grey bars* correspond to date ranges as seen in text (from Särkinen et al. 2013, reproduced with permission from *BMC Evolutionary Biology* 13:214 (2013). doi:10.1186/1471-2148-13-214)

8 Mya (6.6–9/9 Mya). Within the tomato clade in the strict sense (excluding sections *Juglandifolia* and *Lycopersicoides*) species diversification was calculated to have a minimum age of ca. 2 Mya (1.2–2.6 Mya). The cultivated tomato itself belongs to a very recently derived group within the clade and is not a wild species, but instead is a domesticated plant derived from its wild progenitor, *S. pimpinellifolium*, by humans.

Tomatoes Travelling

The origins of crop plants can be difficult to decipher, due at least in part to human transport and use around the world with the globalization that began in the sixteenth century when Europeans first colonized the New World (Mann 2011). Even modern molecular tools can fail to unambiguously resolve origins, especially in groups like tomatoes, where spread has been global and wild species have been extensively used in breeding (Grandillo et al. 2011). How and when *Solanum lycopersicum* was first brought from the Americas to Europe has been debated since the late nineteenth century (de Candolle 1886; Jenkins 1948). The earliest description in the European botanical literature of a tomato dates from the sixteenth century in Pietro Andrea Matthioli's (Latinized as Petrus

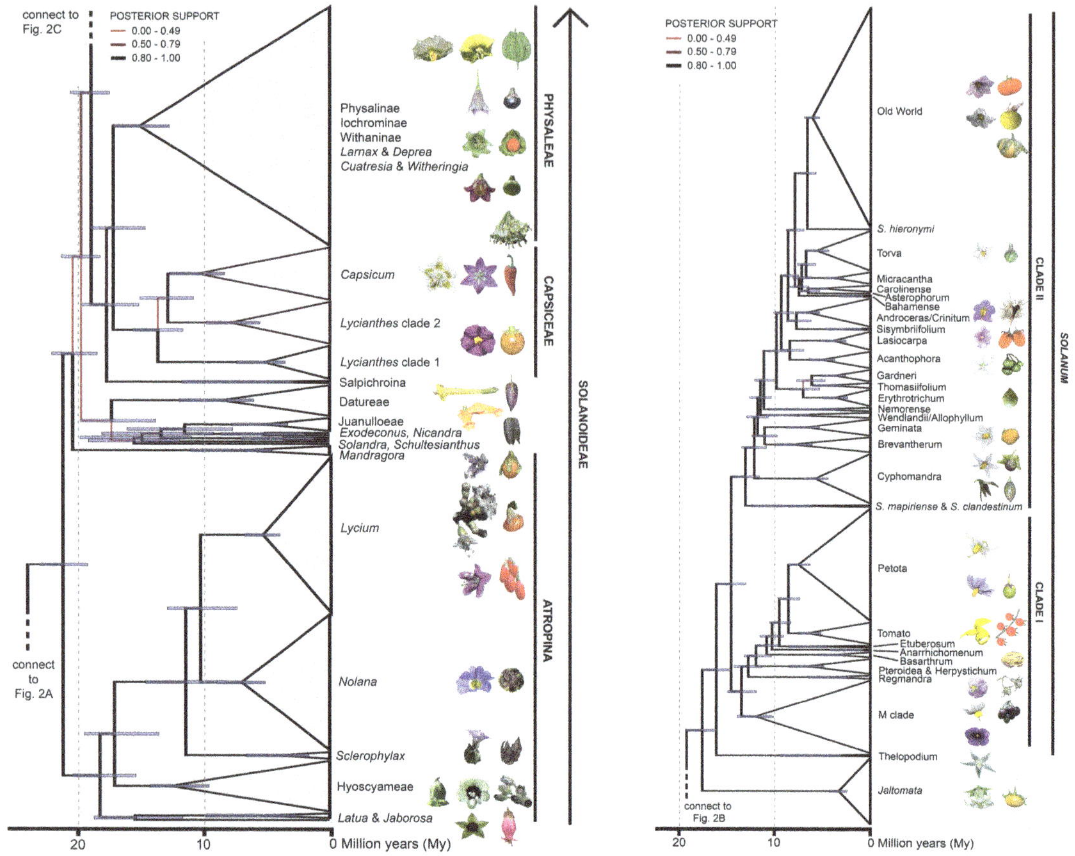

Fig. 2.2 (continued)

Andrea Matthiolus and sometimes also written as Mattioli) Italian language commentary upon the work of the first century Greek botanist Dioscorides of Anazarbos (Mattioli 1544). Tomatoes were classified and identified by comparison with plants already known in Europe and from classical Greek references, and, following this tradition, Mattioli (1544) described tomatoes in his section "Della Mandragorae," (On Mandrakes) as: "Portansi à i tempi nostri d'un'altra spetie in Italia stiacciante come le mele rose, and fatte a spicci, de colour prima verdi and come son mature, di color d'oro, lequali pur si mangiano nel medesmo modo" (Another species has been brought to Italy in our time, flattened like the "mele rose" [variety of apple] and segmented, green at first and when ripe of a golden colour, which is eaten in the same manner). Most probably the oldest illustration of tomatoes is a watercolour part of the unpublished manuscript of Leonard Fuchs (see frontispiece of Peralta et al. 2008a, b), and it is considered a "chimera" since represent in one plant fruits of different shapes and colours (round, flat, segmented, red and yellow) and even green fruits with stripes that might correspond to a wild species. This painting demonstrates that various different types of tomatoes (perhaps even wild species) were known in Europe by mid-sixteenth century. The earliest published illustration of a tomato is a rather crude woodcut of a plant with eight-parted flowers and fascinated fruits in Dodoens' herbal (1554) published in the Netherlands. Contemporaneous published illustrations of tomatoes in the sixteenth and seventeenth century literature (see Fig. 2.3) all depict plants with large, fascinated flowers and multi-locular fruit, clearly showing that tomatoes came to Europe not as

Fig. 2.3 An early wood cut illustration of *Solanum lycopersicum* (Mattioli 1590), showing the fasciated flowers and large multi-locular fruits present in early European tomatoes. *Source* Reproduced with permission of the Library of the Natural History Museum, London

small-fruited wild species, but as domesticated, large-fruited plants. These early introductions were said to have yellow (Mattioli 1544; Besler 1613) or red (Besler 1613) fruits.

de Candolle (1886) suggested the tomato was introduced from Peru for both historical and botanical reasons, and subsequent workers on the group (Müller 1940; Luckwill 1943). Jenkins (1948) suggested that Mexico was the area from which the plants were introduced to Europe, based mostly on linguistic (the Nahuatl name for *S. lycopersicum* is 'jitomatl', very like tomato) evidence and the lack of archaeological or linguistic evidence for any domestication in South America. Peralta and Spooner (2007) considered the origins for the cultivated tomato to be uncertain, and concluded that evidence is inconclusive regarding either a Mexican or a Peruvian initial site of domestication. Recent work with high density molecular markers has helped to shed light on some aspects of the story (see below).

Small-fruited cherry tomatoes were considered to be the wild progenitors of *S. lycopersicum* (de Candolle 1886; Müller 1940; Luckwill 1943; Rick and Holle 1990); these small-fruited plants are otherwise morphologically nested within the variation of the cultivated tomato and they are often seen growing in what appear to be wild conditions. Nesbitt and Tanksley (2001), however, suggested that many of these plants with small fruits were the results of admixtures with the wild species, *S. pimpinellifolium*. Molecular analyses of SNPs in a large collection of small-fruited tomatoes (Ranc et al. 2008) showed that cherry-type tomatoes were a complex mixture of *S. pimpinellifolium* and *S. lycopersicum* and did not form a distinct, recognizable group either based on morphology or molecules. Blanca et al. (2013) used the SOLCap platform to analyze a different set of small and large-fruited tomatoes from both germplasm collections and wild origin. They found that a set of Andean accessions could be distinguished from both *S. pimpinellifolium* and *S. lycopersicum*, but that these plants did not all have small fruits. Accessions from the eastern slopes of the Andes in Ecuador and Peru were suggested to be early cultivars, with Mesoamerican accessions also distinct from those found elsewhere in the world. Blanca et al. (2013) hypothesize that the plants from Ecuador and Peru represent early domesticates, pre-breeding populations, and that the tomato was truly developed as a cultivated plant in Mexico and Mesoamerica after being taken there in pre-Columbian times. European heritage varieties show more molecular similarity to Mesoamerican accessions than to South American ones. The similarity of climate in Mexico and the European Mediterranean may have contributed to the ease of introduction of the tomato post-1520.

Blanca et al. (2013) distinguish these pre-breeding Andean populations at the varietal level as var. *cerasiforme*. This has been traditional in the tomato literature for plants of *S. lycopersicum* with small fruits, but we consider these plants to be the product of domestication, not of evolution by natural selection, and thus should not be named using the *International Code of Nomenclature for algae, fungi, and plants* (McNeill et al. 2012). In addition, Blanca et al. (2013) found that the South American accessions they identified as distinct had a wide range of fruit sizes; the accessions were better distinguished using a panel of morphological characteristics (similar to those used to distinguish *S. pimpinellifolium* and *S. lycopersicum* by Peralta et al. 2008a), thus use of '*cerasiforme*' could cause confusion. We suggest this distinct set of accessions be named according to the *International Code of Nomenclature for Cultivated Plants* (Brickell et al. 2009), as has been done for potatoes (Huamán and Spooner 2002). These conventions for naming pertain to "plants whose origin or selection is primarily due to the intentional actions of mankind" (Brickell et al. 2009). As Blanca et al. (2013) point out, further sampling of South American traditional cultivars is necessary to better understand these patterns. New collecting in the Andes where tomato pre-breeding and early domestication occurred is a priority before this diversity disappears.

Diversity within the cultivated species is likely to be well conserved ex situ; Ross (1998) cited 62,832 accessions of mainly of *S. lycopersium* maintained in gene banks around the world. A wealth of studies using isozymes (Rick and Holle 1990) and molecular markers (Williams and St. Clair 1993; Villand et al. 1998; Blanca et al. 2013) have demonstrated the high genetic diversity of landrace cultivars in South America. Nevertheless, areas close to the origin of tomatoes have not been sufficiently explored to recover these valuable genetic resources. The richness of cultural values in Andean communities is also reflected by their crop diversity, traditional cultivation and culinary practices. Small farmers developed a sustainable agriculture using ancestral land practices that are less aggressive to the environment, select crops adapted to the local conditions and maintain their own seed. Social, economic and ecological factors are affecting the in situ conservation of these genetic resources. Recently, germplasm recuperation efforts have been focused in tomato local landraces or "criollos" in Bolivia (Gonzáles et el. 2011) and Argentina (Peralta et al. 2008b, Fig. 2.4). These landraces were incorporated in the Argentinean Vegetable Crop Germplasm Bank System (Clausen et al. 2008, http://inta.gob.ar/documentos/red-de-bancos-y-colecciones-de-germoplasma/), evaluated in the field for agronomic and fruit quality traits and their potential use in breeding programmes (Peralta et al. 2008b). Traditional tomato varieties are characterized by their fruit qualities, mainly metabolites (Asprelli et al. 2016), antioxidants (Di Paola Naranjo et al. 2016a, b) and organic volatiles (Cortina et al. 2016), and typical flavour that consumers appreciate and now demand, although their seeds are not longer available. Recovery and return of these locally adapted varieties to their original communities will contribute to their sustainable maintenance. In basic research, the value of these Andean accessions has been demonstrated in their contribution to understanding the role of epigenetics in the

Fig. 2.4 Fruits from three tomato landraces from Argentina. "Platense": plurilocular, round, flattened and segmented; "Corazón de Buey": plurilocular, heart shape, slighly segmented; and "Largo": 2 or 3 locules, elongated. These landraces are cultivated for their quality traits (flavor, color, aroma) by local farmers in rural Argentina

determination of relevant agronomic traits (Quadrana et al. 2014). Additional collections and characterization of South American traditional cultivars are necessary not only for understanding diversity patters and evolutionary relationships, but also to reveal the domestication history and elucidate the genetics of agronomic and quality traits. Recuperation, conservation and uses of local landraces, particularly those from South and Central America, in tomato breeding is essential to incorporate valuable traits, such as fruit flavour and nutritional and health beneficial components, that humans have selected for over the course of improvement of tomatoes in local situations.

Summary

The cultivated tomato, *Solanum lycopersicum*, is a member of the large and diverse genus Solanum of the derived Asterid family Solanaceae. It belongs to a group of 13 closely related species all of which occur in arid habitats on the west coast of South America. The tomatoes are sister to the potatoes, and began to diversify only very recently, after the rise of the Andes and the development of the arid western deserts. Tomatoes were probably brought to Europe by the Spanish from Mesoamerica, and thence distributed worldwide. Traditional, early cultivars from the eastern slopes of the Andes in Ecuador and Peru are distinct from other cultivated populations but harbour a great diversity of fruit size and are not only small-fruited. Further collecting of feral populations and local varieties from South America will contribute to elucidate the diversity and origins of the cultivated tomato, as well as to reveal the genetics of agronomic and quality traits. Efforts to conserve the variation in *S. lycopersicum* itself, and not only related wild species, in its area of origin are a priority. Tomato landraces, selected and adapted to their local environments, are promising genetic sources to incorporate valuable traits in cultivated varieties.

Acknowledgments We thank Mathilde Causse for asking us to write this chapter, and all our solanaceous colleagues for years of fruitful research together, especially David M. Spooner and Lynn Bohs. SK thanks the National Science Foundation (USA) for funding under the Planetary Biodiversity Inventory programme (PBI Solanum: a worldwide treatment, DEB-0316614), the European Commission for funding under the FP6 Integrated Project EU-SOL (PL 016214) and SYNTHESYS programme (http://www.synthesys.info/ which is financed by European Community Research Infrastructure Action under the FP6 "Structuring the European Research Area" Programme); IE thanks María Sance, Pablo Asprelli, Marilu Makuch, Leonardo Togno, Inés Lorello, Patricia Occhiuto, Estela Valle, Marina Insani, Claudio Galmarini, Ramón Asis and Fernando Carrari for all their efforts to recover and valorize genetic resources in Argentina, also to CONICET, Universidad Nacional de Cuyo, INTA and Agencia Nacional de Promoción Científica y Tecnológica (ANPCyT) for funding under FONCYT (PICTO 08-12903 and PICTR 01942).

References

Alvarez AE, van de Wiel CCM, Smulders MJM, Vosman B (2001) Use of microsatellites to evaluate genetic diversity and species relationships in the genus *Lycopersicon*. Theor Appl Genet 103:1283–1292

APG III (2009) An update of the Angiosperm Phylogeny Group classification for the orders and families of flowering plants: APG III. Bot J Linn Soc 161:105–121

Asprelli PD, Sance M, Insani M, Asis R, Valle EM, Carrari F, Galmarini CR, Peralta IE (2016) Agronomic performance and fruit nutritional quality of an Andean tomato collection. Acta Horticulturae (in press)

Barboza GE, Knapp S, Särkinen TE (2013) Grupo VII. Moreloide. In: Anton AM, Zuloaga FO (eds), Barboza GE (coord) Flora Argentina vol 13, Solanaceae. IOBDA- IMBIV, CONICET, Buenos Aires, pp 231–264

Baum DA, Smith SD (2012) Tree thinking: an introduction to phylogenetic biology. Roberts and Co., Greenwood

Bell CD, Soltis DE, Soltis PS (2010) The age and diversification of the angiosperms re-visited. Am J Bot 97:1296–1303

Bentham G, Hooker JD (1873) Solanaceae. Genera Plant 2:882–913

Besler B (1613) Hortus eystettensis. Published by the author, Nuremberg

Blanca J, Cañizares J, Cordero L, Pascual L, Diez MJ, Nuez F (2013) Variation revealed by SNP genotyping and morphology provides insight into the origin of the tomato. PLoS ONE 7(10):e48198

Bohs L (1995) Transfer of *Cyphomandra* (Solanaceae) and its species to *Solanum*. Taxon 44:583–587

Bohs L (2005) Major clades in *Solanum* based on ndhF sequences. In: Keating RC, Hollowell VC, Croat TB (eds) A Festschrift for William G. D'Arcy: the legacy of a taxonomist. Monographs in Systematic Botany from the Missouri Botanical Garden, vol 104. Missouri Botanical Garden Press, St. Louis, pp 27–49

Bohs L, Olmstead RG (1997) Phylogenetic relationships in *Solanum* (Solanaceae) based on ndhF sequences. Syst Bot 22:5–17

Bohs L, Olmstead RG (1999) *Solanum* phylogeny inferred from chloroplast DNA sequence data. In: Nee M, Symon DE, Lester RN, Jessop JP (eds) Solanaceae IV: advances in biology and utilization. Royal Botanic Gardens, Kew, pp 97–110

Bohs L, Olmstead RG (2001) A reassessment of *Normania* and *Triguera* (Solanaceae). Plant Syst Evol 228:33–48

Bretó MP, Asins MJ, Carbonell EA (1993) Genetic variability in *Lycopersicon* species and their genetic relationships. Theor Appl Genet 86:113–120

Brickell CD, Alexander C, David JC, Hetterscheid WLA, Leslie AC, Malecot V, Jin XB, Cubey JJ (2009) International code of nomenclature for cultivated plants (ICNCP or Cultivated Plant Code) incorporating the rules and recommendations for naming plants in cultivation, 8th edn. Adopted by the International Union of Biological Sciences International Commission for the Nomenclature of Cultivated Plants. Regnum Vegetabile 151; Scripta Horticulturae 10. International Society for Horticultural Science, Leuven, Belgium

Caicedo AL, Schaal BA (2004) Population structure and phylogeography of *Solanum pimpinellifolium* inferred from a nuclear gene. Mol Ecol 13:1871–1882

Canady MA, Ji Y, Chetelat RT (2006) Homeologous recombination in *Solanum lycopersicodes* introgression lines of cultivated tomato. Genetics 174:1775–1778

Causse M, Desplat N, Pascual L, Le Paslier M-C, Sauvage C, Bauchet G, Bérard A, Bounon R, Tchoumakov M, Brunel D, Bouchet J-P (2013) Whole genome resequencing in tomato reveals variation associated with introgression and breeding events. BMC Genom 14:791

Chetelat RT, Ji Y (2007) Cytogenetics and evolution. In: Razdan MK, Mattoo AK (eds) Genetic improvement of solanaceous crops, vol 2, Tomato. Science, Enfield, pp 77–112

Chetelat RT, Meglic Y (2000) Molecular mapping of chromosome segments introgressed from *Solanum lycopersicoides* into cultivated tomato (*Lycopersicon esculentum*). Theor Appl Genet 100:232–241

Chetelat RT, Rick CM, Cisneros P, Alpert KB, DeVerna JW (1998) Identification, transmission, and cytological behavior of *Solanum lycopersicoides* Dun. monosomic alien addition lines in tomato (*Lycopersicon esculentum* Mill.). Genome 41:40–50

Child A (1990) A synopsis of *Solanum* subgenus *Potatoe* (G. Don) D'Arcy section *Tuberarium* (Dunal) Bitter (s.l.). Feddes Rep 101:209–235

Clausen AM, Ferrer ME, Formica ME (2008) Situación de los recursos Fitogenéticos en la Argentina. II Informe Nacional 1996–2006. Publicaciones Regionales. Ediciones INTA

Correll DS (1958) A new species and some nomenclatural changes in *Solanum* section *Tuberarium*. Madroño 14:232–236

Correll DS (1962) The potato and its wild relatives. Contr Texas Res Found Bot Studies 4:1–606

Cortina PR, Asis R, Peralta IE, Asprelli PD, Santiago AN (2016) Determination of volatile organic compounds in Andean tomato landraces by headspace solid phase microextraction-gas chromatography-mass spectrometry. J Braz Chem Soc 1–12. http://jbcs.sbq.org.br/imagebank/pdf/160126AR.pdf

D'Arcy WG (1972) Solanaceae studies II: typification of subdivisions of *Solanum*. Ann Mo Bot Gard 59:262–278

D'Arcy WG (1987) The circumscription of Lycopersicon. Solanaceae Newsl 2:60–61

D'Arcy WG (1991) The Solanaceae since 1976, with a review of its biogeography. In: Hawkes JG, Lester RN, Nee M, Estrada N (eds) Solanaceae III: taxonomy, chemistry, evolution. Royal Botanic Gardens, Kew, pp 75–137

Darwin SC, Knapp S, Peralta IE (2003) Taxonomy of tomatoes in the Galapagos Islands: native and introduced species of *Solanum* section *Lycopersicon* (Solanaceae). Syst Biodivers 12:29–53

Davis J, Yu D, Evans W, Gokirmak T, Chetelat RT, Stotz HU (2009) Mapping of loci from *Solanum lycopersicoides* conferring resistance or susceptibility to *Botrytis cinerea* in tomato. Theor Appl Genet 119:305–314

de Candolle ALPP (1886) Origin of cultivated plants, 2nd ed. D. Appleton, New York (1959 reprint; Hafner Publishing Company, New York)

de Tournefort JP (1694) Éléments de Botanique. Imprimerie Royale, Paris

Di Paola Naranjo RD, Otaiza S, Saragusti A2, Baroni V, Carranza Adel V, Peralta IE, Valle EM, Carrari F, Asis R (2016a) Hydrophilic antioxidants from Andean tomato landraces assessed by their bioactivities in vitro and in vivo. Food Chem 206:146–155. doi:10.1016/j.foodchem.2016.03.027

Di Paola Naranjo RD, Otaiza S, Saragusti AC, Baroni V, Carranza AV, Peralta IE, Valle EM, Carrari F, Asis R (2016b) Data on polyphenols and biological activity analyses of an Andean tomato collection and their relationships with tomato traits and geographical origin. Data Brief 7:1258–1268. doi:10.1016/j.dib.2016.04.005

Dunal MF (1813) Histoire naturelle, médicale et économique des *Solanum* et des genres qui ont été confundus avec eux. France, Montpellier

Dunal MF (1852) Solanaceae. In: De Candolle AP (ed) Prodromus systematis naturalis regni vegetabilis, vol 13, pp 1–450

Fosberg FR (1987) New nomenclatural combinations for Galápagos plant species. Phytologia 62:181–183

Fridman E, Carrari F, Liu YS, Fernie AR, Zamir D (2004) Zooming in on a quantitative trait for tomato yield using interspecific introgressions. Science 305:1786–1789

Frodin D (2004) History and concepts of big plant genera. Taxon 53:753–776

Grandillo S, Chetelat R, Knapp S, Spooner D, Peralta, I, Cammareri M, Perez O, Tripodi P, Termolino P,

Chuisano ML, Ercolano MR, Frusciante L, Monti L, Pignone D (2011) 9. *Solanum* sect. *Lycopersicon*. In: Kole C (ed) Wild crop relatives: genomics and breeding resources. Volume 5—vegetables. Springer, Heidelberg, pp 129–216

Gonzáles J et al (2011) Catálogo de poblaciones de tomate nativo e introducido en Bolivia. Impresiones Poligraf, Bolivia

Hawkes JG (1990) The potato: evolution biodiversity and genetic resources. Belhaven, London

Heine H (1976) Flora de la Nouvelle Caledonie, vol 7. Museum National D'Histoire Naturelle, Paris

Huamán Z, Spooner DM (2002) Reclassification of landrace populations of cultivated potatoes (*Solanum* sect. *Petota*). Am J Bot 89:947–965

Hunziker AT (1979) South American Solanaceae: a synoptic survey. In: Hawkes JG, Lester RN, Skelding AD (eds) The biology and taxonomy of Solanaceae. Academic, London, pp 49–85

Hunziker AT (2001) Genera Solanacearum, the genera of Solanaceae illustrated arranged according to a new system. ARG Gantner, Ruggell

Jenkins JA (1948) The origin of the cultivated tomato. Econ Bot 2:379–392

Jussieu AL (1789) Genera plantarum. Herissant V & Barrios T, Paris

Knapp S (2002) *Solanum* section *Geminata*. Fl Neotrop 84:1–405

Knapp S, Bohs L, Nee M, Spooner DM (2004) Solanaceae: a model for linking genomics and biodiversity. Comp Funct Genom 5:285–291

Koenig D, Jimenez-Gómez JM, Kimura S, Fulop D, Chitwood DH, Headland LR, Kumar R, Covington MF, Kumar Devisetty U, Tat AV, Tohge T, Bolger A, Schneeberger K, Ossowski S, Lanz C, Xiong G, Taylor-Teeples, M, Brady SM, Pauly M, Weigel D, Usadel B, Fernie AR, Peng J, Sinha N, Maloof JN (2013) Comparative transcriptomics reveals patterns of selection in domesticated and wild tomato. Proc Natl Acad Sci USA 110(28):E2655–E2662. doi:10.1073/pnas.1309606110

Linnaeus C (1753) Species plantarum, 1st edn. L. Salvius, Stockholm

Lippman Z, Tanksley SD (2001) Dissecting the genetic pathway to extreme fruit size in tomato using a cross between the small-fruited wild species *Lycopersicon pimpinellifolium* and *L. esculentum* var. Giant Heirloom. Genetics 158:413–422

Lucatti AF, van Heusden AW, de Vos RCH, Visser RGF, Vosman B (2013) Differences in insect resistance between tomato species endemic to the Galapagos islands. BMC Evol Biol 13:175

Luckwill LC (1943) The genus *Lycopersicon*: an historical, biological, and taxonomical survey of the wild and cultivated tomatoes. Aberdeen Univ Stud 120:1–44

Macbride JF (1962) Solanaceae. In: Flora of Peru. Field Mus Nat Hist Bot Ser 13:3–267

Mann CC (2011) 1493: Uncovering the New World Columbus created. Alfred A. Knopf, New York

Marshall JA, Knapp S, Davey MR, Power JB, Cocking EC, Bennett MD, Cox AV (2001) Molecular systematics of *Solanum* section *Lycopersicum* (*Lycopersicon*) using the nuclear ITS rDNA region. Theor Appl Genet 103:1216–1222

Mattioli PA (1544) Di Pedacio Dioscoride Anazarbeo libri cinque della historia, et materia medicinale trodotti in lingua uolgare Italiana. N. de Bascarini, Venice

Mattioli PA (1590) Kreutterbuch deß hochgelehrten unnd weitberühmten Herrn D. Petri Andreae Matthioli. Johann Feyerabend für Peter Fischer & Heinrich Tack, Frankfurt

McClean PE, Hanson MR (1986) Mitochondrial DNA sequence divergence among *Lycopersicon* and related *Solanum* species. Genetics 112:649–667

McNeill J, Barrie FR, Buck WR, Demoulin V, Greuter W, Hawksworth DL, Herendeen PS, Knapp S, Marhold K, Prado J, Prud'homme van Reine WF, Smith GF, Wiersema JH, Turland NJ (2012) International Code of Nomenclature for algae, fungi, and plants (Melbourne Code). Regnum Vegetabile 154. Koelz Scientific Books, Königstein, Germany

Miller P (1731) The Gardener's dictionary, 1st edn. Published for the author, London

Miller P (1754) The Gardener's dictionary, Abridged 4th edn. Published for the author, London

Miller P. (1807) The gardener's and botanist's dictionary, posthumous edition, ed. Thomas Martyn. F.C. & J. Rivington, London

Miller JC, Tanksley SD (1990) RFLP analysis of phylogenetic relationships and genetic variation in the genus *Lycopersicon*. Theor Appl Genet 80:437–448

Mueller LA, Tanksley SD, Giovannoni JJ, van Eck J, Stack S, Choi D, Kim BD, Chen M, Cheng Z, Li C, Ling H, Xue Y, Seymour G, Bishop G, Bryan G, Sharma R, Khurana J, Tyagi A, Chattopadhyay D, Singh NK, Stiekema W, Lindhout P, Jesse T, Lankhorst RK, Bouzayen M, Shibata D, Tabata S, Granell A, Botella MA, Giuliano G, Frusciante L, Causse M, Zamir D (2005) The tomato sequencing project, the first cornerstone of the international Solanaceae project (SOL). Comp Funct Genom 6(3):153–158

Müller CH (1940) A revision of the genus *Lycopersicon*. USDA Misc Publ 382:1–28

Nee M (1999) A synopsis of *Solanum* in the New World. In: Nee M, Symon DE, Lester RN, Jessop JP (eds) Solanaceae IV: advances in biology and utilization. Royal Botanic Gardens, Kew, pp 285–333

Nesbitt TC, Tanksley SD (2002) Comparative sequencing in the genus *Lycopersicon*: implications for the evolution of fruit size in the domestication of cultivated tomatoes. Genetics 162:365–379

Olmstead RG, Palmer JD (1997) Implications for phylogeny, classification, and biogeography of *Solanum* from cpDNA restriction site variation. Syst Bot 22:19–29

Olmstead RG, Sweere JA, Spangler RE, Bohs L, Palmer JD (1999) Phylogeny and provisional classification of the Solanaceae based on chloroplast DNA. In: Nee M, Symon DE, Lester RN, Jessop JP (eds) Solanaceae IV: advances in biology and utilization. Royal Botanic Gardens, Kew, pp 111–137

Palmer JD, Zamir D (1982) Chloroplast DNA evolution and phylogenetic relationships in Lycopersicon. Proc Natl Acad Sci USA 79:5006–5010

Peralta IE, Spooner DM (2001) Granule-Bound Starch Synthase (GBSSI) gene phylogeny of wild tomatoes (*Solanum* L. section *Lycopersicon* [Mill.] Wettst. subsection *Lycopersicon*). Am J Bot 88:1888–1902

Peralta IE, Spooner DM (2005) Morphological characterization and Relationships of wild tomatoes (Solanum L. Section Lycopersicon [Mill.] Wettst. Subsection Lycopersicon). Monogr Syst Bot Mo Bot Gard 104:227–257

Peralta IE, Spooner DM (2007) History, origin and early cultivation of tomato (Solanaceae). In: Razdan MK, Mattoo AK (eds) Genetic improvement of Solanaceous crops, vol 2, tomato. Science, Enfield, pp 1–27

Peralta IE, Knapp S, Spooner DM (2005) New species of wild tomatoes (*Solanum* section *Lycopersicon*: Solanaceae) from Northern Peru. Syst Bot 30(2):424–434

Peralta IE, Knapp S, Spooner DM (2006) Nomenclature for wild and cultivated tomatoes. Feature article. Rep Tomato Genet Coop 56:6–12

Peralta IE, Spooner DM, Knapp S (2008a) Taxonomy of wild tomatoes and their relatives (*Solanum* sections *Lycopersicoides*, *Juglandifolia*, *Lycopersicon*; Solanaceae). Syst Bot Monogr 84:1–186

Peralta IE, Makuch M, García Lampasona S, Occhiuto PN, Asprelli PD, Lorello IM, Togno L (2008b) Catálogo de poblaciones criollas de pimiento, tomate y zapallo colectadas en valles andinos de la Argentina. Editorial INTA, Mendoza

Pertuzé RA, Ji Y, Chetelat RT (2002) Comparative linkage map of the *Solanum lycopersicoides* and *S. sitiens* genomes and their differentiation from tomato. Genome 45:1003–1012

Quadrana L, Almeida J, Asís R, Duffy T, Dominguez PG, Bermúdez L, ContiG, Corrêa da Silva JV, Colot V, Asurmendi S, Fernie AR, Rossi M, Peralta I, Carrari F (2014) Natural occurring epialleles determine vitamin E accumulation in tomato fruits. Nat Comm 5:4027. doi:10.1038/ncomms5027

Ranc N, Muños S, Santoni S, Causse M (2008) A clarified position for *Solanum lycopersicum* var. *cerasiforme* in the evolutionary history of tomatoes (Solanaceae). BMC Plant Biol 8:130

Rick CM (1979) Biosystematic studies in *Lycopersicon* and closely related species of *Solanum*. In: Hawkes JG, Lester RN, Skelding AD (eds) The biology and taxonomy of Solanaceae, Linn Soc Symp Ser 7. Academic, New York, pp 667–677

Rick CM (1988) Tomato-like nightshades: affinities, autoecology, and breeders opportunities. Econ Bot 42:145–154

Rick CM, Chetelat RT (1995) Utilization of related wild species for tomato improvement. Acta Hortic 412: 21–38

Rick CM, Holle M (1990) Andean *Lycopersicon esculentum* var *cerasiforme* genetic variation and its evolutionary significance. Econ Bot 43(Suppl. 3):69–78

Rick CM, Laterrot H, Philouze J (1990) A revised key for the *Lycopersicon* species. Tomato Genet Coop Rep 40:31

Rodriguez F, Wu F, Ané C, Tanksley S, Spooner DM (2010) Do potatoes and tomatoes have a single evolutionary history, and what proportion of the genome supports this history? BMC Evol Biol 9:191

Ross RJ (1998) Review paper: global genetic resources of vegetables. Plant Var Seeds 11:39–60

Särkinen T, Bohs L, Olmstead RG, Knapp S (2013) A phylogenetic framework for the evolutionary study of the nightshades (Solanaceae): a dated 1000-tip tree. BMC Evol Biol 13:214

Schauer N, Semel Y, Roessner U, Gur A, Balbo I, Carrari F, Pleban T, Perez-Melis A, Bruedigam C, Kopka J, Willmitzer L, Zamir D, Fernie AR (2006) Comprehensive metabolic profiling and phenotyping of interspecific introgression lines for tomato improvement. Nat Biotechnol 24:447–454

Seithe A (1962) Die Haararten der Gattung *Solanum* L. und ihre taxonomische Verwertung. Bot Jahrb Syst 81:261–336

Seymour G, Ostergaard L, Chapman NH, Knapp S, Martin C (2013) Fruit ripening and development. Ann Rev Plant Biol 64:219–241

Smith SD, Peralta IE (2002) Ecogeographic surveys as tools for analyzing potential reproductive isolating mechanisms: an example using *Solanum juglandifolium* Dunal, *S. ochranthum* Dunal, *S. lycopersicoides* Dunal, and *S. sitiens* I.M. Johnston. Taxon 51:341–349

Spooner DM, Anderson GJ, Jansen RK (1993) Chloroplast DNA evidence for the interrelationships of tomatoes, potatoes, and pepinos (Solanaceae). Am J Bot 80:676–688

Spooner DM, Peralta IE, Knapp S (2005) Comparison of AFLPs to other markers for phylogenetic inference in wild tomatoes [*Solanum* L. section *Lycopersicon* (Mill.) Wettst. subsection *Lycopersicon*]. Taxon 54:43–61

Stack SM, Covey PA, Anderson LK, Bedinger PA (2009) Cytogenetic characterization of species hybrids in the tomato clade. Tomato Genet Coop Rep 59:57–61

Stommel JR (2001) USDA 97L63, 97L66 and 97L97: tomato breeding lines with high fruit beta-carotene content. HortScience 36:387–388

Symon DE (1981) The Solanaceous genera *Browallia*, *Capsicum*, *Cestrum*, *Cyphomandra*, *Hyoscyamus*, *Lycopersicon*, *Nierembergia*, *Physalis*, *Petunia*, *Salpichroa*, *Withania*, naturalized in Australia. J Adelaide Bot Gard 3:133–166

Symon DE (1985) The Solanaceae of New Guinea. J Adelaide Bot Gard 8:1–177

Taylor IB (1986) Biosystematics of the tomato. In: Atherton JG, Rudich J (eds) The tomato crop: a scientific basis for improvement. Chapman and Hall, London, pp 1–34

Tomato Genome Consortium (2012) The tomato genome sequence provides insights into fleshy fruit evolution. Nature 485:635–641

Van der Knaap E, Lippman ZB, Tanksley SD (2002) Extremely elongated tomato fruit controlled by four quantitative trait loci with epistatic interactions. Theor Appl Genet 104:241–247

Villand J, Skroch PW, Lai T, Hanson P, Kuo CG, Nienhuis J (1998) Genetic variation among tomato accessions from primary and secondary centers of diversity. Crop Sci 38:1339–1347

Whalen MD (1979) Taxonomy of *Solanum* section *Androceras*. Gentes Herb 11:359–426

Williams CE, St. Clair DA (1993) Phenetic relationships and levels of variability detected by restriction fragment length polymorphism and random amplified polymorphic DNA analysis of cultivated and wild accessions of *Lycopersicum esculentum*. Genome 36:619–630

Wu F, Mueller LA, Crouzillat D, Petiard V, Tanksley SD (2006) Combining bioinformatics and phylogenetics to identify large sets of single copy, orthologous genes (COSII) for comparative, evolutionary and systematics studies: a test case in the Euasterid plant clade. Genetics 174:1407–1420

Gene Mapping in Tomato

3

Mathilde Causse and Silvana Grandillo

Abstract

Tomato is a model species for genetic analyses since a long time. Many mutations controlled by a single gene were discovered and the underlying genes were mapped first on the tomato genetic map. Most of these genes are involved in fruit colour and shape, in plant growth and architecture and in disease resistances. With the construction of high-density molecular genetic maps, many genes were located on the genome and subsequently several of them were fine-mapped and further identified by positional cloning. Today with the availability of the tomato genome sequence these genes are physically located on the genome and the identification of new ones is being considerably accelerated. The alignment of the physical and genetic maps allowed the identification of hot spots of recombination and of large regions where recombination is almost suppressed, whatever the progeny studied. The impact of this heterogeneity in recombination is discussed.

Keywords

Tomato · Gene mapping · Mutations · Resistance · Fruit quality

Introduction

Tomato has been a model species for genetic analyses for years. The diversity of its fruit colour, shape and size has interested geneticists since the early work of genetic mapping. Butler (1952) proposed one of the first genetic maps including more than 50 loci corresponding to phenotypic mutations. Nevertheless, until the discovery of molecular markers in the late 1980s, the location of mutations on genetic maps was not really precise as it was impossible to

M. Causse (✉)
UR 1052 Génétique et Amélioration des Fruits et Légumes, INRA, CS60094, 84250 Montfavet, France
e-mail: mathilde.causse@avignon.inra.fr

S. Grandillo
Research Division Portici, Italian National Research Council, Institute of Bioscience and BioResources (CNR-IBBR), Via Università 133, 80055 Portici, Napoli, Italy

M. Causse et al. (eds.), *The Tomato Genome*, Compendium of Plant Genomes,
DOI 10.1007/978-3-662-53389-5_3

simultaneously map many loci. Molecular markers have enabled biologists to construct saturated linkage maps of the genome and to systematically localize mutations of interest on these maps. Over years, more and more markers were discovered and the genotyping cost decreased. Following isozymes, the first DNA markers, based on the detection of Restriction Fragment Length Polymorphisms (RFLP), allowed the construction of a reference map of the tomato genome (Tanksley et al. 1992). With more than 1000 loci, spread on the 12 chromosomes, this map allowed the precise localization of several mutations and of a few genes of interest. New mutations or genes of interest were subsequently mapped using either F_2 populations or pairs of near isogenic lines differing only in the region of the interesting gene (Laterrot 1996). Bulks of individuals were later used (following the Bulk Segregant Analysis method), together with markers based on PCR amplification of the DNA (RAPD or AFLP markers). Following the identification of PCR markers linked to the gene of interest, specific PCR markers were set up, simplifying the genotyping step for breeders. Nevertheless, PCR markers such as RAPD or AFLP are dominant and map for the most part close to the centromeres, reducing their potential efficiency for gene mapping in tomato (Grandillo and Tanksley 1996; Haanstra et al. 1999; Saliba-Colombani et al. 2000). Markers based on the variation in the number of small sequence repeats (microsatellites or SSR) were then discovered and mapped on the reference map or used for the construction of new maps (He et al. 2002; Liu et al. 2005). To increase the number of markers available and to use the microsynteny observed with the *Arabidopsis thaliana* genome, Fulton et al. (2002) proposed the use of Conserved Ortholog Sequences (COS) as markers.

The polymorphism revealed by RFLP markers among cultivated accessions was very low and only a few markers were polymorphic and thus useful for mapping genes in such genetic background (Saliba-Colombani et al. 2000). Interspecific progenies were much more polymorphic and maps based on progenies derived from crosses with every wild species related to tomato were constructed (Labate et al. 2007). A population of introgression lines derived from a cross with a *Solanum pennellii* accession (Eshed and Zamir 1995) was particularly useful to discover new genes and quantitative trait loci (QTL) involved in fruit colour, size and plant traits (Zamir 2001).

More recently, several tomato accessions were used to sequence fragments of expressed sequences and identify Expressed Sequence Tags (ESTs), allowing the first Single Nucleotide Polymorphism (SNP) markers to be discovered and mapped (Labate and Baldo 2005; Sim et al. 2009). With the access to the tomato genome sequence (Tomato Genome Consortium 2012), the increased throughput of sequencing and the advances in Next Generation Sequencing technologies, it has been possible to discover thousands of SNPs through RNA sequencing (RNAseq). The SolCAP consortium developed a SNP array carrying more than 8000 SNPs chosen to reveal polymorphisms among cultivated accessions (Sim et al. 2012). Another SNP array was developed by Víquez-Zamora et al. (2013). Today, thanks to the tomato genome sequence availability, several projects of resequencing whole genomes of tomato accessions allowed the discovery of several millions of SNP (Causse et al. 2013; Aflitos et al. 2014; Lin et al. 2014) and the construction of genetic maps at the intraspecific level is now possible (Shirasawa et al. 2010). Large SNP arrays permit the rapid mapping of new loci of interest (Viquez-Zamora et al. 2014).

Genes and Loci Involved in Morphological and Fruit Characteristics

Among the major mutations used in tomato, the self-pruning (*sp*) mutation was discovered about 100 years ago and confers the determinate growth behaviour. It was largely used in processing tomato for field grown production. The tomato *SELF-PRUNING* (*SP*) gene is the homolog of the *Antirrhinum majus CENTRORADIALIS* (*CEN*) and *Arabidopsis thaliana TERMINAL FLOWER1* (*TFL1*) genes (Pnueli et al. 1998).

Many mutations in genes related to the carotenoid pathway were identified and correspond to specific fruit colours (Hirschberg 2001). Among them the *B/ogc* locus has been shown to correspond to two mutations in the same gene responsible for either yellow or dark red colour of the fruit (Ronen et al. 2000). Recently the gene conferring the uniform ripening (*u*) phenotype was cloned and shown to correspond to a Golden 2-like (GLK) transcription factor, which determines the chlorophyll accumulation and distribution in developing fruit (Powell et al. 2012). The *y* locus, responsible for the pink fruit colour (due to a colourless peel which lacks the yellow flavonoid pigment naringenin chalcone), was also cloned. It corresponds to a MYB transcription factor (Adato et al. 2009; Ballester et al. 2010). Several alleles and their polymorphisms were identified at the *y* locus, thanks to the recent resequencing of more than 300 tomato accessions (Lin et al. 2014). Several mutations confer a long shelf life to the fruit. The most widely used, *rin* (for ripening inhibitor) corresponds to a deletion in a MADS BOX transcription factor (Vrebalov et al. 2002). Another important discovery was the mutation at the *Cnr* locus (Colourless non-ripening), which was one of the first epiallele discovered in tomato (Manning et al. 2006). Table 3.1 lists the genes involved in morphological and fruit mutations.

Disease Resistance Genes

Tomato is susceptible to many pathogens and all the resistance genes (R) were discovered in wild relatives. Many tomato disease resistance genes were mapped and characterized (Table 3.2). Since the first positionally cloned R gene (*Pto*, by Martin et al. 1993), more than 20 genes were cloned and characterized. Their structure and evolution was analyzed and the great conservation among genes conferring resistance to different types of pathogens revealed. The majority of R genes cloned so far encode proteins with a nucleotide-binding site (NBS) and a leucine-rich repeat (LRR) region (Ellis et al. 2000).

Mutant Collections

Many natural mutations were discovered in tomato. The Tomato Genetic Resources Center (TGRC, Davis, California, USA) collection encompasses more than 1000 monogenic mutants at over 600 loci, including spontaneous and induced mutations affecting many aspects of plant development and morphology, disease resistance genes, protein marker stocks, and other traits of economic importance (Chetelat 2005). Genetic data on individual stocks, including phenotypes, images, chromosome locations, etc. are available at the TGRC website (http://tgrc.ucdavis.edu/).

An additional series of provisional (i.e. less well-characterized) mutants is also available. The Hebrew University of Jerusalem developed an isogenic mutant library in the genetic background of cv. M82 (http://zamir.sgn.cornell.edu/mutants/index.html). A total of 13,000 M_2 families, generated by ethylmethane sulfonate (EMS) and fast-neutron mutagenesis, were phenotypically analyzed and catalogued into at least 3417 mutations (Menda et al. 2004). This series of mutations includes many previously described mutant phenotypes as well as many novel mutants, and multiple alleles per locus. Screening this collection allowed the discovery of interesting alleles which interact with the *SP* gene and whose mutation modify its expression and may allow optimization of crop productivity (Park et al. 2014). Other collections of mutants are available (Okabe et al. 2011). Together these mutant collections provide important tools for analyses of gene function either through forward or reverse genetic approaches (Chap. 5).

Recombination Heterogeneity

Many genes/mutations were mapped on a genetic map but not yet cloned (Table 3.3). The recent availability of the tomato genome sequence confirmed earlier observations that recombination is unevenly distributed along chromosomes and that large pieces of the chromosomes around the centromeres do not recombine at all

Table 3.1 Cloned genes with a phenotyped mutant mapped on the tomato genome assembly

ITAG gene model	Gene symbol	locus_name	Chromosome	Start	End	Phenotypic descriptors	References
Solyc01g008930	au	AUREA phytochromobilin synthase	1	2,948,574	2,955,890	Phytochrome-deficient	Muramoto et al. (2005)
Solyc01g056340	hp-2	De-etiolated 1	1	46,495,644	46,516,174	High pigment	Mustilli et al. (1999)
Solyc01g059870	phyB1	Apophytochrome B1	1	61,760,931	61,767,869	Red light reception	Weller et al. (2001)
Solyc01g079620	y	Colourless epidermis*	1	71,255,600	71,258,882	Pink epidermis	Ballester et al. (2010)
Solyc01g100490	chln	Chloronerva	1	82,262,052	82,263,287	Chlorophyll deficiency	Ling et al. (1999)
Solyc01g104340	gr	Green ripe	1	84,508,287	84,509,191	Reduces ethylene sensitivity in fruit	Barry and Giovannoni (2006)
Solyc02g021650	hp-1	UV damaged DNA binding protein 1	2	14,069,796	14,090,192	High pigment fruit pericarp	Lieberman et al. (2004); Liu et al. (2004)
Solyc02g077390	s	Compound inflorescence	2	36,913,957	36,915,889	Inflorescence branching	Lippman et al. (2008)
Solyc02g077920	Cnr	Colourless non-ripening	2	37,323,107	37,320,931	Inhibition of ripening	Manning et al. (2006)
Solyc02g080250	Wo	Wooly	2	39,094,298	39,095,666	High trichome density	Yang et al. (2011)
Solyc02g081120	Me	Knotted 2	2	39,767,063	39,773,953	Leaf complexity	Parnis et al. (1997)
Solyc02g081670	an	Anantha	2	40,120,235	40,121,602	Compound inflorescence, aborted flowers	Lippman et al. (2008)
Solyc02g089160	d	Dwarf	2	45,622,114	45,624,672	Dwarf plant, dpy	Bishop et al. (1996)
Solyc02g090890	hp-3	Zeaxanthin epoxidase	2	46,947,557	46,953,158	High pigment in fruits	Thompson et al. (2000)
Solyc03g007960	wf	Beta-carotene hydroxylase-2	3	2,447,949	2,450,014	White flower	Galpaz et al. (2006)
Solyc03g031860	r	Phytoene synthase 1	3	8,606,749	8,610,050	Yellow fruit	Fray and Grierson (1993)
Solyc03g063100	sft	Single flower truss	3	30,564,833	30,568,648	Single flower truss	Molinero-Rosales et al. (2004)

(continued)

Table 3.1 (continued)

ITAG gene model	Gene symbol	locus_name	Chromosome	Start	End	Phenotypic descriptors	References
Solyc03g083910	sucr	Sucrose accumulator	3	47,401,871	47,397,595	Accumulates predominantly sucrose in mature fruit, rather than glucose and fructose	Sato et al. (1993)
Solyc03g118160	fa	Falsiflora	3	61,162,449	61,164,404	Leafy inflorescence	Molinero-Rosales et al. (1999)
Solyc03g119060	div	Divaricata	3	61,827,331	61,831,038	Small squarrose plant with intercostally yellowish leaves and ventrally purple	van der Biezen et al. (1996)
Solyc03g119770	SlBrc1a	Branched1a	3	62,381,910	62,383,042	Shoot branching	Martín-Trillo et al. (2011)
Solyc04g051510	cu-3	Curl-3	4	49,870,634	49,874,257	Dwarf	Montoya et al. (2002)
Solyc04g074180	cry1a	Cryptochrome 1A	4	57,772,528	57,777,909	Blue light reception	Weller et al. (2001)
Solyc04g076850	e	Entire	4	59,354,677	59,358,365	Reduced leaf complexity	Zhang et al. (2007)
Solyc04g082520	cwp1	Cuticular water permeability 1	4	63,765,366	63,766,988	Microfissure/dehydration of fruits	Hovav et al. (2007)
Solyc05g005020	gwd	Alpha-glucan water dikinase	5	32,905	50,320	Starch excess phenotype and reduced pollen germination	Nashilevitz et al. (2009)
Solyc05g009380	lyr	Lyrate	5	3,536,207	3,540,567	Reduced leaf complexity	David-Schwartz et al. (2009)
Solyc05g012020	rin	Ripening inhibitor	5	5,217,073	5,230,708	Never ripening	Vrebalov et al. (2002)
Solyc05g012020	mc	Macrocalyx	5	5,217,073	5,230,708	Large sepals	Vrebalov et al. (2002)
Solyc05g053410	phyB2	Apophytochrome B2	5	62,648,223	62,653,411	Red light reception	Weller et al. (2001)
Solyc06g051550	fe	fe inefficient	6	31,547,361	31,549,010	Iron deficiency	Ling et al. (2002)
Solyc06g069240	SlBrc1b	Branched1b	6	39,396,681	39,398,027	Shoot branching	Martín-Trillo et al. (2011)
Solyc06g074240	B	Beta-carotene	6	42,288,127	42,289,623	Increased fruit beta-carotene	Ronen et al. (2000)
Solyc06g074350	sp	Self-pruning	6	42,361,623	42,363,883	Determinate plant habit	Pnueli et al. (1998)
Solyc06g074910	C	Potato leaf	6	42,804,036	42,806,196	Simple leaves	Busch et al. (2011)
Solyc07g056570	not	Notabilis	7	61,684,846	61,686,663	ABA deficiency. Wilty	Burbidge et al. (1999)

(continued)

Table 3.1 (continued)

ITAG gene model	Gene symbol	locus_name	Chromosome	Start	End	Phenotypic descriptors	References
Solyc07g062680	La	Lanceolate	7	62,593,583	62,594,785	Simple leaves	Ori et al. (2007)
Solyc07g062840	gob	Goblet	7	62,710,395	62,710,928	Shoot apical meristem terminates but occasionally partially recovers	Berger et al. (2009)
Solyc07g066250	ls	Lateral suppresser	7	64,958,148	64,959,434	Few or no axillary branches; corolla suppressed; partially male sterile	Schumacher et al. (1999)
Solyc07g066480	flacca	Flacca	7	65,118,760	65,130,514	Wilty	Sagi et al. (2002)
Solyc08g080090	Gr	Green flesh	8	60,582,066	60,579,438	Green fruit flesh	Barry et al. (2008)
Solyc09g075440	Nr	Never ripe	9	62,631,866	62,639,953	Not ripening	Wilkinson et al. (1995)
Solyc10g044670	phyA	Apophytochrome A	10	22,854,459	22,859,333	Far red light insensitive	Weller et al. (2001)
Solyc10g081650	t	Carotenoid isomerase	10	62,006,972	62,011,520	Orange fruit flesh	Isaacson et al. (2002)
Solyc10g081470	L-2	Lutescent-2	10	61,858,478	61,851,435	Altered chloroplast development and delayed ripening	Barry et al. (2012)
Solyc10g008160	*u*	Fruit ripening *(uniform ripening)*	10	2,293,088	2,295,824	Increased chlorophyl content	Powell et al. (2012)
Solyc11g010570	j	Jointless	11	3,640,857	3,645,766	No pedicel abscission zone	Mao et al. (2000)
Solyc11g011260	pro	Procera	11	4,303,769	4,305,535	Suppressed axillary bud development and altered branching architecture	Bassel et al. (2008)
Solyc11g011990	ghost	Plastid terminal oxidase	11	4,937,989	4,942,674	White seedlings	Josse et al. (2000)
Solyc11g069030	bl	Blind	11	50,686,745	50,688,284	Stem terminating in first inflorescence; midribs may develop adventitious shoots	Schmitz et al. (2002)
Solyc12g008980	Del	Del	12	2,285,372	2,290,327	Orange fruit	Ronen et al. (1999)
Solyc12g009470	Yg2	Heme oxygenase 1	12	2,726,504	2,729,459	Phytochrome-deficient	Terry and Kendrick (1996)
*	ht-a	HT-A	12	47,894,134	47,896,625	Self incompatibility gene from *S. peruvianum*	Kondo et al. (2002)
*	ht-b	HT-B	12	47,896,590	47,896,869	Self incompatibility gene from *S. peruvianum*	Kondo et al. (2002)

*indicate loci that do not have a corresponding Solyc gene

Table 3.2 Disease resistance genes cloned

ITAG gene model	Gene symbol	locus_name	Chromosome	Start	End	Phenotypic descriptors	References
*	cf-9	Cladosporium fulvum resistance-9	1	3,972,616	3,975,213	Defense responses to specific races of *Cladosporium fulvum*	Jones et al. (1994)
Solyc02g062560	tm-1	Tobacco mosaic virus resistance-1	2	28,856,001	28,864,037	TMV resistance	Ishibashi et al. (2007)
Solyc03g005870	pot	Eukaryotic translation initiation factor 4E (eIF4E)	3	590,087	59,328	Potyvirus immunity	Piron et al. (2010)
Solyc03g114600	asc	Alternaria stem canker resistance	3	58,592,692	58,594,435	Alternaria stem canker resistance	Mesbah et al. (1999)
*	Hero	Heterodera rostochiensis resistance	4	1,795,425	1,799,656	Cyst nematode resistance	Ernst et al. (2002)
Solyc05g013280	pfr	Pseudomonas resistance	5	6,379,823	6,385,909	Pseudomonas resistance	Salmeron et al. (1996)
Solyc05g013290	Fen	Fenthion	5	6,387,418	6,388,380	Confers sensitivity to the insecticide Fenthion	Martin et al. (1994)
Solyc05g013300	pto	Pseudomonas syringae pv tomato resis.	5	6,404,160	6,405,095	Resistance to bacterial speck disease	Martin et al. (1994)
Solyc05g052620	coi1	Coronatine-insensitive 1	5	61,968,745	61,974,317	Jasmonate signalling in *Pseudomonas syringae* response	Katsir et al. (2008)
Solyc05g007850	bs-4	Bacterial spot disease resistance protein 4	5	62,136,023	62,140,081	Mediates HR against *Xanthomonas campestris pv. vesicatoria*	Schornack et al. (2004) and Ballvora et al. (2001)
Solyc06g008270	Cf2.2	Cladosporium fulvum resistance 2.2	6	2,139,526	2,142,150	*Cladosporium fulvum* resistance	Dixon et al. (1996)
Solyc06g008300	cf2	Cladosporium fulvum resistance 2.1	6	2,164,746	2,161,344	*Cladosporium fulvum* resistance	Dixon et al. (1996)

(continued)

Table 3.2 (continued)

ITAG gene model	Gene symbol	locus_name	Chromosome	Start	End	Phenotypic descriptors	References
*	Mi-1.1	Mi-1.1	6	2,327,468	2,331,298	Unknown	Bhattarai et al. (2007)
*	Mi-1.2	Mi-1.2	6	2,354,799	2,358,629	*Meloidogyne incognita* resistance	Kaloshian et al. (1998)
Solyc06g051170, Solyc06g051180, and Solyc06g051190	Ty-1/Ty-3	TYLCV	6	30,862,817	30,879,542	Tomato yellow leaf curl virus	Verlaan et al. (2013)
Solyc09g005080	Ve2	Verticillium wilt disease resistance	9	48,645	52,064	verticillium wilt disease resistance	Kawchuk et al. (2001)
Solyc09g005090	Ve1	Verticillium wilt disease resistance	9	55,478	58,639	Verticillium wilt disease resistance	Kawchuk et al. (2001)
*	Tm-2	Tobacco mosaic virus resistance-2	9	13,621,396	13,623,981	TMV resistance	Lanfermeijer et al. (2003)
Solyc09g098130	Sw-5	Spotted wilt resistance-5	9	67,301,675	67,305,412	Spotted wilt resistance	Brommonschenkel et al. (2000)
Solyc09g092280-Solyc09g092310	Ph-3	Phytophtora infestans	9L	66,764,694	66,795,551	Late blight resistance	Zhang et al. (2013, 2014)
Solyc11g071430	I-2	Immunity to Fusarium wilt race 2	11	51,992,917	51,996,706	Fusarium resistance	Ori et al. (1997)

*indicate loci that do not have a corresponding Solyc gene

Table 3.3 Genes mapped on a genetic map with a phenotype but not yet cloned

Gene symbol	Phenotypic descriptors	Chromosome	References
S	Self incompatibility	1	Tanksley and Loaiza-Figueroa (1985) and Rivers et al. (1993)
ms-10	Male sterility	2	Tanksley et al. (1992)
af	Anthocyanin free	5	Rick 1980 (cited by Tanksley et al. 1992)
tf	Trifoliate	5	Rick 1980 (cited by Tanksley et al. 1992)
ae	Entirely anthocyaninless	8	Rick 1980 (cited by Tanksley et al. 1992)
h	Hairs absent	10	Rick 1980 (cited by Tanksley et al. 1992)
ag	Anthocyanin gainer	10	Rick 1980 (cited by Tanksley et al. 1992)
hl	Hairless	11	Rick 1980 (cited by Tanksley et al. 1992)
a	Anthocianinless	11	Rick 1980 (cited by Tanksley et al. 1992)
alb	Albescent	12	Rick 1980 (cited by Tanksley et al. 1992)
alc	Fruit ripening *(alcobaca)*	10	Kinzer et al. (1990)
nor	Fruit ripening *(non-ripening)*	10	Moore et al. (2002)
j-2	Jointless	12	Budiman et al. (2004)
Disease resistance genes mapped but not yet cloned			
Cf-4	Cladosporium fulvum (leaf mold)	1	Thomas et al. (1997)
Cf-1	Cladosporium fulvum (leaf mold)	1	Jones et al. (1993)
rx-1, rx-2,	Hypersensitive reaction	1	Yu et al. (1995)
Cf-ECP2, Cf-ECP3	Cladosporium fulvum (leaf mold)	1	Haanstra et al. (1999) and Yuan et al. (2002)
Cf-ECP5	Cladosporium fulvum (leaf mold)	1	Haanstra et al. (2000)
Cf-ECP1, Cf-ECP4	Cladosporium fulvum (leaf mold)	1	Soumpourou et al. (2007)
I-5,	*Fusarium oxysporum f. sp.lycopersici* (race 2)	2	Sela-Buurlage et al. (2001)
Xv4	*Xanthomonas campestris pv. vesicatoria (race T3*	3	Astua-Monge et al. (2000)
pot-1	*Potato virus Y* (PVY) and *Tobacco etch virus* (TEV)	3S	Parrella et al. (2002a, b) and Ruffel et al. (2005)
py-1	*Pyrenochaeta lycopersici (corky root)*	3S	Doganlar et al. (1998)
ol-2	*Oidium neolycopersici (Powdery Mildew)*	4C	De Giovanni et al. (2004) and Bai et al. (2008)
rx-3	*Xanthomonas campestris* pv. *vesicatoria*	5	Yu et al. (1995)
Ol-1	*Oidium neolycopersici (Powdery Mildew)*	6L	Huang et al. (2000a) and Bai et al. (2005)
Ol-3	*Oidium neolycopersici (Powdery Mildew)*	6L	Huang et al. (2000b) and Bai et al. (2005)
Ol-4	*Oidium neolycopersici (Powdery Mildew)*	6	Bai et al. (2004, 2005)
Ol-5	*Oidium neolycopersici (Powdery Mildew)*	6L	Bai et al. (2005)
Am	Alfalfa Mosaic Virus (AMV)	6S	Parrella et al. (2004)

(continued)

Table 3.3 (continued)

Gene symbol	Phenotypic descriptors	Chromosome	References
Cf-5	Cladosporium fulvum (leaf mold)	6S	Dixon et al. (1998)
Mi-9	*Meloydogine* spp. nematode (root-knot)	6S	Jablonska et al. (2007)
I-3	*Fusarium oxysporum f. sp.lycopersici (race 3) (fusarium wilt)*	7L	Hemming et al. (2004) and Lim et al. (2008)
I-1	*Fusarium oxysporum f. sp.lycopersici* (race 1) (fusarium wilt)	7	Sarfatti et al. (1991) and Scott et al. (2004)
Frl	*Fusarium oxysporum* f. sp. *radici-lycopersici (root rot)*	9	Vakalounakis et al. (1997)
I-6	*Fusarium oxysporum f. sp.lycopersici (race 2) (fusarium wilt)*	10	Sela-Buurlage et al. (2001)
Ph-2	*Phytophtora infestans (late blight)*	10L	Moreau et al. (1998)
al	*Fusarium oxysporum* f. sp. *lycopersici* (race 1) (fusarium wilt)	11S	Scott et al. (2004)
Sm	*Stemphyllium spp. (grey leaf spot)*	11	Behare et al. (1991)
Cmr	Cucumber Mosaic Virus (CMV)	12	Stamova and Chetelat (2000)
Lv	*Leveillula taurica*	12C	Chunwongse et al. (1994, 1997)
Mi-3, Mi-5	*Meloidogyne* spp (nematode)	12S	Yaghoobi et al. (1995)

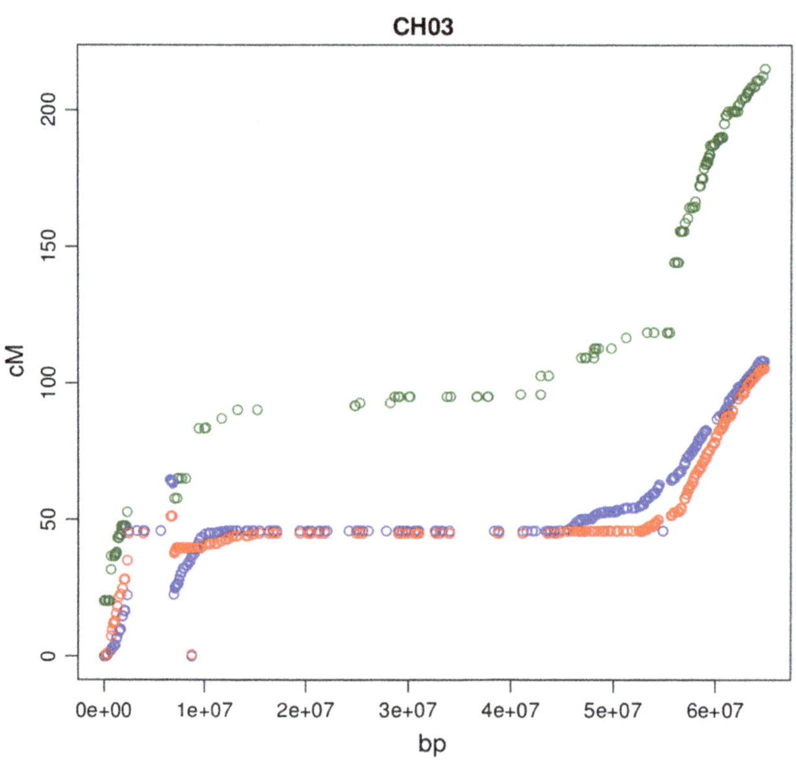

Fig. 3.1 Relationships between physical and genetic distances: example of chromosome 3

(Sim et al. 2012; Tomato Genome Consortium 2012). If the recombination frequencies may vary from one progeny to the other (Fig. 3.1), these regions do not recombine more in any. The ratio of kb per cM thus greatly varies hampering the characterization of some mutations due to the lack of recombination. Hopefully, these regions of low recombination also correspond to regions with lower gene density.

Many genes involved in morphological traits or disease resistances remain to be characterized. The high-quality genome sequence and millions of SNPs available today constitute unique resources to rapidly identify new genes of interest. High throughput genotyping technologies combined to the information on gene annotation and expression in various tissues should make the task much easier.

References

Adato A, Mandel T, Mintz-Oron S et al (2009) Fruit-surface flavonoid accumulation in tomato is controlled by a SlMYB12-regulated transcriptional network. PLoS Genet 5:e1000777

Aflitos SA, Schijlen EGWM, de Jong JHSGM et al (2014) Exploring genetic variation in the tomato (*Solanum* section *Lycopersicon*) clade by whole-genome sequencing. Plant J 80:136–148

Astua-Monge G, Minsavage GV, Stall RE et al (2000) *Xv4-vrxv4*: a new gene-for-gene interaction identified between *Xanthomonas campestris* pv. *vesicatoria* Race T3 and the wild tomato relative *Lycopersicon pennellii*. Mol Plant Microbe Interact 13:1346–1355

Bai Y, van der Hulst R, Huang CC et al (2004) Mapping *Ol-4*, a gene conferring resistance to *Oidium neolycopersici* and originating from *Lycopersicon peruvianum* LA2172, requires muliallelic, single-locus markers. Theor Appl Genet 109:1215–1223

Bai Y, van der Hulst R, Bonnema G et al (2005) Tomato defense to *Oidium neolycopersici*: dominant *Ol* genes confer isolate-dependent resistance via a different mechanism than recessive *ol-2*. Mol Plant Microbe Interact 18(4):354–362

Bai Y, Pavan S, Zheng Z et al (2008) Naturally occurring broad-spectrum powdery mildew resistance in a Central American tomato accession is caused by loss of *Mlo* function. Mol Plant Microbe Interact 21(1):30–39

Ballester AR, Molthoff J, de Vos R et al (2010) Biochemical and molecular analysis of pink tomatoes: deregulated expression of the gene encoding transcription factor SlMYB12 leads to pink tomato fruit color. Plant Phys 152:71–84

Ballvora A, Pierre M, van den Ackerveken G et al (2001) Genetic mapping and functional analysis of the tomato *Bs4* locus governing recognition of the *Xanthomonas campestris pv. vesicatoria* AvrBs4 protein. Mol Plant Microbe Interact 14:629–638

Barry CS, Giovannoni JJ (2006) Ripening in the tomato *Green-ripe* mutant is inhibited by ectopic expression of a protein that disrupts ethylene signaling. Proc Natl Acad Sci USA 103(20):7923–7928

Barry CS, McQuinn RP, Chung MY et al (2008) Amino acid substitutions in homologs of the STAY-GREEN protein are responsible for the *green-flesh* and *chlorophyll retainer* mutations of tomato and pepper. Plant Physiol 147(1):179–187

Barry CS, Aldridge GM, Herzog G et al (2012) Altered chloroplast development and delayed fruit ripening caused by mutations in a zinc metalloprotease at the *lutescent2* locus of tomato. Plant Physiol 159 (3):1086–1098

Bassel GW, Mullen RT, Bewley JD (2008) *procera* is a putative DELLA mutant in tomato (*Solanum lycopersicum*): effects on the seed and vegetative plant. J Exp Bot 59(3):585–593

Behare J, Laterrot H, Sarfatti M et al (1991) Restriction fragment length polymorphisms mapping of the Stemphylium resistance gene in tomato. Mol Plant Microbe Interact 4:489–492

Berger Y, Harpaz-Saad S, Brand A, Melnik H, Sirding N, Alvarez JP, Zinder M, Samach A, Eshed Y, Ori N (2009) The NAC-domain transcription factor GOBLET specifies leaflet boundaries in compound tomato leaves. Development 136(5):823–832

Bhattarai KK, Li Q, Liu Y, Dinesh-Kumar SP, Kaloshian I (2007) The *Mi-1*-mediated pest resistance requires *Hsp90* and *Sgt1*. Plant Physiol 144(1):312–323

Bishop GJ, Harrison K, Jones JD (1996) The tomato *Dwarf* gene isolated by heterologous transposon tagging encodes the first member of a new cytochrome P450 family. Plant Cell 8(6):959–969

Brommonschenkel SH, Frary A, Tanksley SD (2000) The broadspectrum tospovirus resistance gene *Sw-5* of tomato is a homolog of the root-knot nematode resistance gene *Mi*. Mol Plant Microbe Interact 13:1130–1138

Budiman MA, Chang S-B, Lee S et al (2004) Localization of *jointless*-2 gene in the centromeric region of tomato chromosome 12 based on high resolution genetic and physical mapping. Theor Appl Genet 108:190–196

Burbidge A, Grieve TM, Jackson A et al (1999) Characterization of the ABA-deficient tomato mutant notabilis and its relationship with maize *Vp14*. Plant J 17(4):427–431

Busch BL, Schmitz G, Rossmann S, Piron F, Ding J, Bendahmane A, Theres K (2011) Shoot branching and leaf dissection in tomato are regulated by homologous gene modules. Plant Cell 23(10):3595–3609

Butler L (1952) The linkage map of the tomato. J Heredity 43:25–35

Causse M, Desplat N, Pascual L, Le Paslier MC, Sauvage C, Bauchet G et al (2013) Whole genome resequencing in tomato reveals variation associated with introgression and breeding events. BMC Genom 14:791

Chetelat RT (2005) Revised list of monogenic stocks. Rep Tomato Genet Coop 55:48–69

Chunwongse J, Bunn TB, Crossman C et al (1994) Chromosomal localization and molecular-marker tagging of the powdery mildew resistance gene (Lv) in tomato. Theor Appl Genet 89:76–79

Chunwongse S, Doganlar C, Crossman JJ et al (1997) High-resolution genetic map of the *Lv* resistance locus in tomato. Theor Appl Genet 95:220–223

David-Schwartz R, Koenig D, Sinha N (2009) LYRATE is a key regulator of leaflet initiation and lamina outgrowth in tomato. Plant Cell 21:3093–3104

De Giovanni C, Dell'Orco P, Bruno A et al (2004) Identification of PCR-based markers (RAPD, AFLP) linked to a novel powdery mildew resistance gene (*ol-2*) in tomato. Plant Sci 166:41–48

Dixon MS, Jones DA, Keddie JS et al (1996) The tomato *Cf-2* disease resistance locus comprises two functional genes encoding leucine-rich repeat proteins. Cell 84:451–459

Dixon MS, Hatzixanthis K, Jones DA, Harrison K, Jones JD (1998) The tomato *Cf-5* disease resistance gene and six homologs show pronounced allelic variation in leucine-rich repeat copy number. Plant Cell 10(11):1915–1925

Doganlar S, Dodson J, Gabor B et al (1998) Molecular mapping of the *py-1* gene for resistance to corky root rot (*Pyrenochaeta lycopersici*) in tomato. Theor Appl Genet 97:784–788

Ellis J, Dodds P, Pryor T (2000) Structure, function and evolution of plant disease resistance genes. Curr Opin Plant Biol 3:278–284

Ernst K, Kumar A, Kriseleit D et al (2002) The broad-spectrum potato cyst nematode resistance gene (*Hero*) from tomato is the only member of a large gene family of NBS-LRR genes with an unusual amino acid repeat in the LRR region. Plant J 31(2):127–136

Eshed Y, Zamir D (1995) An introgression line population of *Lycopersicon pennellii* in the cultivated tomato enables the identification and fine mapping of yield associated QTL. Genetics 141:1147–1162

Fray RG, Grierson D (1993) Identification and genetic analysis of normal and mutant phytoene synthase genes of tomato by sequencing, complementation and co-suppression. Plant Mol Biol 22(4):589–602

Fulton TM, van der Hoeven R, Eanetta NT et al (2002) Identification, analysis, and utilization of conserved ortholog set markers for comparative genomics in higher plants. Plant Cell 14:1457–1467

Galpaz N, Ronen G, Khalfa Z, Zamir D et al (2006) A chromoplast-specific carotenoid biosynthesis pathway is revealed by cloning of the tomato *white-flower* locus. Plant Cell 18(8):1947–1960

Grandillo S, Tanksley SD (1996) Genetic analysis of RFLPs, GATA microsatellites and RAPDs in a cross between *L. esculentum* and *L. pimpinellifolium*. Theor Appl Genet 92:957–965

Haanstra JPW, Wye C, Verbakel H et al (1999) An integrated high-density RFLP-AFLP map of tomato based on two *Lycopersicon esculentum* × *L. pennellii* F2 populations. Theor Appl Genet 99:254–271. doi:10.1007/s001220051231

Haanstra JPW, Meijer-Dekens F, Laugé R et al (2000) Mapping strategy for resistance genes against *Cladosporium fulvum* on the short arm of chromosome 1 of tomato: *Cf-ECP5* near the *Hcr9* milky way cluster. Theor Appl Genet 101:661–668

He C, Poysa V, Yu K (2002) Development and characterization of simple sequence repeat (SSR) markers and their use in determining relationships among *Lycopersicon esculentum* cultivars. Theor Appl Genet 106:363–373

Hemming MN, Basuki S, McGrath DJ et al (2004) Fine mapping of the tomato *I-3* gene for fusarium wilt resistance and elimination of a co-segregating resistance gene analogue as a candidate for *I-3*. Theor Appl Genet 109(2):409–418

Hirschberg J (2001) Carotenoid biosynthesis in flowering plants. Curr Opin Plant Biol 4:210–218

Hovav R, Chehanovsky N, Moy M et al (2007) The identification of a gene (*Cwp1*), silenced during Solanum evolution, which causes cuticle microfissuring and dehydration when expressed in tomato fruit. Plant J 52(4):627–639

Huang CC, Cui YY, Weng CR et al (2000a) Development of diagnostic PCR markers closely linked to the tomato powdery mildew resistance gene *Ol-1* on chromosome 6 of tomato. Theor Appl Genet 101:918–924

Huang CC, Hoefs-Van De Putte PM, Haanstra-Van Der Meer JG et al (2000b) Characterization and mapping of resistance to *Oidium lycopersicum* in two *Lycopersicon hirsutum* accessions: evidence for close linkage of two *Ol*-genes on chromosome 6 of tomato. Heredity 85:511–520

Isaacson T, Ronen G, Zamir D et al (2002) Cloning of *tangerine* from tomato reveals a carotenoid isomerase essential for the production of beta-carotene and xanthophylls in plants. Plant Cell 14(2):333–342

Ishibashi K, Masuda K, Naito S et al (2007) An inhibitor of viral RNA replication is encoded by a plant resistance gene. Proc Natl Acad Sci USA 104:13833–13838

Jablonska B, Ammiraju JSS, Bhattarai KK et al (2007) The *Mi-9* gene from *Solanum arcanum* conferring heat-stable resistance to root-knot nematodes is a homolog of *Mi-1*. Plant Physiol 143:1044–1054

Jones DA, Dickinson MJ, Balint-Kurti PJ et al (1993) Two complex resistance loci revealed in tomato by classical and RFLP mapping of the *Cf-2*, *Cf-4*, *Cf-5*, and *Cf-9* genes for resistance to *Cladosporium fulvum*. Mol Plant Microbe Interact 6:348–357

Jones DA, Thomas CM, Hammond-Kosack KE et al (1994) Isolation of the tomato Cf-9 gene for resistance to *Cladosporium fulvum* by transposon tagging. Science 266:789–793

Josse EM, Simkin AJ, Gaffé J et al (2000) A plastid terminal oxidase associated with carotenoid desaturation during chromoplast differentiation. Plant Physiol 123(4):1427–1436

Kaloshian I, Yaghoobi J, Liharska T et al (1998) Genetic and physical localization of the root-knot nematode resistance locus *Mi* in tomato. Mol Gen Genet 257 (3):376–385

Katsir L, Schilmiller AL, Staswick PE et al (2008) COI1 is a critical component of a receptor for jasmonate and the bacterial virulence factor coronatine. Proc Natl Acad Sci USA 105(19):7100–7105

Kawchuk LM, Hachey J, Lynch DR et al (2001) Tomato *Ve* disease resistance genes encode cell surface-like receptors. Proc Natl Acad Sci USA 98(11):6511–6515

Kinzer SM, Schwager SJ, Mutschler MA (1990) Mapping of ripening-related or -specific cDNA clones of tomato (*Lycopersicon esculentum*). Theor Appl Genet 79:489–496

Kondo K, Yamamoto M, Matton DPY et al (2002) Cultivated tomato has defects in both S-RNase and HT genes required for stylar function of self-incompatibility. Plant J 29(5):627–636

Labate JA, Baldo AM (2005) Tomato SNP discovery by EST mining and resequencing. Mol Breed 16:343–349

Labate JA, Grandillo S, Fulton T et al (2007) Tomato. In: Kole C (ed) Genome mapping and molecular breeding in plants, vol 5, Vegetables. Springer, Berlin, pp 11–135

Lanfermeijer FC, Dijkhuis J, Sturre MJ et al (2003) Cloning and characterization of the durable tomato mosaic virus resistance gene *Tm-2(2)* from *Lycopersicon esculentum*. Plant Mol Biol 52(5):1037–1049

Laterrot H (1996) Stock List. Rep Tom Genet Coop 46:34

Lieberman M, Segev O, Gilboa N et al (2004) The tomato homolog of the gene encoding UV-damaged DNA binding protein 1 (DDB1) underlined as the gene that causes the *high pigment-1* mutant phenotype. Theor Appl Genet 108(8):1574–1581

Lim GT, Wang GP, Hemming MN et al (2008) High resolution genetic and physical mapping of the *I-3* region of tomato chromosome *7* reveals almost continuous microsynteny with grape chromosome *12* but interspersed microsynteny with duplications on *Arabidopsis* chromosomes *1*, *2* and *3*. Theor Appl Genet 118(1):57–75

Lin T, Zhu G, Zhang J et al (2014) Genomic analyses provide insights into the history of tomato breeding. Nat Genet 46:1220–1226

Ling HQ, Koch G, Bäumlein H et al (1999) Map-based cloning of *chloronerva*, a gene involved in iron uptake of higher plants encoding nicotianamine synthase. Proc Natl Acad Sci USA 96(12):7098–7103

Ling HQ, Bauer P, Bereczky Z et al (2002) The tomato *fer* gene encoding a bHLH protein controls iron-uptake responses in roots. Proc Natl Acad Sci USA 99 (21):13938–13943

Lippman ZB, Cohen O, Alvarez JP et al (2008) The making of a compound inflorescence in tomato and related nightshades. PLoS Biol 11:e288

Liu Y, Roof S, Ye Z et al (2004) Manipulation of light signal transduction as a means of modifying fruit nutritional quality in tomato. Proc Nat Acad Sci USA 101:9897–9902

Liu Y, Chen H, Wei Y et al (2005) Construction of a genetic map and localization of QTLs for yield traits in tomato by SSR markers. Prog Nat Sci 15:793–797

Manning K, Tör M, Poole M et al (2006) A naturally occurring epigenetic mutation in a gene encoding an SBP-box transcription factor inhibits tomato fruit ripening. Nat Genet 38(8):948–952

Mao L, Begum D, Chuang HW et al (2000) *JOINTLESS* is a MADS-box gene controlling tomato flower abscission zone development. Nature 406 (6798):910–913

Martin GB, Brommonschenkel S, Chunwongse J et al (1993) Map-based cloning of a protein-kinase gene conferring disease resistance in tomato. Science 262:1432–1436

Martin GB, Frary A, Wu T et al (1994) A member of the tomato Pto gene family confers sensitivity to fenthion resulting in rapid cell death. Plant Cell 6(11):1543–1552

Martín-Trillo M, Grandío EG, Serra F et al (2011) Role of tomato *BRANCHED1*-like genes in the control of shoot branching. Plant J 67(4):701–714

Menda N, Semel Y, Peled D et al (2004) *In silico* screening of a saturated mutation library of tomato. Plant J 38:861–872

Mesbah LA, Kneppers TJ, Takken FL et al (1999) Genetic and physical analysis of a YAC contig spanning the fungal disease resistance locus *Asc* of tomato (*Lycopersicon esculentum*). Mol Gen Genet 261(1):50–57

Molinero-Rosales N, Jamilena M, Zurita S et al (1999) *FALSIFLORA*, the tomato orthologue of *FLORICAULA* and *LEAFY*, controls flowering time and floral meristem identity. Plant J 20(6):685–693

Molinero-Rosales N, Latorre A, Jamilena M et al (2004) *SINGLE FLOWER TRUSS* regulates the transition and maintenance of flowering in tomato. Planta 218 (3):427–434

Montoya T, Nomura T, Farrar K et al (2002) Cloning the tomato curl3 gene highlights the putative dual role of the leucine-rich repeat receptor kinase tBRI1/SR160 in plant steroid hormone and peptide hormone signaling. Plant Cell 14(12):3163–3176

Moore S, Vrebalov J, Payton P et al (2002) Use of genomics tools to isolate key ripening genes and analyse fruit maturation in tomato. J Exp Bot 53 (377):2023–2030

Moreau P, Thoquet P, Olivier J et al (1998) Genetic Mapping of *Ph-2*, a single locus controlling partial resistance to Phytophthora *infestans* in tomato. Mol Plant Microbe Interact 11(4):259–269

Muramoto T, Kami C, Kataoka H et al (2005) The tomato photomorphogenetic mutant, *aurea*, is deficient in phytochromobilin synthase for phytochrome chromophore biosynthesis. Plant Cell Physiol 46(4):661–665

Mustilli AC, Fenzi F, Ciliento R et al (1999) Phenotype of the tomato *high pigment-2* mutant is caused by a mutation in the tomato homolog of *DEETIOLATED1*. Plant Cell 11:145–157

Nashilevitz S, Melamed-Bessudo C, Aharoni A et al (2009) The *legwd* mutant uncovers the role of starch phosphorylation in pollen development and germination in tomato. Plant J 57(1):1–13

Okabe Y, Asamizu E, Saito T et al (2011) Tomato TILLING technology: development of a reverse genetics tool for the efficient isolation of mutants from Micro-Tom mutant libraries. Plant Cell Physiol 52(11):1994–2005

Ori N, Eshed Y, Paran I et al (1997) The *I2C* family from the wilt disease resistance locus *I2* belongs to the nucleotide binding, leucine-rich repeat superfamily of plant resistance genes. Plant Cell 9(4):521–532

Ori N, Cohen AR, Etzioni A et al (2007) Regulation of *LANCEOLATE* by miR319 is required for compound-leaf development in tomato. Nat Genet 39(6):787–791

Park SJ, Jiang K, Tal L et al (2014) Optimization of crop productivity in tomato using induced mutations in the florigen pathway. Nat Genet 46(12):1337–1342

Parnis A, Cohen O, Gutfinger T et al (1997) The dominant developmental mutants of tomato, *Mouse-ear* and *Curl*, are associated with distinct modes of abnormal transcriptional regulation of a Knotted gene. Plant Cell 9(12):2143–2158

Parrella G, Ruffel S, Moretti A et al (2002a) Recessive resistance genes against potyviruses are localized in colinear genomic regions of the tomato (*Lycopersicon* spp.) and pepper (*Capsicum* spp.) genomes. Theor Appl Genet 105(6-7):855–861

Parrella G, Ruffel S, Moretti A et al (2002b) Recessive resistance genes against potyviruses are localized in colinear genomic regions of the tomato (*Lycopersicon* spp.) and pepper (*Capsicum* spp.) genomes. Theor Appl Genet 105(6-7):855–861

Parrella G, Moretti A, Gognalons P et al (2004) The *Am* gene controlling resistance to *Alfalfa mosaic virus* in tomato is located in the cluster of dominant resistance genes on chromosome 6. Phytopathology 94:345–350

Piron F, Nicolaï M, Minoïa S et al (2010) An induced mutation in tomato eIF4E leads to immunity to two potyviruses. PLoS ONE 5(6):e11313

Pnueli L, Carmel-Goren L, Hareven D et al (1998) The *SELF-PRUNING* gene of tomato regulates vegetative to reproductive switching of sympodial meristems and is the ortholog of *CEN* and *TFL1*. Development 125:1979–1989

Powell AL, Nguyen CV, Hill T et al (2012) *Uniform ripening* encodes a Golden 2-like transcription factor regulating tomato fruit chloroplast development. Science 336(6089):1711–1715

Rivers BA, Bernatzky R, Robinson SJ et al (1993) Molecular diversity at the self-incompatibility locus is a salient feature in natural populations of wild tomato (*Lycopersicon peruvianum*). Mol Gen Genet 238 (3):419–427

Ronen GL, Cohen M, Zamir D et al (1999) Regulation of carotenoid biosynthesis during tomato fruit development: expression of the gene for lycopene epsilon-cyclase is down-regulated during ripening and is elevated in the mutant *Delta*. Plant J 17:341–351

Ronen G, Carmel-Goren L, Zamir D et al (2000) An alternative pathway to β-carotene formation in plant chromoplasts discovered by map-based cloning of *Beta* and *old-gold* color mutations in tomato. Proc Nat Acad Sci USA 97:11102–11107

Ruffel S, Gallois JL, Lesage ML et al (2005) The recessive potyvirus resistance gene *pot-1* is the tomato orthologue of the pepper *pvr2-eIF4E* gene. Mol Gen Genomics 274:346–353

Sagi M, Scazzocchio C, Fluhr R (2002) The absence of molybdenum cofactor sulfuration is the primary cause of the *flacca* phenotype in tomato plants. Plant J 31 (3):305–317

Saliba-Colombani V, Causse M, Gervais L et al (2000) Efficiency of RFLP, RAPD, and AFLP markers for the construction of an intraspecific map of the tomato genome. Genome 43:29–40

Salmeron J, Oldroyd G, Rommens C et al (1996) Tomato *Prf* is a member of the leucine-rich repeat class of plant disease resistance genes and lies embedded within the *Pto* kinase gene cluster. Cell 86:123–133

Sarfatti M, Abu-Abied M, Katan J et al (1991) RFLP mapping of *I1*, a new locus in tomato conferring resistance against *Fusarium oxysporum* f. sp. *lycopersici* race 1. Theor Appl Genet 82:22–26

Sato T, Iwatsubo T, Takahashi M et al (1993) Intercellular localization of acid invertase in tomato fruit and molecular cloning of a cDNA for the enzyme. Plant Cell Physiol 34(2):263–269

Schmitz G, Tillmann E, Carriero F et al (2002) The tomato *Blind* gene encodes a MYB transcription factor that controls the formation of lateral meristems. Proc Natl Acad Sci USA 99(2):1064–1069

Schornack S, Ballvora A, Gürlebeck D et al (2004) The tomato resistance protein Bs4 is a predicted non-nuclear TIR-NB-LRR protein that mediates defense responses to severely truncated derivatives of AvrBs4 and overexpressed AvrBs3. Plant J 37 (1):46–60 *Erratum* in: Plant J 37(5):787

Schumacher K, Schmitt T, Rossberg M et al (1999) The Lateral suppressor (*Ls*) gene of tomato encodes a new member of the VHIID protein family. Proc Natl Acad Sci USA 96(1):290–295

Scott JW, Agrama HA, Jones JP (2004) RFLP-based analysis of recombination among resistance genes to *Fusarium* wilt races 1, 2, and 3 in tomato. J Am Soc Hortic Sci 129:394–400

Sela-Buurlage MB, Budai-Hadrian O, Pan Q, Zamir D, Fluhr R (2001) Genome-wide dissection of Fusarium resistance in tomato reveals multiple complex loci. Mol Gen Genet 265:1104–1111

Shirasawa K, Isobe S, Hirakawa H et al (2010) SNP discovery and linkage map construction in cultivated tomato. DNA Res 17(6):381–391

Sim SC, Robbins MD, Chilcott C et al (2009) Oligonucleotide array discovery of polymorphisms in cultivated tomato (*Solanum lycopersicum* L.) reveals patterns of SNP variation associated with breeding. BMC Genom 10:466

Sim S-C, Durstewitz G, Plieske J et al (2012) Development of a large SNP genotyping array and generation of high-density genetic maps in tomato. PLoS ONE 7 (7):e40563

Soumpourou E, Iakovidis M, Chartrain L et al (2007) The *Solanum pimpinellifolium Cf-ECP1* and *Cf-ECP4* genes for resistance to *Cladosporium fulvum* are located at the *Milky Way* locus on the short arm of chromosome 1. Theor Appl Genet 115:1127–1136

Stamova BS, Chetelat RT (2000) Inheritance and genetic mapping of cucumber mosaic virus resistance introgressed from *Lycopersicon chilense* into tomato. Theor Appl Genet 101:527–537

Tanksley SD, Loaiza-Figueroa F (1985) Gametophytic self-incompatibility is controlled by a single major locus on chromosome 1 in *Lycopersicon peruvianum*. Genetics 82:5093–5096

Tanksley SD, Ganal MW, Prince JP et al (1992) High density molecular linkage maps of the tomato and potato genomes. Genetics 132:1141–1160

Terry MJ, Kendrick RE (1996) The *aurea* and *yellow-green-2* mutants of tomato are deficient in phytochrome chromophore synthesis. J Biol Chem 271(35):21681–21686

Thomas CM, Jones DA, Parniske M et al (1997) Characterization of the tomato *Cf-4* gene for resistance to *Cladosporium fulvum* identifies sequences that determine recognitional specificity in *Cf-4* and *Cf-9*. Plant Cell 9:2209–2224

Thompson AJ, Jackson AC, Parker RA et al (2000) Abscisic acid biosynthesis in tomato: regulation of zeaxanthin epoxidase and 9-cis-epoxycarotenoid dioxygenase mRNAs by light/dark cycles, water stress and abscisic acid. Plant Mol Biol 42(6):833–845

Tomato Genome Consortium (2012) The tomato genome sequence provides insights into fleshy fruit evolution. Nature 485(7400):635–641

Vakalounakis DJ, Laterrot H, Moretti A et al (1997) Linkage between *Frl* (*Fusarium oxysporum* f.sp. *radicis-lycopersici* resistance) and *Tm-2* (tobacco mosaic virus resistance-2) loci in tomato (*Lycopersicon esculentum*). Ann Appl Biol 130:319–323

van der Biezen EA, Brandwagt BF, van Leeuwen W et al (1996) Identification and isolation of the *FEEBLY* gene from tomato by transposon tagging. Mol Gen Genet 251(3):267–280

Verlaan MG, Hutton SF, Ibrahem RM et al (2013) The tomato yellow leaf curl virus resistance genes *Ty-1* and *Ty-3* are allelic and code for DFDGD-Class RNA-dependent RNA polymerases. PLoS Genet 9(3): e1003399

Víquez-Zamora M, Vosman B, van de Geest H et al (2013) Tomato breeding in the genomics era: insights from a SNP array. BMC Genom 14:354

Viquez-Zamora AM, Caro Rios CM, Finkers R et al (2014) Mapping in the era of sequencing: high density genotyping and its application for mapping TYLCV resistance in *Solanum pimpinellifolium*. BMC Genom 15:1152. doi:10.1186/1471-2164-15-1152

Vrebalov J, Ruezinsky D, Padmanabhan V et al (2002) A MADS-box gene necessary for fruit ripening at the tomato *ripening-inhibitor* (*rin*) locus. Science 296 (5566):343–346

Weller JL, Perrotta G, Schreuder ME et al (2001) Genetic dissection of blue-light sensing in tomato using mutants deficient in cryptochrome 1 and phytochromes A, B1 and B2. Plant J 25(4):427–440

Wilkinson JQ, Lanahan MB, Yen HC et al (1995) An ethylene-inducible component of signal transduction encoded by *never-ripe*. Science 270(5243):1807–1809

Yaghoobi J, Kaloshian I, Wen Y et al (1995) Mapping a new nematode resistance locus in *Lycopersicon peruvianum*. Theor Appl Genet 91:457–464

Yang C, Li H, Zhang J et al (2011) A regulatory gene induces trichome formation and embryo lethality in tomato. Proc Natl Acad Sci USA 108(29):11836–11841

Yu ZH, Wang JF, Stall RE et al (1995) Genomic localization of tomato genes that control a hypersensitive reaction to *Xanthomonas campestris* pv. *vesicatoria* (Doidge) dye. Genetics 141(2):675–682

Yuan YN, Haanstra J, Lindhout P et al (2002) The *Cladopsorium fulvum* resistance gene *Cf-ECP3* is part of the *Orion* cluster on the short arm of chromosome 1. Mol Breed 10:45–50

Zamir D (2001) Improving plant breeding with exotic genetic libraries. Nat Rev Genet 2:983–989

Zhang C, Liu L, Wang X et al (2014) The *Ph-3* gene from *Solanum pimpinellifolium* encodes CC-NBS-LRR protein conferring resistance to *Phytophthora infestans*. Theor Appl Genet 127(6):1353–1364

Zhang C, Liu L, Zheng Z et al (2013) Fine mapping of the *Ph-3* gene conferring resistance to late blight (*Phytophthora infestans*) in tomato. Theor Appl Genet 126 (10):2643–2653

Zhang J, Chen R, Xiao J et al (2007) A single-base deletion mutation in *SlIAA9* gene causes tomato (*Solanum lycopersicum*) *entire* mutant. J Plant Res 120(6):671–678

Molecular Mapping of Quantitative Trait Loci in Tomato

4

Silvana Grandillo and Maria Cammareri

Abstract

A major objective in modern biology is deciphering the genetic and molecular bases of natural phenotypic variation. Over the past three decades, the tomato clade (*Solanum* sect. *Lycopersicon*) has been a model system not only for the identification and positional cloning of quantitative trait loci (QTL), but also for the development of new molecular breeding strategies aimed at a more efficient exploration and exploitation of the rich biodiversity stored in wild germplasm for hundreds of biologically and agronomically relevant quantitative traits. The numerous QTL mapping studies conducted so far have resulted in the detection of several thousands of QTL. Despite this wealth of genetic information, the molecular bases have been revealed for only a handful of major QTL. The release of the tomato genome sequences, along with the rapid development of cost-effective next-generation sequencing (NGS) technologies, new mapping resources, and the evergrowing "omic" platforms, are holding the promise to reverse this trend. This deluge of genomic resources are undoubtedly reshaping QTL analyses also in this crop, allowing a reexamination of the variation and inheritance of complex traits at the intraspecific level, increasing the spectrum of potentially valuable alleles available for breeding. In this framework, precision phenotyping, advanced bioinformatics tools, as well as public phenotype "warehousing" databases are foreseen as the necessary tools to boost our understanding of the genetic and molecular architecture of quantitative traits, and to guarantee sustainable crop improvements in the face of an evergrowing human population and changing climates.

S. Grandillo (✉) · M. Cammareri
Research Division Portici, Italian National Research Council, Institute of Bioscience and BioResources (CNR-IBBR), Via Università 133, 80055 Portici, Naples, Italy
e-mail: grandill@unina.it; silvana.grandillo@ibbr.cnr.it

M. Causse et al. (eds.), *The Tomato Genome*, Compendium of Plant Genomes,
DOI 10.1007/978-3-662-53389-5_4

Keywords
Tomato · QTL · Association mapping · Introgression lines · Wild relatives

Abbreviations

AB	Advanced backcross
AM	Association mapping
BC	Backcross
BIL	Backcross inbred line
cM	CentiMorgans
COSII	Conserverd ortholog set II
GWAS	Genome-wide Association Studies
IL	Introgression line
ILH	Introgression line hybrid
LD	Linkage disequilibrium
MAF	Minor frequency slleles
MAS	Marker-assisted selection
MLMM	Multilocus mixed model
NGS	Next-generation sequencing
NIL	Near isogenic line
PCR	Polymerase chain reaction
QTL	Quantitative trait loci
QTN	Quantitative trait nucleotide
RFLP	Restriction fragment length polymorphism
RIL	Recombinant inbred line
RNAi	RNA interference
RS	Reproductive stage
SG	Seed germination
SGe	Selective genotyping
SGN	SOL genomics network
SNP	Single nucleotide polymorphism

Introduction

The phenotypic variation of many traits of agricultural and evolutionary importance is of quantitative nature, and results from the combined action of multiple segregating loci that may interact with each other as well as with the environment, making the dissection of the genetic architecture and molecular basis of these traits a notoriously challenging endeavor (Falconer 1989). Before the advent of molecular markers, the genetics of complex traits was studied in general terms by "quantitative genetics" (Mather 1949), and no information was available about the number and location of the underlying genes, termed polygenes by Mather (1941).

The theoretical landmarks for mapping polygenes were set already in 1923 when Sax reported the association of seed size in bean (a quantitatively inherited trait) with seed-coat pigmentation (a discrete monogenic trait).

Subsequently, Thoday (1961) elaborated the basic approach for using marker genes in segregating populations to systematically map and characterize individual polygenes, and Geldermann (1975) introduced the term quantitative trait locus (QTL) to describe a genetic locus where functionally different alleles segregate and cause significant effects on a polygenic trait. However, the application of Thoday's idea had to wait until the 1980s when isozyme markers started to be applied as a general tool for QTL analyses in tomato (*Solanum lycopersicum*) (Tanksley et al. 1982; Vallejos and Tanksley 1983; Weller et al. 1988) and in maize (Edwards et al. 1987).

Numerous factors influence the power of detecting QTL, including the heritability of the trait, gene action, the type of mapping population, marker coverage, the number and individual effects of QTL, as well as the distance between marker loci and QTL affecting the trait (Tanksley 1993; Mackay et al. 2009). Early tomato QTL mapping studies mainly applied morphological and isozyme markers in F_2 and backcross (e.g., BC_1) populations. Although several quantitative plant and fruit characteristics were analyzed, the number of informative isozyme markers was not sufficient to adequately scan the entire tomato genome for QTL, and it was therefore difficult to precisely estimate QTL positions (Tanksley et al. 1982; Vallejos and Tanksley 1983; Weller et al. 1988). The constraint of limited marker availability was subsequently overcome with the development of DNA-based genetic markers, the first of which were restriction fragment length polymorphisms (RFLPs) (Botstein et al. 1980; Bernatzky and Tanksley 1986). In 1988 Paterson and collaborators reported their pioneering study in which a complete RFLP linkage map, including 63 RFLPs, along with appropriate statistical procedures, were used in an interspecific tomato BC_1 population to map and characterize QTL, thus demonstrating that complex traits could be dissected into single Mendelian factors. Thereafter, the number of RFLP markers available for tomato genetics has increased to approximately 1000 (Tanksley et al. 1992). Meanwhile, QTL mapping in tomato has flourished and has been applied to hundreds of traits of agronomical and biological interest (Tables 4.1, 4.2, 4.3, and 4.4; reviewed by Foolad 2007; Labate et al. 2007; Grandillo et al. 2011, 2013; Grandillo 2013). To this end, different segregating populations and mapping strategies have been used.

An essential requirement for QTL mapping populations is the existence of sufficient polymorphism at marker loci and in genes underlying the trait(s) of interest. Due to several genetic bottlenecks occurred during tomato domestication and breeding, and similarly to other self-pollinated crops, the genotypic diversity within cultivated germplasm is very narrow (Miller and Tanksley 1990; Blanca et al. 2012). This limitation has led tomato geneticists and breeders to also harness the rich genetic variation stored in unadapted germplasm for the development of mapping populations and for breeding (Rick 1982; Bai and Lindhout 2007). As a result, most tomato QTL mapping experiments conducted thus far have used distant crosses between cultivated germplasm and related wild species, although several successful examples of *S. lycopersicum* intraspecific QTL studies have also been reported (Tables 4.1, 4.2, 4.3, and 4.4; Causse et al. 2001, 2007; Saliba-Colombani et al. 2001; reviewed by Foolad 2007; Labate et al. 2007; Grandillo et al. 2011, 2013).

Similarly to other autogamous species, primary segregating populations such as F_2 or early backcross (BC) progenies have been widely used for tomato QTL mapping. However, over time a more variegated repertoire of population structures has been employed including recombinant inbred (RI) populations, advanced backcross (AB) populations, backcross inbred lines (BILs), and introgression lines (ILs) (Tables 4.1, 4.2, 4.3, and 4.4). As for marker technology, following a wide use of RFLP markers, PCR-based markers have gained ground and, in many cases, RFLP maps have been integrated with several types of PCR markers (reviewed by Grandillo et al. 2011, 2013). Although the large majority of known marker systems have found applications in tomato, yet most of them are too laborious and low throughput to meet the requirements of the

Table 4.1 Summary of QTL mapping studies for disease (viral, bacterial, fungal) resistance in tomato

Type of resistance	Source of resistance/paternal parent_mapping population	No. QTL[a]	References[b]
Viral			
Tomato yellow leaf curl virus (TYLCV)	*S. pimpinellifolium* hirsute INRA $\_F_4$	1	Chagué et al. (1997)
	S. chilense LA1932; LA2779; LA1938/Tyking_3F_2s	2; 2; 1	Agrama and Scott (2006)
	S. peruvianum breeding line_F_2	5	Anbinder et al. (2009)
	S. lycopersicum FLA456	4	Kadirvel et al. (2013)
Tomato mottle virus (ToMoV)	*S. chilense* LA1932_F_2	2	Griffiths and Scott (2001)
	S. chilense LA1932; LA2779; LA1938/Tyking_3F_2s	2; 2; 1	Agrama and Scott (2006)
Bacterial			
Bacterial canker (*Clavibacter michiganensis* subsp. *michiganensis*)	*S. arcanum* LA2157_3BC_1s (intraspecific)	5	Sandbrink et al. (1995)
	S. arcanum LA2157_F_2 (interspecific)	3	van Heusden et al. (1999)
	S. habrochaites LA0407_BILs	2	Kabelka et al. (2002), (Coaker et al. 2002; Coaker and Francis 2004)
Bacterial spot (*Xanthomonas sp.*)	*S. lycopersicum* cv. Hawaii 7998 (H7998) (recurrent parent)_BC_1 (interspecific)	3	Yu et al. (1995)
	S. lycopersicum cv. Hawaii 7998 (H7998)_F_2 (intraspecific), AB, BILs	2	Yang et al. (2005)
	S. lycopersicum "cerasiforme" PI 114490_IBC	2	Hutton et al. (2010, 2014)
Bacterial wilt (*Ralstonia solanacearum*) (different races and phylotypes)	*S. lycopersicum* "cerasiforme" (L285)_F_2	3	Danesh et al. (1994)
	S. lycopersicum cv. Hawaii 7996 (H7996)_F_2; $F_{2:3}$; F_3; F_3; F_3, RILs (F_8); RILs (interspecific)	4; 6; 2; 4; 4; 2	Thoquet et al. (1996a, b), Mangin et al. (1999), Wang et al. (2000), Carmeille et al. (2006), Wang et al. (2013)
Fungal			
Anthracnose (*Colletotrichum coccodes*)	*S. lycopersicum* line 115-4 _F_2 (intraspecific)	Several	Stommel and Zhang (2001)
Black mold (*Alternaria alternata*)	*S. cheesmaniae* LA0422_BC_1S_2; BC_1S_3	5	Robert et al. (2001)
Early blight (*Alternaria solani*)	*S. habrochaites* PI 126445_BC_1, BC_1S_1; BC_1 (SGe)[c]	11; 13; 7	Foolad et al. (2002), Zhang et al. (2003a)
	S. arcanum LA2157_F_2, F_3	6	Chaerani et al. (2007)
Gray mold (*Botrytis cinerea*)	*S. habrochaites* LYC4_F_2, BC_2S_1; ILs	3; 10	Finkers et al. (2007a, b)
	S. lycopersicoides LA2951_ILs	7	Davis et al. (2009)

(continued)

Table 4.1 (continued)

Type of resistance	Source of resistance/paternal parent_mapping population	No. QTL[a]	References[b]
Late blight (*Phytophtora infestans*)	*S. habrochaites* LA2099_BC_1s; NILs; (sub-NILs)	15; 18; 3; (2 complex loci)	Brouwer et al. (2004), Brouwer and St. Clair (2004), (Johnson et al. 2012; Haggard et al. 2013)
	S. pennellii LA0716_F_2, ILs	1	Smart et al. (2007)
	S. habrochaites LA1777_ILs, BILs	5	Li et al. (2011b)
	S. pimpinellifolium L3708_$F_{2:3}$	2	Chen et al. (2014)
Powdery mildew (*Oidium lycopersici*)	*S. neorickii* G1.1601_$F_{2:3}$ (BC_2, BC_2S_1, BC_2S_2)	3 (2 fine-mapped)	Bai et al. (2003), (Faino et al. 2012)

[a]A semicolon separates the QTL identified in each population
[b]Related follow-up studies are indicated in parentheses
[c]Selective genotyping

Table 4.2 Summary of QTL mapping studies for pest resistance-related traits in tomato

Type of pest resistance/pest resistance-related traits	Source of resistance/paternal parent_mapping population	No. QTL[a]	Reference[b]
2-Tridecanone	*S. habrochaites* LA0407_F_2	5	Zamir et al. (1984)
	S. habrochaites PI 134417_F_2	3	Nienhuis et al. (1987)
Greenhouse whitefly (*Trialeurodes vaporariorum*) oviposition rate & glandular trichome densities	*S. habrochaites* (CGN1.1561)_F_2	3 & 2	Maliepaard et al. (1995)
Acylsugars level	*S. pennellii* LA0716 _F_2	5	Mutschler et al. (1996), (Lawson et al. 1997)
Acylsugars level, trichome density, percentage acylglucoses, leaf area	*S. pennellii* LA1912 _F_2 (intraspecific)	13	Blauth et al. (1998)
Acylsugars composition	*S. pennellii* LA1912 _F_2 (intraspecific)	6	Blauth et al. (1999)
Acylsugars level & resistance to silverleaf whitefly (*Bemisia tabaci*)	*S. pennellii* LA0716 _BC_1F_1	5 & 2	Leckie et al. (2012)
Acylsugars level & composition	*S. pennellii* LA0716 _BC_1F_1; BC_1F_2	3	Leckie et al. (2013)
Sesquiterpenes	*S. habrochaites* LA1777_IL_s	ns[c]	Van der Hoeven et al. (2000)
Sweetpotato whitefly (*Bemisia tabaci*)	*S. habrochaites* LA1777_F_2	4	Momotaz et al. (2010)
Trichome specialized metabolites	*S. pennelli* LA0716_ILs	ns[c]	Schilmiller et al. (2010, 2012)
Whitefly & type IV trichome characteristics, metabolic profiling	*S. galapagense*_$F_{2:3}$	2 (1 major & 1 minor)	Firdaus et al. (2013)
Two-spotted spider mite (*Tetranychus urticae* Koch)	*S. pimpinellifolium* TO-937_F_4, F_8-RILs	2	Salinas et al. (2013)

[a]A semicolon separates the QTL identified in each population; "&" separates the QTL identified for different traits
[b]Related follow-up studies are indicated in parentheses
[c]ns = the number of QTL was not specified

Table 4.3 Summary of QTL mapping studies for abiotic stress/tolerance resistance in tomato

Stress/tolerance resistance (developmental stage/main specific traits)[a]	Source of tolerance/paternal parent_mapping population	No. QTL[b]	References[c]
Cold			
Cold (VG)	*S. habrochaites*_BC_1	3	Vallejos and Tanksley (1983)
Cold (SG)	*S. pimpinellifolium* LA0722_BC_1S_1	3	Foolad et al. (1998b), (Foolad et al. 1999)
Cold (VG/SW, RAU)	*S. habrochaites* LA1778_BC_1; (NILs, sub-NILs)	9 (1 fine mapped)	Truco et al. (2000), (Goodstal et al. 2005)
Cold/transcriptional profiling	*S. habrochaites* LA1777_ILs, BILs	1	Liu et al. (2012)
Drought			
Drought (VG/WUE)	*S. pennellii*_F_3, BC_1S_1	3	Martin et al. (1989)
Drought (SG)	*S. pimpinellifolium* LA0722_BC_1S_1	4	Foolad et al. (2003)
Drought (VG/WUE)	*S. pennellii* LA0716_ILs, subILs	6 (1 fine mapped)	Xu et al. (2008)
Drought/transcriptional profiling	*S. pennellii* LA0716_ILs	2 (major), 5 (minor)	Gong et al. (2010)
Heat			
Heat (RS/FRN, FST, FN, FW, BX, SN)	*S. esculentum L.*, CL5915-93D4-1-0-3 (heat-tolerant inbred line)	21	Lin et al. (2010)
Nutrient			
Nutrient (VG/SZ, SDG, RA)	*S. pimpinellifolium* CGN 15528 _RIL	62	Khan et al. (2012)
Salt			
Salt (VG/Na^+, Cl^-, K^+ accumulation)	*S. pennellii* LA0716_F_2	6	Zamir and Tal (1987)
Salt (SG)	*S. pennellii* LA0716_F_2 (SGe)[d]	5	Foolad and Jones (1993)
Salt (RS/TW, FN, FW)	*S. pimpinellifolium* L1_F_2	6; 12	Bretó et al. (1994), Monforte et al. (1996)
Salt (RS/TW, FN, FW, EA)	*S. pimpinellifolium* (L1 and L5) and *S. cheesmaniae* L2_F_2	31; 43	Monforte et al. (1997a, b)
Salt (SG)	*S. pennellii* LA0716_F_2 (SGe)[d]	8; 8	Foolad et al. (1997), Foolad and Chen (1998)
Salt (SG)	*S. pimpinellifolium* LA0722 _BC_1S_1	7	Foolad et al. (1998a), (Foolad 1999a)
Salt (VG&RS/TW, FN, FW, EA, PHT, ID)	*S. cheesmaniae* L2_F_2 and subpopulations	8	Monforte et al. (1999)
Salt (VG)	*S. pimpinellifolium* LA0722_BC_1S_1; BC_1 (SGe)[d]	5; 5	Foolad and Chen (1999), (Foolad 1999b, review; Foolad et al. 2001)
Salt (SG&VG)	*S. pimpinellifolium* LA0722_F_9-RILs	9 & 8	Zhang et al. (2003b)
Salt (VG&RS/TW, FN, FW, FRW, NFL, DTF, DRW, DFR, Cl^-)	*S. pimpinellifolium* and *S. cheesmaniae* L2_F_7-RILs	12; 23[f]	Villalta et al. (2007)

(continued)

Table 4.3 (continued)

Stress/tolerance resistance (developmental stage/main specific traits)[a]	Source of tolerance/paternal parent_mapping population	No. QTL[b]	References[c]
Salt (VG/DSW, DLW, LA, K^+ and Na^+ concentration)	*S. pimpinellifolium* and *S. cheesmaniae* L2_2 F_8-RILs	18; 25	Villalta et al. (2008), (Asins et al. 2013)
Salt (RS/FW, FN, TW, LNC, TN, LA, DLW)	*S. galapagense* (L2) and *S. pimpinellifolium* (L5)_F_9-RILs[g]	8	Estañ et al. (2009), (Asins et al. 2010)
Salt (VG/Growth traits (PHT, STEM, LNO, DLW, DRW); Antioxidant content/activity AOX, PHE, FLA, SOD, CAT, APX, POX)	*S. pennellii* LA0716_ILs	125[h]	Frary et al. (2010)
Salt (VG/Growth traits (PHT, STEM, LNO, DLW, FLW, FRW, DRW); K^+, Na^+ and Ca^{2+} concentration)	*S. pennellii* LA0716_ILs	311	Frary et al. (2011)
Salt (VG)	*S. pennellii* LA0716_ILs; *S. lycopersicoides* LA2951_ILs	4; 6	Li et al. (2011a)
Salt (SG) & blossom end	*S. pennellii* LA0716_IL8-3; subILs	ns[i]	Uozumi et al. (2012)
Multiple stresses			
Cold, drought, salt, oxidative, nonstress	*S. pimpinellifolium* LA0722_BC_1S_1; BC_1 (SGe)[d]	14	Foolad et al. (2007)
Salt, osmotic, oxidative & cold (SG), seed quality	*S. pimpinellifolium*_RIL	Numerous (ns)[i]	Kazmi et al. (2012)

[a]*AOX* antioxidant activity, *APX* ascorbate peroxidase activity, *BX* Brix, *CAT* catalase activity, *DFR* days to fruiting, *DLW* dry leaf weight, *DRW* dry root weight, *DSW* dry stem weight, *DTF* flowering time, *EA* earliness, *FLA* total flavonoid content, *FLW* fresh leaf weight, *FN* fruit number, *FRN* flower number, *FRW* fresh root weight, *FST* flower set, *FW* fruit weight, *ID* internodal distance, *LA* leaf area, *LNC* leaf sodium concentration, *LNO* leaf number, *NFL* number of flowers per inflorescence, *PHE* total phenolic content, *PHT* plant height, *POX* glutathione peroxidase activity, *RA* root architecture, *RAU* root ammonium uptake, *RS* reproductive stage, *SG* seed germination, *SDG* seedling growth, *SN* seed number per fruit, *SOD* superoxide dismutase, *SW* shoot wilting, *STEM* stem diameter, *SZ* seed size, *TN* transported sodium, *TW* total fruit weight, *VG* vegetative growth, *WUE* water use efficiency

[b]A semicolon separates the QTL identified in each population; "&" separates the QTL identified for different traits

[c]Related follow-up studies are indicated in parentheses

[d]SGe = selective genotyping

[f]QTL detected for the six traits FW, FN, TW, Cl^-, SF, NL, under both control and high salinity conditions

[g]Both populations used as rootstocks

[h]Number of loci detected for antioxidant content under control and salt conditions

[i]ns = the number of QTL was not specified

genomics era (Víquez-Zamora et al. 2013). These drawbacks are now being circumvented by next-generation sequencing (NGS) projects, which are offering new possibilities to significantly increase genotyping throughput, as well as by the availability of high-throughput Single Nucleotide Polymorphisms (SNPs) arrays that have allowed massive parallel whole genome screening of genotypes (Sim et al. 2012; Víquez-Zamora et al. 2013). In addition, thanks to the recently published whole genome sequences of tomato (Tomato Genome Consortium 2012), next-generation resequencing approaches can be applied also in related germplasm (Causse et al. 2013; Aflitos et al. 2014).

The numerous QTL mapping studies conducted in tomato over the past three decades have provided information about the genetic architecture of complex traits, i.e., estimated number of QTL and magnitude of their estimated additive, dominance, and epistatic effects in multiple environments. These efforts have resulted in the detection of thousands of QTL, many of which are of potential interest for tomato breeding, and

Table 4.4 Summary of QTL mapping studies for plant, flower, fruit and yield traits in tomato

Traits[a]	Wild/paternal parent_mapping population	No. QTL[b]	References[c]
Fruit and seed weight, stigma exsertion, leaf ratio	*S. pennellii* LA0716_BC_1	21	Tanksley et al. (1982)
Brix	*S. chmielewskii* LA1028_(BC_5S_5), derived F_2	ns[d]	Osborn et al. (1987)
Fruit quality (fine-mapping)	*S. chmielewskii* LA1028_BC_1, BC_2F_2 (subILs)	15	Paterson et al. (1988), (Paterson et al. 1990)
Fruit quality, earliness, leaf and plant morphology, yield-related, reproductive	*S. pimpinellifolium* CIAS27_F_2	85[e]	Weller et al. (1988)
Fruit quality	*S. galapagense* LA0483_F_2, F_3	29	Paterson et al. (1991)
Earliness, leaf and plant morphology	*S. pennellii* LA0716_F_2	74	de Vicente and Tanksley (1993)
Fruit quality, yield	*S. chmielewskii*_BILs (BC_2F_5)	ns[d]	Azanza et al. (1994)
Earliness, fruit weight	*S. lycopersicum* IVT KT_1 (breeding line containing *S. pimpinellifolium* and *S. neorickii* introgressions)_F_2	3	Lindhout et al. (1994)
Yield and fruit quality-related	*S. pennellii* LA0716_ILs, ILHs (subILs)	104 (including ***fw2.2***, ***Brx9-2-5***)	Eshed and Zamir (1994, 1995), (Alpert et al. 1995; Alpert and Tanksley 1996; Eshed and Zamir 1996; Eshed et al. 1996; **Frary et al.** 2000; **Fridman et al.** 2000, 2002, 2004; Gur and Zamir 2004; Baxter et al. 2005; Cong and Tanksley 2006)
Fruit weight, soluble solids, seed weight	*S. galapagense* LA0483_F_8-RILs	73[e]	(Paran et al. 1995), Goldman et al. (1995)
Fruit quality, flower and plant morphology, earliness, seed weight and number	*S. pimpinellifolium* LA1589_BC_1	54 (including *fw2.2*, *fs8.1*)	Grandillo and Tanksley (1996a, b), (Alpert et al. 1995; Grandillo et al. 1996, 1999; **Frary et al.** 2000; Ku et al. 2000)
Fruit quality, yield related, earliness	*S. pimpinellifolium* LA1589_BC_2/BC_2F_1/BC_3, QTL-NILs	87	Tanksley et al. (1996)
Flower morphology, SI, UI	*S. habrochaites* LA1777_BC_1	23	Bernacchi and Tanksley (1997)
Fruit quality, yield related, earliness, growth, stigma exsertion	*S. arcanum* LA1708_BC_3/BC_4	166	Fulton et al. (1997)
Plant, fruit, leaf morphology	*S. galapagense* LA0483_RILs	41[e]	Paran et al. (1997)
Fruit quality, yield related, earliness, cover, HA	*S. habrochaites* LA1777_BC_2/BC_3	121	Bernacchi et al. (1998a)
Fruit quality, yield related, earliness, cover	*S. habrochaites* LA1777; *S. pimpinellifolium* LA1589_NILs	25	Bernacchi et al. (1998b)
Fruit quality	*S. pimpinellifolium* LA0722_BC_1/BC_1S_1	59	Chen et al. (1999)

(continued)

Table 4.4 (continued)

Traits[a]	Wild/paternal parent_mapping population	No. QTL[b]	References[c]
Fruit shape (pear-shaped)	*S. pimpinellifolium* LA1589; *S. pennelli* LA0716 (IL2-5)_F_2	2 (1 major-***ovate*** & 1 minor)	Ku et al. (1999), (**Liu et al.** 2002; Huang et al. 2013)
Earliness & fruit weight	*S. lycopersicum* (line *"Early cherry"*) _F_2	2 & 3	Doganlar et al. (2000a)
Seed weight	*S. galapagense* LA0483_F_8-RILs; *S. habrochaites* LA1777_BC_2/BC_3; *S. pimpinellifolium* LA1589_BC_1, BC_2F_6-RIL; *S. neorickii* LA2133_BC_2/BC_3; *S. pimpinellifolium* CIAS27_F_2; *S. pennellii* LA0716_BC_1	24 (including ***sw4.1***)	Doganlar et al. (2000b) (review), (**Orsi and Tanksley** 2009)
Fruit quality, yield related, earliness, cover, HA	*S. neorickii* LA2133_BC_2/BC_3	199	Fulton et al. (2000)
Yield and fruit quality related, fine mapping	*S. habrochaites* LA1777_NILs, subNILs	6	Monforte and Tanksley (2000a, b)
Yield and fruit quality related	*S. habrochaites* LA1777_Chr. 4 ILs, subILs; *S. pennellii* LA0716_Chr. IL, subILs; *S. arcanum* LA1708_Chr. 4 IL	15	Monforte et al. (2001)
Fruit quality, including aroma volatiles	*S. lycopersicum* "cerasiforme" F_7-RILs; (NILs)	81 (including ***lc***)	Saliba-Colombani et al. (2001), (Causse et al. 2002; Lecomte et al. 2004a, b; Chaïb et al. 2006, 2007; Causse et al. 2007; Zanor et al. 2009; **Muños et al.** 2011; Aurand et al. 2012)
Sensory attributes	*S. lycopersicum* "cerasiforme"_F_7-RILs	49	Causse et al. (2001), (Causse et al. 2002, Bertin et al. 2003; Lecomte et al. 2004a, b; Chaib et al. 2006, 2007; Causse et al. 2007; Bertin et al. 2009; Zanor et al. 2009)
Fruit size and shape, seed number and weight	*S. pimpinellifolium* LA1589_F_2	30	Lippman and Tanksley (2001)
Fruit shape	*S. pimpinellifolium* LA1589_F_2	1 (***sun***)	van der Knaap and Tanksley (2001), (**Xiao et al.** 2008, 2009; Jiang et al. 2009; Wu et al. 2011; Huang et al. 2013)
Stem vascular morphology	*S. habrochaites* LA0407	1	Coaker et al. (2002)
Fruit quality, yield related, earliness, seed number and weight, growth	*S. pimpinellifolium* LA1589_BILs (BC_2F_6)	71	Doganlar et al. (2002)
Fruit composition	*S. habrochaites* LA1777_AB; *S. arcanum* LA1708_AB; *S. neorickii* LA2133_AB; *S. pimpinellifollium* LA0722_AB	222	Fulton et al. (2002)
Flower morphology, number of flowers per cluster	*S. pimpinellifolium* LA1237 ("selfer") and LA1581 ("outcrosser") F_2	5	Georgiady et al. (2002)

(continued)

Table 4.4 (continued)

Traits[a]	Wild/paternal parent_mapping population	No. QTL[b]	References[c]
Fruit shape	*S. pimpinellifolium* LA1589_F_2	4	van der Knaap et al. (2002)
Fruit volatiles, untrained sensory evaluation	*S. pennellii* LA0716_ILs, subILs	1	Tadmor et al. (2002)
Fruit quality, yield related	*S. chmielewskii* LA1028_IL, subILs; *S. habrochaites* LA1777_IL, subILs	8	Frary et al. (2003)
Leaf morphology	*S. pennellii* LA0716_ILs	30	Holtan and Hake (2003)
Fruit color, carotenoids	*S. pennellii* LA0716_ILs	16	Liu et al. (2003a)
Fruit shape-related, fruit size, number of flower per cluster, seed number per fruit	*S. pimpinellifolium* LA1589_F_2	50 (including ***fw3.2***)	van der Knaap and Tanksley (2003), (Zhang et al. 2012; **Chakrabarti et al.** 2013)
Fasciated (multiloculed) fruit	*S. lycopersicum* cultivars; *S. pennellii* ILs_F_2; *S. pimpinellifolium* LA1589_F_2	4 (including ***fas***)	Barrero and Tanksley (2004), (**Cong et al.** 2008; Huang et al. 2013)
Fruit weight and composition	*S. pennellii* LA0716_ILs	81	Causse et al. (2004)
Stigma exsertion	*S. pennellii* LA0716_IL2-5	1 (complex locus including ***Style2.1***)	Chen and Tanksley (2004), (**Chen et al.** 2007)
Fruit quality, yield related,	*S. pennellii* LA1657_BC_2/BC_2F_1	84	Frary et al. (2004°)
Leaf, petal, sepal morphology	*S. pennellii* LA0716_F_2	36	Frary et al. (2004b)
Fruit quality	*S. lycopersicum* "cerasiforme" (PI 270248)	ns[d]	Georgelis et al. (2004)
Fruit color	*S. habrochaites* LA0407_BILs (BC_2S_5)/F_3, F_4	2	Kabelka et al. (2004)
Fruit quality	*S. habrochaites* LA1777_Chr. 4 ILs, subILs; *S. arcanum* LA1708_Chr. 4 IL, subILs	15	Yates et al. (2004)
Fruit size and composition, Trans. prof.	*S. pennellii* LA0716_ILs	ns[d]	Baxter et al. (2005)
Hybrid incompatibility	*S. habrochaites* LA1777_ILs; BILs	22	Moyle and Graham (2005)
Metabolite profiling	*S. pennellii* LA0716_ILs	20	Overy et al. (2005)
Fruit antioxidants	*S. pennellii* LA0716_ILs	ns[d]	Rousseaux et al. (2005)
Fruit metabolites & yield related	*S. pennelli* LA0716_ILs	889 & 326	Schauer et al. (2006)
Morphology, yield, fitness	*S. pennelli* LA0716_ILs, ILHs	841	Semel et al. (2006)
Fruit aroma volatiles & organic acids	*S. pennellii* LA0716_ILs	25 & 4	Tieman et al. (2006), (Mageroy et al. 2012)
Fruit shape	*S. pimpinellifolium* LA1589_2 F_2s, BC_1	36; 32; 27	Brewer et al. (2007)
Fruit size and composition, including AsA	*S. pennellii* LA0716_ILs; *S. habrochaites* PI24_BC_2S_1; *S. lycopersicum* "cerasiforme" RILs	23	Stevens et al. (2007, 2008)

(continued)

Table 4.4 (continued)

Traits[a]	Wild/paternal parent_mapping population	No. QTL[b]	References[c]
Flowering time	*S. chmielewskii* CH6047_F_2	8	Jiménez-Gómez et al. (2007)
Fruit shape	*S. pimpinellifolium* LA1589_3F_2 s	20; 23; 20	Gonzalo and van der Knaap (2008)
Partenocarpy, stigma exsertion	*S. habrochaites* LYC4_ILs; *S. habrochaites* (IVT-line 1)_BC_5S_1, F_2	4	Gorguet et al. (2008)
Aroma volatiles	*S. habrochaites* LA1777_ILs, BILs	30	Mathieu et al. (2009)
Hybrid incompatibility	*S. pennellii* LA0716_ILs	19	Moyle and Nakazato (2008), (review Bedinger et al. 2011)
Primary metabolites	*S. pennellii* LA0716_ILs, ILHs	332	Schauer et al. (2008); (Kamenetzky et al. 2010)
Ripening-associated ethylene emission	*S. habrochaites* LA1777_ILs, BILs	17	Dal Cin et al. (2009)
Fruit weight and composition under different fruit loads (HL vs. LL)[f]	*S. chmielewskii* LA1840_ILs	103	Prudent et al. (2009, 2010, 2011)
AsA, phenols, soluble solids, trans. prof.	*S. pennellii* LA0716_ILs, IL12-4, IL7-3	3	Di Matteo et al. (2010, 2013), (Sacco et al. 2013)
Pericarp metabolome at two different fruit load conditions	*S. chmielewskii* LA1840_ILs	240 (HL), 128 (LL)[f]	Do et al. (2010)
Plant weight, yield, brix, harvest index, earliness, metabolites	*S. pennellii* LA0716_ILs, HILs, subILs	1	Gur et al. (2010)
Metabolism and yield-related, genomic analysis	*S. pennellii* LA0716_ILs (BINs 1C, 2B, 4I, 7H, 11C)	104	(Eshed and Zamir 1995; Schauer et al. 2006; Tieman et al. 2006), Kamenetzky et al. (2010)
Trichome specialized metabolites	*S. pennellii* LA0716_ILs	ns[d]	Schilmiller et al. (2010, 2012)
Flowering time-related	*S. pimpinellifolium* PI24039_BC_1F_6	12	Sumugat et al. (2010), (Sumugat and Sugiyama 2010)
Transplanting time and root growth-related	*S. pimpinellifolium* PI24039_BC_1F_6	8	Sumugat et al. (2011)
Vitamin E, CGs	*S. pennellii* LA0716_ILs	6	Almeida et al. (2011), (Quadrana et al. 2014)
Yield-related (shoot and root)/grafting	*S. pennellii* LA0716_ILs, HILs	11	Gur et al. (2011)
Fruit weight/fine mapping	*S. pimpinellifolium* LA1589_BC_1F_5	1	Huang and van der Knaap (2011)
Fruit branched-chain amino acids/CGs	*S. pennellii* LA0716_ILs	25	(Schauer et al. 2006, 2008), Kochevenko and Fernie (2011)
Fruit quality, shelf-life	*S. pimpinellifolium* LA722_16 RILs	8[e]	Pratta et al. (2011)
Enzyme activity for central carbon metabolism in fruit pericarp	*S. pennellii* LA0716_ILs, HILs	27	Steinhauser et al. (2011)
In vitro plant regeneration	*S. pennellii* PE-47_BC_1, F_2	6	Trujillo-Moya et al. (2011)

(continued)

Table 4.4 (continued)

Traits[a]	Wild/paternal parent_mapping population	No. QTL[b]	References[c]
Lycopene	*S. pimpinellifolium* LA2093_RIL	2	Ashrafi et al. (2012), (Kinkade and Foolad 2013)
Fruit firmness/fine mapping/CGs	*S. pennelli* LA0716_ILs	1 (complex locus)	Chapman et al. (2012)
Seed quality, abiotic stress versus control	*S. pimpinellifolium* _RIL	Numerous, ns[d]	Kazmi et al. (2012)
Seed size, seedling growth, root architecture (control vs. nutrient stress)	*S. pimpinellifolium* CGN 15528 _RIL	62	Khan et al. (2012)
Fruit texture, cell wall galactose metabolism	*S. lycopersicum* "cerasiforme" ILs; *S. chmielewskii* LA1840_ILs; *S. pennellii* LA0716_ILs	ns[d]	Lahaye et al. (2012, 2013)
Carotenoids/Trans. prof./ CG verification	*S. pennellii* LA0716_ILs	ns[d]	Lee et al. (2012)
Lycopene, ascorbic acid, brix, fruit weight	*S. pimpinellifolium* (S0801)_$F_{2:3}$	15	Sun et al. (2012)
Seed metabolism	*S. pennellii* LA0716_ILs	30	Toubiana et al. (2012)
Competence for adventitious organ formation	*S. pennellii* LA0716_ILs	6 (bins)	Arikita et al. (2013)
Leaf morphology, RNA-Seq	*S. pennellii* LA0716_ILs	1035	Chitwood et al. (2013)
Fruit quality, fruit size/shape, maturity, yield, plant architecture	*S. habrochaites* LA2099_chr. 5 subILs	41	Haggard et al. (2013)
Brix, physiological characterization	*S. pennellii* LA0716_IL8-3	1	Ikeda et al. (2013)
Polyphenols content in plant organs	*S. pennellii* LA0716_IL7-3, IL10-1, IL12-4	ns[d]	Minutolo et al. (2013)
Fruit quality, shelf-life	*S. pimpinellifolium* LA722_BC_1/ BC_1S_1; BC_2	6	Pereira da Costa et al. (2013)
Fruit shape (modifier loci for *OVATE*)	*S. lycopersicum* $F_{2:3}$	2	Rodríguez et al. (2013)
Root morphology and cellular development	*S. pennellii* LA0716_ILs	Numerous, ns[d]	Ron et al. (2013)
Plant height, fruit firmness, yield, shelf-life	*S. lycopersicum* _F_2	9	Yogendra and Ramanjini Gowda (2013)
Rutin content	*S. habrochaites* LA1777_7 ILs	1	Hanson et al. (2014)
Fruit metabolome	*S. pennellii* LA0716_ILs	2820	Perez-Fons et al. (2014)

[a]*AsA* ascorbic acid content, *CG* candidate gene, *HA* horticultural acceptability (differently measured), *HL* high load, *LL* low load, *SI* self incompatibility, *Trans. Prof.* transcriptional profiling, *UI* unilateral incompatibility
[b]A semicolon separates the QTL identified in each population; "&" separates the QTL identified for different traits; cloned QTL are indicated in parenthesis and in bold
[c]Related previous or follow-up studies are indicated in parentheses; the studies that have cloned the QTL are indicated in bold
[d]ns = the number of QTL was not specified
[e]Number of significant marker × trait associations
[f]HL = high load; LL = low load

whose molecular bases still wait to be revealed (Tables 4.1, 4.2, 4.3, and 4.4) (reviewed by Foolad 2007; Labate et al. 2007; Grandillo et al. 2011, 2013; Grandillo 2013; Alseekh et al. 2013).

During these years, the tomato clade (*Solanum* sect. *Lycopersicon*), which encompasses the cultivated tomato (*S. lycopersicum*) and its 12 wild relatives (Peralta et al. 2008), has proven to be a model system not only for the identification (Paterson et al. 1988) and positional cloning of QTL (Frary et al. 2000; Fridman et al. 2000, 2004), but also for the development of new molecular breeding approaches aimed at ensuring a more efficient use of the wealth of genetic variation hold in wild germplasm (Tanksley and Nelson 1996; Tanksley et al. 1996; Tanksley and McCouch 1997; Zamir 2001).

Although the QTL mapping approach has proven to be an undoubtedly powerful method to dissect the genetic architecture of complex traits and for breeding, nevertheless, it suffers from several drawbacks including the restricted allelic variation, the low-resolution mapping, and the time necessary to develop the mapping populations (Korte and Farlow 2013). In order to overcome these limitations and to facilitate the association of phenotypes to genotypes, alternative approaches have been suggested including linkage disequilibrium (LD)-based association analysis, also referred to as association mapping (AM) (Flint-Garcia et al. 2003; Gupta et al. 2005), and next generation genetic-mapping populations such as Multi-parent Advanced Generation Inter-Cross (MAGIC) populations (Cavanagh et al. 2008). Over the last years, the availability of the tomato genome sequences (Tomato Genome Consortium 2012), the related new high-throughput genotyping tools, and the development of new methodological approaches have allowed successful applications of both strategies also in tomato (Sauvage et al. 2014; Pascual et al. 2015). These advances are paving the way for a more efficient exploitation of *S. lycopersicum* germplasm in breeding programs.

The status of QTL mapping in tomato has been the subject of several reviews (Foolad 2007; Labate et al. 2007; Grandillo et al. 2011, 2013; Grandillo 2013), and most of the studies have been summarized and updated in Tables 4.1, 4.2, 4.3, and 4.4. Therefore, also because of space limitations, in this current review we do not attempt to provide a comprehensive discussion of the subject, but rather we focus on a few aspects, highlighting the new opportunities that the tomato genome sequences and the related genomic tools are providing for the genetic and molecular dissection of complex traits and to accelerate the improvement of this important crop.

IL-Based Analysis of Complex Traits and Breeding

Since the first QTL mapping studies conducted in interspecific crosses of tomato, it became evident that the approach allowed a more efficient detection of "cryptic" genetic variants (Tanksley et al. 1982; Weller et al. 1988; de Vicente and Tanksley 1993). This suggested that despite their overall inferior phenotype, unadapted germplasm is likely to be a rich source of agronomically favorable QTL alleles (Tanksley and McCouch 1997). However, in order to increase the efficiency with which natural biodiversity could be mined to improve yield, adaptation and quality of elite germplasm, and thus to bridge the gap between QTL mapping and QTL-based breeding, new concepts and strategies needed to be developed. These new methods should have also allowed circumventing some of the constraints posed by QTL mapping conducted in early biparental segregating generations (F_2, F_3, and BC_1) or in RILs. The high proportion of donor parent alleles that still segregate in these populations, in fact, may result in overshadowing effects of major QTL on the effects of independently segregating minor QTL, as well as in relatively high level of epistatic interactions between donor QTL alleles and other donor genes. Thereby, favorable donor QTL alleles detected in these mapping populations often lose their effects once they are introgressed into the genetic background of elite lines. In addition, in the case of interspecific crosses involving exotic germplasm, QTL analyses might be further complicated by partial or complete sterility problems, since a few genes for sterility may impede population development and/or the obtention of meaningful measurements for agronomical important traits (such as fruit characters).

In order to address these issues, two related molecular breeding strategies, the "Advanced Backcross (AB) QTL analysis" (Tanksley and Nelson 1996; Tanksley et al. 1996) and the "introgression line (IL) populations" or "exotic libraries" (Eshed and Zamir 1994, 1995; Zamir 2001), have been implemented first in tomato, and then in several other crops (Grandillo et al. 2008, 2013; Grandillo 2013). These methods were proposed to more efficiently unlock the genetic potential stored in seed banks and in exotic germplasm for the development of improved varieties, thereby expanding the genetic base of crop species (Tanksley and McCouch 1997; Zamir 2001). Both approaches have allowed the detection of favorable wild QTL alleles for numerous traits of agronomical and biological interest along with the development of ILs or QTL-NILs that can be used in marker-assisted breeding programs (Grandillo et al. 2008; Grandillo 2013). Sets of ILs or QTL-NILs have also been developed from intraspecific crosses (Lecomte et al. 2004a; Chaïb et al. 2006). In some instances, they have been used to verify, stabilize, and fine-map QTL, in the same or in different genetic backgrounds, and therefore only a relatively small proportion of the donor parent genome was represented in the developed ILs (Paterson et al. 1990; Tanksley et al. 1996; Bernacchi et al. 1998b; Monforte and Tanksley 2000b, Monforte et al. 2001; Lecomte et al. 2004b; Chaïb et al. 2006).

In tomato, the AB-QTL analysis method has been applied to six interspecific crosses involving the same *S. lycopersicum* parent (cv. E6203) and six wild species, selected to represent a broad spectrum of the phylogenetic tree: *S. pimpinellifolium* LA1589 (Tanksley et al. 1996), *S. arcanum* LA1708 (Fulton et al. 1997), *S. habrochaites* LA1777 (Bernacchi et al. 1998a, b), *S. neorickii* LA2133 (Fulton et al. 2000), and *S. pennellii* LA1657 (Frary et al. 2004a), *S. chilense* LA1932 (Termolino et al. 2010) (Table 4.4). These populations have been analyzed for numerous horticultural traits important for the tomato processing industry, using replicated field trials in several locations worldwide (Table 4.4). Overall, wild QTL alleles with favorable effects were detected for more than 45 % of traits evaluated across the first five AB populations (reviewed by Grandillo et al. 2008). In addition, the first four AB-QTL populations have also been analyzed for biochemical traits possibly contributing to flavor (Fulton et al. 2002).

Concomitantly, the IL approach was proposed in D. Zamir's laboratory, and the first tomato whole genome IL population was developed which comprised a core set of 50 lines carrying single RFLP-defined homozygous chromosomal segments of the distantly related, wild desert green-fruited species *S. pennellii* LA0716 in the background of the processing inbred cv. M82 (Eshed and Zamir 1994, 1995). Several properties of IL populations contribute to their power in detecting and stabilizing QTL, and they have been widely discussed elsewhere (Zamir 2001; Lippman et al. 2007; Grandillo et al. 2008; Grandillo 2013). Collectively the *S. pennellii* LA0716 ILs represent whole genome coverage of the wild parent in overlapping segments, which define unique "bins" where genes and QTL can be mapped, albeit at an initial average coarse resolution. Another important feature of this IL library is its permanent nature, as it can be maintained by self-pollination, and this aspect allows replicated measurements to be taken across different environments, years, and laboratories (Eshed and Zamir 1995).

The numerous advantages of IL populations for the analyses of complex traits have become manifest since the first experiments conducted with the *S. pennellii* IL library (and, in some cases, also with the correspondent heterozygous lines, HILs) to map and fine-map QTL underlying horticultural yield and fruit quality traits (Eshed and Zamir 1995, 1996; Eshed et al. 1996).

Thenceforth, the *S. pennellii* IL population, and subsequently also its second generation consisting of 76 ILs and subILs (Pan et al. 2000; http://solgenomics.net/), have been publicly available, and have been used to analyze a plethora of biologically and agronomically relevant traits including whole-plant morphology and yield (also heterosis), primary and secondary metabolic composition, fruit color, enzyme activities, leaf, fruit, and root morphology, cellular development, biotic and abiotic stress tolerance, hybrid incompatibility, and gene expression (Tables 4.1, 4.2, 4.3, and 4.4) (Grandillo et al. 2011, 2013; Grandillo 2013), resulting in more than 3069 QTL identified in this population to date (reviewed in Alseekh et al. 2013).

To aid in the discovery of the genes underlying the many QTL described to date, the mapping resolution of the *S. pennellii* LA0716 IL library was improved through the addition of 285 marker-defined subILs, which break up the 37 largest ILs of the initial population—corresponding to approximately 75 % of the genome; and work is going on to generate sublines also for the remaining 25 % of the genome. Seeds for the subILs as well as F_2 seeds for each IL are publically available (Alseekh et al. 2013).

Panels of ILs, deriving from both interspecific as well as intraspecific crosses, represent also a very valuable resource to get more precise estimates of epistatic interactions (Eshed and Zamir 1996; Causse et al. 2007) and of QTL × genotype interactions (Eshed and Zamir 1995; Eshed et al. 1996; Monforte et al. 2001; Gur and Zamir 2004; Lecomte et al. 2004a; Chaïb et al. 2006; Causse et al. 2007). The immortality of IL populations allows taking phenotypic measurements on multiple replicates, which reduces the environmental effects and increases statistical power. By replicating the trials in more than one location and over time, it becomes possible to estimate QTL × environment interactions (Paterson et al. 1991; Eshed et al. 1996; Monforte et al. 2001; Liu et al. 2003b; Gur and Zamir 2004; Rousseaux et al. 2005). In this respect, a unique characteristic of the *S. pennellii* library is that phenotypic data from 45 IL experiments, in which 355 traits were scored in replicated measurements by multiple laboratories, have been deposited in the phenotype warehouse of Phenom Networks (http://phnserver.phenome-networks.com/) (Zamir 2013). The data can be browsed and statistically analyzed online; in alternative, they can be downloaded from the site to be analyzed using alternative statistical softwares. This tool allows comparisons of new data collected from the *S. pennellii* ILs with the results already available on the site.

Another relevant feature of IL biology, especially in the context of interspecific crosses, is the exposure of new transgressive phenotypes, not present in the parental lines. This phenomenon is caused by novel epistatic relationships arising between the donor parent alleles, and the independently evolved molecular networks of the recipient parent (Lippman et al. 2007). A recent example is provided by Chitwood et al. (2013) who have characterized the *S. pennellii* IL library for a suite of vegetative traits, ranging from leaf shape, size, complexity, and serration traits to cellular traits, such as stomatal density and epidermal cell phenotypes. Thus, leading to the identification of 1035 QTL, 826 toward the direction of *S. pennellii* and 209 transgressive, beyond the phenotype of the domesticated parent. Additionally, Shivaprasad et al. (2012) have explored the possible involvement of epigenetics and small silencing RNA in the occurrence of stable transgressive phenotypes observed in the *S. pennellii* LA0716 IL library. Their results indicate that different sRNA-based mechanisms could be implicated in transgressive segregation, and that the transgressive accumulation of miRNA and siRNAs is an indication of the hidden potential of parents that becomes manifest in the hybrids.

The IL approach has also facilitated the exploration of the genetic basis of heterosis (Semel et al. 2006), along with its application for IL-based crop improvement, as shown by the development of a new leading hybrid of processing tomato through marker-assisted pyramiding of three *S. pennellii* introgressions carrying heterotic QTL (Gur and Zamir 2004; Lippman et al. 2007).

One shortcoming of most IL populations is the relatively low map resolutions; nevertheless, each IL can be used as the starting point for high-resolution mapping. In this way, tight linkage of multiple QTL affecting one or more trait(s) can be discerned from pleiotropy (Alpert and Tanksley 1996; Eshed and Zamir 1996; Monforte and Tanksley 2000b; Monforte et al. 2001; Fridman et al. 2002; Frary et al. 2003; Chen and Tanksley 2004; Lecomte et al. 2004b; Stevens et al. 2008; Chapman et al. 2012; Haggard et al. 2013). Moreover, the identification of molecular markers more closely linked to the QTL of interest is the basis for marker selection (MAS) of elite breeding lines carrying individual or a combination of QTL.

Thanks to these properties, the *S. pennellii* ILs have soon demonstrated to be an efficient tool for the positional cloning of QTL (Frary et al. 2000; Fridman et al. 2000, 2004). However, in spite of the successes achieved so far, delimiting a QTL to a single gene or to a quantitative trait nucleotide (QTN) using genetic approaches is still an arduous and labor-intensive task. Therefore, over the years, alternative strategies have been tested to short list candidate genes for target QTL. For example, the *S. pennellii* IL population has been used to explore the potential of the "candidate gene approach" to identify candidate genes for QTL affecting tomato fruit color (Liu et al. 2003b), tomato fruit size, and composition (Causse et al. 2004), as well as fruit AsA content (Stevens et al. 2008), and vitamin E (Almeida et al. 2011). While no colocation was initially found between candidate genes and fruit color QTL (Liu et al. 2003b), several putative associations were observed in the other three studies.

Natural genetic variation stored in IL populations can also facilitate the integration of multiple cutting-edge "omic" platforms (genomic, transcriptomic, proteomic, and/or metabolomic) and large physiological data sets, along with statistical network analysis, allowing multi-faceted systems-level analysis of integrated developmental networks, and the identification of candidate genes underlying complex traits (Schauer et al. 2006, 2008; Lippman et al. 2007). These approaches can help identifying previously uncharacterized networks or pathways, in addition to candidate regulators of such pathways (Saito and Matsuda 2010). The availability of a full-genome sequence can further facilitate reducing the list of genes in the QTL interval, since the analysis of the annotation might indicate a more likely candidate. In tomato, numerous studies have already demonstrated the power of these approaches to gain insights into the genetic basis of compositional quality in tomato fruit (Schauer et al. 2006, 2008), of seed "primary" metabolism (Toubiana et al. 2012), or for the analysis of "secondary" metabolism (Schilmiller et al. 2010, 2012), as well to unfold interorgan correlations (Toubiana et al. 2012). Furthermore, Morgan et al. (2013) have showed that detailed biochemical characterization of the *S. pennellii* IL library can provide useful information to guide metabolic engineering strategies aimed at increasing health-related compounds of tomato fruit. Recently, Lee et al. (2012) used a systems-based approach combining transcriptomic analysis (based on the TOM2 oligonucleotide array) and metabolic data to identify key genes regulating tomato fruit ripening and carotenoid accumulation. Altogether, these examples suggest that with the continued development of genetic and "omic" tools, more detailed systems-level analyses will be possible, increasing the efficiency in discovery, candidate gene identification and cloning of target QTL.

Considering the numerous successful applications of the *S. pennellii* LA0716 IL library, in order to accelerate the rate of progress of introgression breeding, Zamir (2001) proposed to invest in the establishment of a genetic infrastructure of "exotic libraries." Along this line, for tomato, besides the *S. pennellii* LA0716 library, additional populations of ILs and BILs, covering different fractions of the wild species genomes, have been developed and/or further refined for other wild tomato relatives including *S. habrochaites* LA1777 (Monforte and Tanksley 2000a; Tripodi et al. 2010; Grandillo et al. 2014; S. Grandillo et al., unpublished results), *S. habrochaites* LA0407 (Finkers et al. 2007b), *S. chmielewskii* LA1840 (Prudent et al. 2009), *S. neorickii* LA2133 (Fulton et al. 2000; D. Zamir personal communication), *S. pimpinellifolium* LA1589 (Doganlar et al. 2002; D. Zamir personal communication), *S. pimpinellifolium* TO-937 (Barrantes et al. 2014) and the wild tomato-like nightshade *S. lycopersicoides* LA2951 (Chetelat and Meglic 2000; Canady et al. 2005). Some of these populations have already been used to identify QTL for several traits (Tables 4.1, 4.2, 4.3, and 4.4). For instance, the *S. chimielewskii* LA1840 ILs have been used to explore the effect of different fruit loads on QTL detection (Prudent et al. 2009, 2010, 2011; Do et al. 2010; Kromdijk et al. 2014).

In order to facilitate marker-assisted breeding based on these wild species resources, and to facilitate comparisons between function maps of tomato and potato, some of the above-mentioned IL/BIL populations have been anchored to the potato genome using a common set of ~120 COSII markers (Wu et al. 2006; Tripodi et al. 2010; S. Grandillo et al. unpublished results). The multispecies IL platform includes ILs and BILs derived from the *S. neorickii* LA2133 AB population (Fulton et al. 2000; D. Zamir personal communication), a new set of *S. habrochaites* LA1777 ILs (Grandillo et al. 2014), the *S. chmielewskii* LA1840 IL population and the *S. pennellii* LA0716 ILs and subILs (Alseekh et al. 2013). These genetic resources expose highly divergent phenotypes, providing a rich segregation for whole genome naturally selected genetic variation affecting yield, morphological, and biochemical traits, thus allowing multiallelic effects to be captured.

The production of such congenic and permanent resources, however, is quite an arduous and time-consuming task, which can take several years. The development of new high-throughput molecular platforms that allow automated genotyping is making IL development a much more efficient and precise process (Severin et al. 2010; Xu et al. 2010; Schmalenbach et al. 2011). Dense genetic maps, in fact, allow high-resolution localization of the introgressed segments, which is essential if one has to select ILs carrying single and small marker-defined segments for genome-wide coverage of the donor parent genome. Furthermore, IL populations genotyped at very high resolution should facilitate rapid and precise localization of QTL and subsequent identification of the underlying genes. In this respect, the *S. pennellii* LA0716 IL library has been genotyped using the high-density "SolCAP" SNP array (Sim et al. 2012), as well as using a diversity arrays technology (DArT) platform, which has resulted, on average, in tenfold increase of the number of markers available for each IL (Van Schalkwyk et al. 2012). Additionally, Chitwood et al. (2013) have genotyped the *S. pennellii* library at ultra-high density, using two complementary approaches, RNA-Seq and RESCAN, which have resulted in a precise definition of the boundaries of each IL at both the genomic and transcriptomic levels. The combination of these data with the recently completed tomato genome has also allowed the exact gene content of each IL to be determined, which should aid the molecular characterization of QTL as well as breeding efforts.

The recent availability of the genome sequences of the parents for some of the IL populations described above is further enhancing the potential of these congenic and permanent genetic resources. In order to support QTL analyses in the *S. pennellii* IL library, following on from the release of the genome sequence for tomato (*S. lycopersicum* cv Heinz) and of a draft sequence of *S. pimpinellifolium* (Tomato Genome Consortium 2012), Bolger et al. (2014) have recently released the genome sequences for the M82 cultivar and *S. pennellii* LA0716. Anchoring the *S. pennellii* genome to the genetic map has allowed the identification of candidate genes for stress tolerance traits; in addition, the study has provided evidence for the role of transposable elements in the evolution of these traits (Bolger et al. 2014). These results demonstrate the power of sequencing the parental lines of permanent genetic populations that have been extensively phenotyped. It is worth noting, that within the SOL-100 sequencing project (http://solgenomics.net/organism/sol100/view), sequences are becoming available for most of the parents of the tomato IL libraries described above, which will further enhance the value of these genetic resources.

Association Mapping and Next-generation Populations

QTL analysis conducted in biparental mapping populations, using the linkage mapping approach, has proven to be an effective tool to identify the genetic basis of complex traits in plants, including tomato. The approach, in fact, has several advantages, such as the lack of structure in the mapping population, the presence of alleles segregating at a balanced frequency, and the possibility to detect rare alleles and epistasis. However, the method is limited by the restricted allelic variation in biparental mapping populations (as only two alleles at a given locus can be studied simultaneously), the low-resolution mapping (generally limited to 10–20 cM) due to the reduced generations of recombination that can lead to extended linkage blocks, and the time-consuming crosses that are necessary for QTL mapping (Zhu et al. 2008).

Linkage disequilibrium (LD)-based association analysis, also known as association mapping (AM), has been proposed as an alternative approach, which can overcome these drawbacks. The approach has been pioneered in human genetics, where it has been exploited broadly to analyze human diseases (Kerem et al. 1989; Corder et al. 1994; reviewed by Visscher et al. 2012). Thanks to the rapid advances in the development of genomic tools and the consequent reduction in costs of genomic technologies, AM is now becoming a popular and powerful strategy also in crop genetics and crop improvement (for review, see Rafalski 2010; Flint-Garcia et al. 2003; Gupta et al. 2005; Zhu et al. 2008; Larsson et al. 2009; Korte and Farlow 2013). Two AM methodologies are in use: candidate gene association and whole genome scan, also called Genome-Wide Association Study (GWAS) (Rafalski 2010).

AM approaches rely on natural patterns of LD (the nonrandom association of alleles at different loci in the population), as they use panels of theoretically unrelated individuals. For crops, the method capitalizes on the wide range of phenotypic variation and historical recombination events accumulated in natural populations and collections of landraces, breeding materials, and varieties to infer marker-phenotype associations (reviewed by Flint-Garcia et al. 2003; Rafalski 2010; Korte and Farlow 2013). This allows reducing research time, to sample a broader genetic diversity, and to take advantage of a much greater genetic resolution, due to a larger number of recombination events. By contrast, the AM approach requires a thorough understanding of both the genetic structure and the extent of LD of the collection studied (Flint-Garcia et al. 2003; Myles et al. 2009). The decay of LD has been shown to differ dramatically between species, and generally LD is higher in selfing species like cultivated tomato and rice, than in outcrossing species; however, it can vary significantly even within a species, and among loci within a population, sometimes caused by positive selection (Flint-Garcia et al. 2003; Myles et al. 2009; Robbins et al. 2011). The rate of LD decay influences the resolution with which a QTL can be mapped, the number and density of markers, as well as the experimental design needed to perform an association analysis (Myles et al. 2009). AM approaches can result in increased resolution compared to linkage mapping populations, as long as enough markers are provided; and, in an ideal scenario, they can lead to the identification of the causative polymorphism(s) of a QTL. Because of domestication, crops are liable not only to higher levels of LD, but also to population structure (the presence of subgroups with unequal distribution of alleles in the population studied), and cryptic relatedness (the presence of close relatives in a sample of unrelated individuals) that all need to be taken into account in statistical analyses (Ranc et al. 2012; Korte and Farlow 2013). To handle the confounding effect of background loci that may be present throughout the genome due to LD, and thus to address the problem of high LD in GWA scans, Segura et al. (2012) proposed a multilocus mixed model (MLMM). In addition, several statistical methods have been suggested to reduce the risk of detecting spurious false-positive or false-negative associations in GWA studies due to population structure and cryptic relatedness (Flint-Garcia et al. 2003; Mitchell-Olds 2010).

Despite the advantages of AM in terms of higher resolution, allelic richness and speed, pitfalls do exist, and hence linkage mapping is considered a valuable complementary approach (Larsson et al. 2013). For this reason, the two strategies are often applied together to mitigate each other flaws, for example to validate the associations identified by AM, thus reducing spurious associations (Flint-Garcia et al. 2003; Larsson et al. 2013).

In tomato, a few association studies have been conducted to dissect morphophysical and fruit traits. Nesbitt and Tanksley (2002) used a collection of 39 cherry tomato accessions to identify associations between fruit size and genomic sequence of the *fw2.2* region, which controls fruit weight (Frary et al. 2000). However, the small collection used prevented from finding any significant association. Subsequently, Mazzucato et al. (2008) investigated associations between 29 simple sequence repeat (SSR) markers and 15 morphophysiological traits in a collection of 50 tomato landraces. Recent association studies, which have included cherry tomato accessions (*S. lycopersicum* "cerasiforme"), have shown the potential of this genetic material to identify QTL by GWAS in tomato (Ranc et al. 2012; Xu et al. 2013). In particular, Ranc et al. (2012) carried out a pilot study to define the optimal conditions, including the marker density needed, to perform GWAS in the tomato by using an association panel of 90 tomato accessions (63 *S. lycopersicum* "cerasiforme"—cherry type, 17 *S. lycopersicum*—large fruited, 10 *S. pimpinellifolium*), focusing on chromosome 2, on which several clusters of QTL for fruit morphology and quality traits had been previously mapped (Causse et al. 2002). In another recent study, Xu et al. (2013) used low-density genome-wide-distributed SNP markers (SNPlexTM assay of 192 SNPs) on a large collection of 188 tomato accessions (44 heirloom and vintage cultivars (*S. lycopersicum*), 127 *S. lycopersicum* "cerasiforme" (cherry tomato) and 17 *S. pimpinellifolium* accessions) phenotyped for ten fruit quality traits. The results highlighted that GWAS in tomato should be easier with the group of *S. lycopersicum* "cerasiforme" accessions, characterized by an admixture structure (their genomes being mosaics of *S. lycopesicum* and the closely related wild species *S. pimpinellifolium*) as they exhibited higher minor frequency alleles (MAF) on average than cultivated group, lower LD and a less structured pattern. In spite of a high level of LD found in the collection at the whole genome level, a mixed linear model allowed the identification of several associations between SNP markers and fruit traits. However, the SNP density was still too low to identify SNPs in candidate genes.

Over the last years, the release of the tomato genome sequences (Tomato Genome Consortium 2012) and derived genomic tools such as a high-density SNP genotyping array (Sim et al. 2012) have offered new opportunities for GWAS in this crop. Shirasawa et al. (2013) analyzed a large collection of 663 tomato accessions with approximately 1300 SNPs obtained from resequencing analysis. Although, GWAS identified SNPs that were significantly associated with the measured agronomical traits, yet, the study investigated a limited number of traits (eight) with low precision on the association collection. More recently, Sauvage et al. (2014) have successfully applied high-resolution GWA using a MLMM as a general method for mapping complex traits in structured populations, to decipher the genetic architecture of tomato fruit composition traits. For this purpose, a core collection of 163 tomato accessions composed of *S. lycopersicum*, *S. lycopersicum* "cerasiforme," and *S. pimpinellifolium* was genotyped with 5995 SNP markers spread over the whole genome. GWAS was conducted on a large set of metabolic traits that showed stability over 2 years, and the analysis allowed the identification of promising candidate loci underlying traits such as fruit malate and citrate levels.

Although, AM has rarely been used to identify the molecular bases of QTL in tomato, recently it has been successfully applied to identify QTNs responsible for locule number differences between *S. lycopersicum* "cerasiforme" and *S. lycopersicum* Muños et al. (2011). Furthermore, a combined approach was pursued by Chakrabarti et al. (2013) to clone the tomato fruit mass QTL *fw3.2*; in this case, association mapping followed by segregation analysis allowed to circumvent the low rate of LD decay found around the *fw3.2* locus, and to identify a SNP in the promoter of the *SlKLUH* gene.

In order to overcome many of the shortcomings of both traditional biparental QTL mapping and AM approaches, a new generation of genetic-mapping populations, including Multi-parent Advanced Generation Inter-Cross (MAGIC) populations, have been proposed (Cavanagh et al. 2008). These next-generation populations combine the controlled crosses of QTL mapping with multiple parents and several generations of intermating to provide increased recombination and mapping resolution and to expand (albeit up to a certain point) allelic richness within the mapping population. The first tomato MAGIC population has been recently developed by Pascual et al. (2015) intercrossing eight resequenced *S. lycopersicum* founder lines, which had been selected to cover a wide range of genetic diversity. The study has shown the potential of this tomato MAGIC population for a better exploitation of intraspecific genetic variation, QTL mapping and for the identification of causal polymorphisms.

From QTL to QTN and Epialleles

A fundamental question in modern biology is identifying the causative genes and the genetic changes underlying complex traits. Whereas much progress has been made in detecting QTL, the molecular cloning of the underlying genes is lagging behind.

In tomato, map-based strategies, using higher resolution near-isogenic lines derived from the *S. pennellii* LA0716 ILs, were successfully applied for cloning the first-ever QTL: *fw2.2* (fruit weight) (Frary et al. 2000; Cong et al. 2002) and *Brix9-2-5* (sugar yield, or Brix) (Fridman et al. 2000, 2004). Both are major QTL, as natural genetic variation at *fw2.2* alone can change the size of fruit by up to 30 % (Frary et al. 2000), while *Brix9-2-5* can increase sugars by as much as 25 % (Fridman et al. 2000, 2004). The gene underlying *fw2.2* encodes a negative regulator of cell division, member of the *Cell Number Regulator* (*CNR*) family, and controls tomato fruit mass as well as organ size in other species, e.g., maize (Guo et al. 2010; Guo and Simmons 2011) and nitrogen-fixing nodule number (Libault et al. 2010). While modest changes in transcript quantity and in the timing of gene expression were correlated with natural variation at *fw2.2*, on the other hand, altered enzyme activity, as a result of a single nucleotide change in a cell wall invertase gene, *LIN5*, leading to a single amino acid change in the corresponding protein in an area very close to the substrate-binding site of the enzyme, was found to be the cause for the variation between the cultivated and wild species alleles at *Brix9-2-5* (Fridman et al. 2004). A comparative association study between the nucleotide polymorphism and activity of *LIN5* conducted in a set of ILs derived from additional tomato species led to the identification of the causative quantitative trait nucleotide (QTN) (Fridman et al. 2004). These first two studies demonstrated that IL-based Mendelian segregation is a very efficient way to partition continuous variation for complex traits into discrete molecular components. Furthermore, these QTL were the first among many showing that, similarly to the variation found for numerous genes that control monogenic traits, variation in QTL alleles in plants can be identified in both coding and regulatory regions of single genes (Paran and Zamir 2003; Salvi and Tuberosa 2005).

Because of domestication and selection, tomato cultivars show a wide variation in fruit morphology (size and shape) that is under the control of a large number of QTL (Grandillo et al. 1999; Tanksley 2004; van der Knaap et al. 2014). Wild and semi-wild forms of tomato carry small fruit that might weigh only a few grams and that are usually round and bilocular. By contrast, fruit from modern tomato varieties may contain many locules (up to 10 or more) and weigh up to 1 kg, and come in a wide variety of shapes that have been recently classified in eight shape categories (flat, ellipsoid, rectangular, oxheart, heart, long, obovoid, and round) using the software program Tomato Analyzer (Brewer et al. 2006, 2007; Rodriguez et al. 2010, 2011). Among the numerous fruit mass QTL identified in tomato, six loci [*fruit weight1.1* (*fw1.1*), *fw2.2*, *fw2.3*, *fw3.1*/*fw3.2*, *fw4.1*, and *fw9.1*] are postulated to be major QTL; whereas major fruit shape QTL include *ovate*, *locule number* (*lc*), *sun*, *fs8.1* and *fasciated* (*f* or *fas*) (Grandillo et al. 1999; Tanksley 2004; Chakrabarti et al. 2013; van der Knaap et al. 2014).

Following the positional cloning of *fw2.2*, significant efforts have been invested in deciphering the molecular basis of tomato fruit morphology. The results obtained so far from the map-based cloning of six tomato fruit shape and weight genes demonstrate that inversions, duplications, as well as SNPs in promoters and coding regions control the phenotypic diversity of the tomato fruit (reviewed by Monforte et al. 2014; Van der Knaap et al. 2014). The cloning of *fw2.2* revealed that one of the earliest steps in the evolution of larger tomato fruit was caused by a heterochronic regulatory mutation in a cell cycle–control gene, as more cells were observed in large compared with small fruits (Frary et al. 2000; Cong et al. 2002). More recently, Chakrabarti et al. (2013) have reported the fine mapping and cloning of a second major tomato fruit mass QTL, *fw3.2*, encoding the ortholog of KLUH, SlKLUH, a P450 enzyme of the CYP78A subfamily. A combination of association mapping followed by segregation analysis, and transgenic studies allowed the identification of a likely regulatory SNP in the promoter of the gene that was highly associated with fruit mass. The increase in fruit mass resulted from the production of extra cell layers in the pericarp, taking place after fertilization, which implies that *SlKLUH* affects cell division.

Changes in *fw2.2* and other cell cycle related genes, however, cannot explain the extreme fruit size observed in modern tomato cultivars. Rather, the development of extreme fruit size has been associated to several QTL affecting locule number, which can influence both fruit size and shape. Two of these QTL, *fas* (chromosome 11) and *lc* (chromosome 2), and their epistatic interactions, explain most of the phenotypic variation (Lippman and Tanksley 2001; Barrero and Tanksley 2004). Both QTL affect organ (carpel) number rather than size, but *fas* exerts the larger effect; in addition, both QTL influence flat fruit shape (Lippman and Tanksley 2001; Barrero and Tanksley 2004; Barrero et al. 2006; Rodriguez et al. 2011). Besides *fas* and *lc*, other two major fruit shape QTL, whose molecular bases have been deciphered, are *ovate* (chromosome 2) and *sun* (chromosome 7), and both influence fruit elongation (Tanksley 2004; Rodriguez et al. 2011).

Positional cloning of *ovate* was achieved using segregating populations derived from *S. pennellii* ILs (Liu et al. 2002). The gene encodes a protein in the Ovate Family Protein (OFP) that is thought to negatively regulate transcription of target genes (Liu et al. 2002; van der Knaap et al. 2014), and a premature stop codon in *OVATE* controls fruit elongation. The *OVATE* gene affects fruit shape well before anthesis, and the increase in fruit elongation is caused by cell proliferation in the proximal region of the developing ovary (van der Knaap and Tanksley 2001; Monforte et al. 2014; van der Knaap et al. 2014).

The same *S. pennellii* IL-based strategy was adopted to clone the gene underlying the *fas* QTL, which was found to encode a YABBY-like transcription factor; a mutation in *FAS* leads to an increase in locule number which affects both fruit shape (flattened fruit) and fruit mass (larger fruit) (Lippman and Tanksley 2001; Cong et al. 2008). Initially, the mutation was postulated to be caused by a large insertion in the first intron of *YABBY* (Cong et al. 2008); however, a reexamination of the nature of the genome rearrangement at the *fas* locus demonstrated that the mutation is due to a 294-kb inversion disrupting the *YABBY* gene (Huang and van der Knaap 2011).

For the cloning of the other two major fruit shape QTL, *sun* and *lc*, the *S. pennellii* IL resource could not be used. For *sun*, the obstacle was given by its map position, as this locus was localized inside a paracentric inversion within the *S. pennellii* genome (van der Knaap et al. 2004).

For *lc*, the limitation derived from its weaker effect on fruit locules compared with that of *fas*, and it was, therefore, necessary to overcome all genetic background effects.

Positional cloning of *sun* revealed that the gene underlying this QTL encodes a member of the IQ domain family (Xiao et al. 2008). The elongated fruit phenotype is caused by an unusual interchromosomal 24.7-kb gene duplication event mediated by the long-terminal repeat retrotransposon *Rider*, which results in a much higher expression of *SUN* throughout floral and fruit development and an extremely elongated fruit (Xiao et al. 2008; Jiang et al. 2009; Wu et al. 2011). Although fruit shape patterning mediated by *SUN* is most likely established before anthesis, yet, the most significant fruit shape changes take place after fertilization, during the cell division stage of fruit development (van der Knaap and Tanksley 2001; Xiao et al. 2009).

More recently, the *lc* QTL was positionally cloned using a combination of map-based cloning to identify the locus region (a sequence of 1600 bp) between a putative ortholog of *WUSCHEL* (*WUS*), which encodes a homeodomain protein that regulates stem cell fate in plants, and a WD40 motif containing protein, and association mapping to refine its molecular characterization, which consisted of two SNPs located approximately 1080-bp downstream of the stop codon of *WUS* (Muños et al. 2011). Subtle changes in the expression of *SlWUS* are likely the cause of the increased number of locules determined by *lc* (van der Knaap et al. 2014). It has also been suggested that the *lc* mutation might cause a loss-of-function regulatory element which would allow a higher expression of *SlWUS*, resulting in maintenance of a larger stem cell population and hence in increased locule numbers (van der Knaap et al. 2014).

Map-based cloning approaches have also been used to decipher the molecular basis of other two major QTL in tomato: *style length 2.1* (*Style 2.1*) (Chen et al. 2007), controlling a key floral attribute associated with the evolution of autogamy in cultivated tomatoes, and *seed weight 4.1* (*sw4.1*) (Orsi and Tanksley 2009). Mapping studies had demonstrated that most of the structural changes that accompanied the evolutionary transition from cross-pollinating to self-pollinating flowers could be explained by a single major QTL on chromosome 2, designated *stigma exertion 2.1* (*se2.1*) (Bernacchi and Tanksley 1997; Fulton et al. 1997). Fine mapping has shown that *se2.1* was a complex locus composed of at least five closely linked genes: three controlling stamen length, one conditioning anther dehiscence, and a fifth one, which accounted for the greatest change in stigma exertion, controlling style length (*Style 2.1*) (Chen and Tanksley 2004). Positional cloning of *Style2.1* revealed that this gene encodes a putative transcription factor that regulates cell elongation in developing styles and that the transition from allogamy to autogamy was caused by a mutation in the *Style2.1* promoter that leads to downregulation of *Style2.1* expression during flower development (Chen et al. 2007).

The numerous QTL mapping studies conducted for tomato seed size in several interspecific crosses have revealed over 20 QTL accounting for most seed size variation; among these, the major QTL *Sw4.1*, mapping on chromosome 4, constantly explained a large fraction (up to 25 %) of the total phenotypic variation in segregating populations (Table 4.4) (reviewed by Doganlar et al. 2000b). For this reason, *Sw4.1* was selected for map-based cloning, and using a combination of genetic, developmental, molecular, and transgenic techniques Orsi and Tanksley (2009) identified a gene encoding an *ABC* transporter gene as the cause of the *Sw4.1* QTL. This gene exerts its control on seed size via gene expression in the developing zygote.

Despite the successes achieved so far, delimiting a QTL to a single gene using genetic approaches is still a technically demanding and daunting undertaking, largely limited to loci exerting large effects upon quantitative variation. In order to enhance the rate of QTL cloning, integrated strategies, which combine near-isogenic line mapping with "omic" analyses (transcriptome or genomic resequencing, metabolome and/or proteome) can be pursued (Wayne and McIntyre 2002). These approaches represent efficient tools for exploring the functional relationship between genotype and phenotype, as they facilitate filtering through candidate genes in a QTL interval. In line with this, Lee et al. (2012) applied ripe fruit transcriptional and metabolic profiling to the *S. pennellii* LA0716 exotic library. Correlation analyses allowed mining for candidate genes, and the ethylene response factor SlERF6 was identified as a valuable target for RNAi analysis, which showed that SlERF6 plays a central role in tomato ripening integrating the ethylene and carotenoid synthesis pathways. This study demonstrated the utility of systems-based analysis to identify genes controlling complex biochemical traits in tomato.

More recently, Quadrana et al. (2014), have identified the gene underlying a major tomato vitmine E (VTE) QTL (*mQTL9-2-6*), which encodes a 2-methyl-6-phytylquinol methyltransferase (namely VTE3(1)). Using a combination of reverse genetic approaches, expression analyses, siRNA profiling and DNA methylation assays, the authors demonstrated that *mQTL9-2-6* is an expression QTL associated with differential methylation of a SINE retrotransposon located in the promoter region of *VTE3(1)*. In addition, different epialleles affecting *VTE3(1)* expression and consequently VTE content in fruits were observed because of spontaneous reversions of promoter DNA methylation. These findings demonstrate that epigenetics can affect quantitative phenotypes of agronomic interest.

Conclusions and Perspectives

We have reviewed more than three decades of research conducted in tomato to dissect the genetic and molecular bases of quantitative traits. Over these years, the tomato clade (*Solanum* sect. *Lycopersicon*) has been at the forefront not only for the localization, characterization, and positional cloning of QTL, but also for the development of new molecular breeding strategies, namely the "AB-QTL" and the "IL libraries," aimed at a more efficient exploitation of the wealth of genetic variation stored in unadapted germplam. The last 20 years of research conducted on the founder *S. pennellii* LA0716 IL library have demonstrated the power of these congenic and permanent resources for the genetic and molecular analyses of QTL, for exploring the genetic bases of heterosis, and for the related practical outcomes, which have resulted in the development of a leading hybrid variety.

The numerous QTL mapping studies conducted in tomato so far have allowed the identification of thousands of QTL many of which are of potential interest for the improvement of this crop. However, despite this richness of genetic information, only a few major QTL have been isolated to date. In order to reverse this trend the tomato research community is capitalizing on the ever growing genetic and "omic" tools, which, in turn, are building on the recently released tomato genome sequences (Tomato Genome Consortium 2012). In this respect, the application of integrated approaches are allowing more detailed systems-level analyses which hold the promise of enhancing our understanding of the functional relationship between genotypes and complex phenotypes (Schauer et al. 2006, 2008; Lee et al. 2012; Chitwood et al. 2013; Pascual et al. 2013).

In addition, the availability of the tomato genome sequences (Tomato Genome Consortium 2012) along with the advent of new cost-effective, high-throughput genotyping, and sequencing technologies are opening new avenues for a reexamination of the variation and inheritance of quantitative traits at the intraspecific level (Pascual et al. 2015; Sauvage et al. 2014). AM approaches can be viewed as complementary to AB-QTL and IL populations as they represent an additional tool for exploring and exploiting extant functional diversity available for each crop species on a much larger scale (Zhu et al. 2008). Furthermore, within the SOL-100 sequencing project (http://solgenomics.net/organism/sol100/view), sequences are becoming available for most of the parents of the tomato IL/BIL populations developed so far. This, in principle, should allow traits to be mapped to known sequence variation, which, in turn, should provide a major advancement in the identification of valuable alleles, further increasing the value of these genetic resources (Bolger et al. 2014). In view of the rapid developments in sequencing technology, it is also foreseen that methods that make use of whole genome sequencing-based technique, such as QTL-seq, will also accelerate crop improvement in a cost-effective way (Takagi et al. 2013).

In order to facilitate the identification of candidate genes and thus help elucidating the molecular basis of quantitative phenotypes, several bioinformatic tools are being developed (Tecle et al. 2010; Chibon et al. 2012). Notably, the Sol Genomics Network (SGN, http://solgenomics.net) has implemented a new QTL module, solQTL, which allows researchers to upload their raw genotype and phenotype QTL data to SGN, perform QTL analysis and dynamically cross-link to relevant genetic, expression and genome annotations, using a user-friendly web interface.

The constant improvements of molecular platforms, the development of new types of genetic resources, along with progresses in bioinformatics and in tools for functionally testing candidate genes are expected to rapidly enhance our ability in unveiling the molecular basis of QTL other than those with a major effect.

In spite of all these technological advances, QTL mapping in biparental populations will probably remain the method of choice for the analysis of epistatic interactions and when rare alleles are involved, especially those with moderate effects (Rafalski 2010). Regardless of the mapping approach used, independent validation of the associations and evaluation of their effects in different genetic backgrounds remain essential aspects of QTL analyses. Furthermore, the role of epigenetics in determining variation in quantitative traits and in phenotypic plasticity needs to be further addressed (Cobb et al. 2013; Quadrana et al. 2014).

Given the wealth of low-cost genomic information, which is rapidly becoming available for most important crop species, phenotyping is emerging as the major bottleneck and funding constraint limiting the power of quantitative traits analyses (Cobb et al. 2013). There is a clear need for precision phenotyping systems able to provide high-quality phenotypic information on the entire collection of genetic factors underlying quantitative phenotypic variation at all levels of biological organization (cells, tissues, organs, and developmental stages) as well as across years, environments, species, and research programs (Chitwood and Sinha 2013; Cobb et al. 2013). Due to the development of high-throughput platforms and image analysis software packages, next-generation phenotyping will require novel data management, access, and storage systems (Cobb et al. 2013). In this framework, public phenotype "warehousing" databases are foreseen as an additional necessary tool to empower our understanding of the genetic and molecular architecture of complex traits (Zamir 2013), and thus to ensure continued advancement in crop improvement aimed at sustainably meeting the demands of a growing human population under changing climates (Godfray et al. 2010).

Acknowledgments The authors thank all the colleagues who provided unpublished information and apologize to those authors whose work we could not highlight because of space limitations. Research in the laboratory of S. Grandillo and M. Cammareri was supported in part by the EUSOL project PL 016214-2, the Italian Ministry of University and Research (MIUR) project GenoPOM-PRO, a dedicated grant from the Italian Ministry of Economy and Finance to the National Research Council for the project "Innovazione e Sviluppo del Mezzogiorno —Conoscenze Integrate per Sostenibilità ed Innovazione del Made in Italy Agroalimentare—Legge n. 191/2009," and the PON R&C 2007–2013 grant financed by the Italian MIUR in cooperation with the European Funds for the Regional Development (FESR).

References

Aflitos S, Schijlen E, Jong H et al (2014) Exploring genetic variation in the tomato (Solanum section Lycopersicon) clade by whole-genome sequencing. Plant J 80(1):136–148

Agrama HA, Scott JW (2006) Quantitative trait loci for Tomato yellow leaf curl virus and Tomato mottle virus resistance in tomato. J Am Soc Hort Sci 131(2): 637–645

Almeida J, Quadrana L, Asís R et al (2011) Genetic dissection of vitamin E biosynthesis in tomato. J Exp Bot 62(11):3781–3798

Alpert K, Tanksley S (1996) High-resolution mapping and isolation of a yeast artificial chromosome contig containing *fw2.2*: a major fruit weight quantitative trait locus in tomato. Proc Natl Acad Sci USA 93:15503–15507

Alpert K, Grandillo S, Tanksley SD (1995) *fw2.2*: a major QTL controlling fruit weight is common to both red- and green-fruited tomato species. Theor Appl Genet 91:994–1000

Alseekh S, Ofner I, Pleban T et al (2013) Resolution by recombination: breaking up *Solanum pennellii* introgressions. Trends Plant Sci 18(10):536–538

Anbinder I, Reuveni M, Azari R et al (2009) Molecular dissection of Tomato leaf curl virus resistance in tomato line TY172 derived from *Solanum peruvianum*. Theor Appl Genet 119(3):519–530

Arikita FN, Azevedo MS, Scotton DC et al (2013) Novel natural genetic variation controlling the competence to form adventitious roots and shoots from the tomato wild relative *Solanum pennellii*. Plant Sci 199–200:121–130

Ashrafi H, Kinkade MP, Merk H et al (2012) Identification of novel quantitative trait loci for increased lycopene content and other fruit quality traits in a tomato recombinant inbred line population. Mol Breed 30(1):549–567

Asins MJ, Bolarín MC, Pérez-Alfocea F et al (2010) Genetic analysis of physiological components of salt tolerance conferred by *Solanum* rootstocks. What is the rootstock doing for the scion? Theor Appl Genet 121(1):105–115

Asins MJ, Villalta I, Aly MM et al (2013) Two closely linked tomato HKT coding genes are positional candidates for the major tomato QTL involved in Na^+/K^+ homeostasis. Plant, Cell Environ 36(6):1171–1191

Aurand R, Faurobert M, Page D et al (2012) Anatomical and biochemical trait network underlying genetic variations in tomato fruit texture. Euphytica 187 (1):99–116

Azanza F, Young TE, Kim D et al (1994) Characterization of the effect of introgressed segments of chromosome 7 and 10 from *Lycopersicon chmielewskii* on tomato soluble solids, pH, and yield. Theor Appl Genet 87:965–972

Bai YL, Lindhout P (2007) Domestication and breeding of tomatoes: what have we gained and what can we gain in the future? Ann Bot 100:1085–1094

Bai Y, Huang CC, van der Hulst R et al (2003) QTLs for tomato powdery mildew resistance (*Oidium lycopersici*) in *Lycopersicon parviflorum* G1.1601 co-localize with two qualitative powdery mildew resistance genes. Mol Plant Microbe Interact 16:169–176

Barrantes W, Fernández-del-Carmen A, López-Casado G et al (2014) Highly efficient genomics-assisted development of a library of introgression lines of *Solanum pimpinellifolium*. Mol Breed 34:1817–1831

Barrero LS, Tanksley SD (2004) Evaluating the genetic basis of multiple-locule fruit in a broad cross section of tomato cultivars. Theor Appl Genet 109:669–679

Barrero LS, Cong B, Wu F et al (2006) Developmental characterization of the *fasciated* locus and mapping of Arabidopsis candidate genes involved in the control of floral meristem size and carpel number in tomato. Genome 49(8):991–1006

Baxter CJ, Sabar M, Quick WP et al (2005) Comparison of changes in fruit gene expression in tomato introgression lines provides evidence of genome-wide transcriptional changes and reveals links to mapped QTLs and described traits. J Exp Bot 56:1591–1604

Bedinger PA, Chetelat RT, McClure B et al (2011) Interspecific reproductive barriers in the tomato clade: opportunities to decipher mechanisms of reproductive isolation. Sex Plant Reprod 24(3):171–187

Bernacchi D, Tanksley SD (1997) An interspecific backcross of *Lycopersicon esculentum* × *L. hirsutum*: linkage analysis and a QTL study of sexual compatibility factors and floral traits. Genetics 147:861–877

Bernacchi D, Beck-Bunn T, Eshed Y et al (1998a) Advanced backcross QTL analysis in tomato. I. Identification of QTLs for traits of agronomic importance from *Lycopersicon hirsutum*. Theor Appl Genet 97:381–397

Bernacchi D, Beck-Bunn T, Emmatty D et al (1998b) Advanced backcross QTL analysis of tomato. II. Evaluation of near-isogenic lines carrying single-donor introgressions for desirable wild QTL-alleles derived from *Lycopersicon hirsutum* and *L. pimpinellifolium*. Theor Appl Genet 97:170–180 and 1191–1196

Bernatzky R, Tanksley S (1986) Toward a saturated linkage map in tomato based on isozymes and random cDNA sequences. Genetics 112:887–898

Bertin N, Borel C, Brunel B et al (2003) Do genetic make-up and growth manipulation affect tomato fruit size by cell number, or cell size and DNA endoreduplication? Ann Bot 92(3):415–424

Bertin N, Causse M, Brunel B et al (2009) Identification of growth processes involved in QTLs for tomato fruit size and composition. J Exp Bot 60(1):237–248

Blanca J, Cañizares J, Cordero L et al (2012) Variation revealed by SNP genotyping and morphology provides insight into the origin of the tomato. PLoS ONE 7:e48198

Blauth SL, Churchill GA, Mutschler MA (1998) Identification of quantitative trait loci associated with acylsugar accumulation using intraspecific populations of the wild tomato, *Lycopersicon pennellii*. Theor Appl Genet 96:458–467

Blauth SL, Steffens JC, Churchill GA et al (1999) Identification of QTLs controlling acylsugar fatty acid composition in an intraspecific population of *Lycopersicon pennellii* (Corr.) D'Arcy. Theor Appl Genet 99:373–381

Bolger A, Scossa F, Bolger ME et al (2014) The genome of the stress-tolerant wild tomato species *Solanum pennellii*. Nat Genet 46(9):1034–1039

Botstein D, White RL, Skolnick M et al (1980) Construction of a genetic linkage map in man using restriction fragment length polymorphism. Am J Hum Genet 32:314–331

Bretó MP, Asins MJ, Carbonell EA (1994) Salt tolerance in *Lycopersicon* species. III. Detection of quantitative trait loci by means of molecular markers. Theor Appl Genet 88:395–401

Brewer MT, Lang L, Fujimura K et al (2006) Development of a controlled vocabulary and software application to analyze fruit shape variation in tomato and other plant species. Plant Physiol 141:15–25

Brewer MT, Moyseenko JB, Monforte AJ et al (2007) Morphological variation in tomato fruit: a comprehensive analysis and identification of loci controlling fruit shape and development. J Exp Bot 58:1339–1349

Brouwer DJ, St. Clair DA (2004) Fine mapping of three quantitative trait loci for late blight resistance in tomato using near isogenic lines (NILs) and sub-NILs. Theor Appl Genet 108:628–638

Brouwer DJ, Jones ES, St. Clair DA (2004) QTL analysis of quantitative resistance to *Phytophthora infestans* (late blight) in tomato and comparison with potato. Genome 47:475–492

Canady MA, Meglic V, Chetelat RT (2005) A library of *Solanum lycopersicoides* introgression lines in cultivated tomato. Genome 48:685–697

Carmeille A, Caranta EC, Dintinger EJ et al (2006) Identification of QTLs for *Ralstonia solanacearum* race 3-phylotype II resistance in tomato. Theor Appl Genet 114:110–121

Causse M, Saliba-Colombani V, Lesschaeve I et al (2001) Genetic analysis of organoleptic quality in fresh market tomato. 2. Mapping QTLs for sensory attributes. Theor Appl Genet 102:273–283

Causse M, Saliba-Colombani V, Lecomte L et al (2002) QTL analysis of fruit quality in fresh market tomato: a few chromosome regions control the variation of sensory and instrumental traits. J Exp Bot 53:2089–2098

Causse M, Duffe P, Gomez MC et al (2004) A genetic map of candidate genes and QTLs involved in tomato fruit size and composition. J Exp Bot 55:1671–1685

Causse M, Chaïb J, Lecomte L et al (2007) Both additivity and epistasis control the genetic variation for fruit quality traits in tomato. Theor Appl Genet 115 (3):429–442

Causse M, Desplat N, Pascual L et al (2013) Whole genome resequencing in tomato reveals variation associated with introgression and breeding events. BMC Genom 14:791. doi:10.1186/1471-2164-14-791

Cavanagh C, Morell M, Mackay I et al (2008) From mutations to MAGIC: resources for gene discovery, validation and delivery in crop plants. Curr Opin Plant Biol 11:215–221

Chaerani R, Smulders MJ, van der Linden CG et al (2007) QTL identification for early blight resistance (*Alternaria solani*) in a *Solanum lycopersicum* x *S. arcanum* cross. Theor Appl Genet 114(3):439–450

Chagué V, Mercier JC, Guénard M et al (1997) Identification of RAPD markers linked to a locus involved in quantitative resistance to TYLCV in tomato by bulked segregant analysis. Theor Appl Genet 95:671–677

Chaïb J, Lecomte L, Buret M et al (2006) Stability over genetic backgrounds, generations and years of quantitative trait locus (QTLs) for organoleptic quality in tomato. Theor Appl Genet 112:934–944

Chaïb J, Devaux MF, Grotte MG et al (2007) Physiological relationships among physical, sensory, and morphological attributes of texture in tomato fruits. J Exp Bot 58(8):1915–1925

Chakrabarti M, Zhang N, Sauvage C et al (2013) A cytochrome P450 regulates a domestication trait in cultivated tomato. Proc Natl Acad Sci USA 10 (42):17125–17130

Chapman NH, Bonnet J, Grivet L et al (2012) High-resolution mapping of a fruit firmness-related quantitative trait locus in tomato reveals epistatic interactions associated with a complex combinatorial locus. Plant Physiol 159(4):1644–1657

Chen AL, Liu CY, Chen CH et al (2014) Reassessment of QTLs for late blight resistance in the tomato accession L3708 using a restriction site associated DNA (RAD) linkage map and highly aggressive isolates of *Phytophthora infestans*. PLoS ONE. doi:10.1371/journal.pone.0096417

Chen FQ, Foolad MR, Hyman J et al (1999) Mapping of QTLs for lycopene and other fruit traits in a *Lycopersicon esculentum* × *L. pimpinellifolium* cross and comparison of QTLs across tomato species. Mol Breed 5:283–299

Chen KY, Tanksley SD (2004) High-resolution mapping and functional analysis of *se2.1*: a major stigma exsertion quantitative trait locus associated with the evolution from allogamy to autogamy in the genus *Lycopersicon*. Genetics 168:1563–1573

Chen KY, Cong B, Wing R et al (2007) Changes in regulation of a transcription factor lead to autogamy in cultivated tomatoes. Science 318:643–645

Chetelat RT, Meglic V (2000) Molecular mapping of chromosome segments introgressed from *Solanum lycopersicoides* into cultivated tomato (*Lycopersicon esculentum*). Theor Appl Genet 100:232–241

Chibon PY, Schoof H, Visser RG et al (2012) Marker2sequence, mine your QTL regions for candidate genes. Bioinformatics 2028(14):1921–1922

Chitwood DH, Sinha NR (2013) A census of cells in time: quantitative genetics meets developmental biology. Curr Opin Plant Biol 16(1):92–99

Chitwood DH, Kumar R, Headland LR (2013) A quantitative genetic basis for leaf morphology in a set of precisely defined tomato introgression lines. Plant Cell 25(7):2465–2481

Coaker GL, Francis DM (2004) Mapping, genetic effects, and epistatic interaction of two bacterial canker resistance QTLs from *Lycopersicon hirsutum*. Theor Appl Genet 108:1047–1055

Coaker GL, Meulia T, Kabelka EA et al (2002) A QTL controlling stem morphology and vascular development in *Lycopersicon esculentum* × *Lycopersicon hirsutum* (Solanaceae) crosses is located on chromosome 2. Am J Bot 89:1859–1866

Cobb JN, Declerck G, Greenberg A et al (2013) Next-generation phenotyping: requirements and strategies for enhancing our understanding of genotype-phenotype relationships and its relevance to crop improvement. Theor Appl Genet 126(4):867–887

Cong B, Tanksley SD (2006) *Fw2.2* and cell cycle control in developing tomato fruit: a possible example of gene co-option in the evolution of a novel organ. Plant Mol Biol 62:867–880

Cong B, Liu J, Tanksley SD (2002) Natural alleles at a tomato fruit size quantitative trait locus differ by heterochronic regulatory mutations. Proc Natl Acad Sci USA 99:13606–13611

Cong B, Barrero LS, Tanksley SD (2008) Regulatory change in YABBY-like transcription factor led to evolution of extreme fruit size during tomato domestication. Nat Genet 40:800–804

Corder EH, Saunders AM, Risch NJ et al (1994) Protective effect of apolipoprotein-E type-2 allele for late-onset Alzheimer disease. Nat Genet 7:180–184

Dal Cin V, Kevany B, Fei Z et al (2009) Identification of *Solanum habrochaites* loci that quantitatively influence tomato fruit ripening-associated ethylene emissions. Theor Appl Genet 119(7):1183–1192

Danesh D, Aarons S, McGill GE et al (1994) Genetic dissection of oligogenic resistance to bacterial wilt in tomato. Mol Plant-Microbe Interact 7:464–471

Davis J, Yu D, Evans W et al (2009) Mapping of loci from *Solanum lycopersicoides* conferring resistance or susceptibility to *Botrytis cinerea* in tomato. Theor Appl Genet 119:305–314

de Vicente MC, Tanksley SD (1993) QTL analysis of transgressive segregation in an interspecific tomato cross. Genetics 134:585–596

Di Matteo A, Sacco A, Anacleria M et al (2010) The ascorbic acid content of tomato fruits is associated with the expression of genes involved in pectin degradation. BMC Plant Biol 10:163

Di Matteo A, Ruggieri V, Sacco A et al (2013) Identification of candidate genes for phenolics accumulation in tomato fruit. Plant Sci 205–206:87–96

Do PT, Prudent M, Sulpice R et al (2010) The influence of fruit load on the tomato pericarp metabolome in a *Solanum chmielewskii* introgression line population. Plant Physiol 154(3):1128–1142

Doganlar S, Tanksley SD, Mutschler MA (2000a) Identification and molecular mapping of loci controlling fruit ripening time in tomato. Theor Appl Genet 100 (2):249–255

Doganlar S, Frary A, Tanksley SD (2000b) The genetic basis of seed-weight variation: tomato as a model system. Theor Appl Genet 100:1267–1273

Doganlar S, Frary A, Ku H-M et al (2002) Mapping quantitative trait loci in inbred backcross lines of *Lycopersicon pimpinellifolium* (LA1589). Genome 45:1189–1202

Edwards MD, Stuber CW, Wendel JF (1987) Molecular-marker facilitated investigations of quantitative-trait loci in maize. I. Numbers, genomic distribution and types of gene action. Genetics 116:113–125

Eshed Y, Zamir D (1994) Introgressions from *Lycopersicon pennellii* can improve the soluble solids yield of tomato hybrids. Theor Appl Genet 88:891–897

Eshed Y, Zamir D (1995) An introgression line population of *Lycopersicon pennellii* in the cultivated tomato enables the identification and fine mapping of yield-associated QTL. Genetics 141:1147–1162

Eshed Y, Zamir D (1996) Less-than-additive epistatic interactions of quantitative trait loci in tomato. Genetics 143:1807–1817

Eshed Y, Gera G, Zamir D (1996) A genome-wide search for wild-species alleles that increase horticultural yield of processing tomatoes. Theor Appl Genet 93:877–886

Estañ MT, Villalta I, Bolarín MC et al (2009) Identification of fruit yield loci controlling the salt tolerance conferred by *Solanum* rootstocks. Theor Appl Genet 118:305–312

Faino L, Azizinia S, Hassanzadeh BH et al (2012) Fine mapping of two major QTLs conferring resistance to powdery mildew in tomato. Euphytica 184(2): 223–234

Falconer DS (1989) Introduction to quantitative genetics, 3rd edn. Longman Scientific & Technical, Essex

Flint-Garcia SA, Thornsberry JM, Buckler ES (2003) Structure of linkage disequilibrium in plants. Annu Rev Plant Biol 54:357–374

Finkers R, van den Berg P, van Berloo R et al (2007a) Three QTLs for *Botrytis cinerea* resistance in tomato. Theor Appl Genet 114(4):585–593

Finkers R, van Heusden AW, Meijer-Dekens F et al (2007b) The construction of a *Solanum habrochaites* LYC4 introgression line population and the identification of QTLs for resistance to *Botrytis cinerea*. Theor Appl Genet 114:1071–1080

Firdaus S, van Heusden AW, Hidayati N et al (2013) Identification and QTL mapping of whitefly resistance components in *Solanum galapagense*. Theor Appl Genet 126(6):1487–1501

Foolad MR (1999a) Comparison of salt tolerance during seed germination and vegetative growth in tomato by QTL mapping. Genome 42:727–734

Foolad MR (1999b) Genetics of salt tolerance and cold tolerance in tomato: quantitative analysis and QTL mapping. Plant Biotechnol 16:55–64

Foolad MR (2007) Genome mapping and molecular breeding of tomato. Int J Plant Genomics. doi:10.1155/2007/64358

Foolad MR, Chen FQ (1998) RAPD markers associated with salt tolerance in an interspecific cross of tomato (*Lycopersicon esculentum* × *L. pennellii*). Plant Cell Rep 17:306–312

Foolad MR, Chen FQ (1999) RFLP mapping of QTLs conferring salt tolerance during vegetative stage in tomato. Theor Appl Genet 99:235–243

Foolad MR, Jones RA (1993) Mapping salt-tolerance genes in tomato (*Lycopersicon esculentum*) using trait-based marker analysis. Theor Appl Genet 87:184–192

Foolad MR, Stoltz T, Dervinis C et al (1997) Mapping QTLs conferring salt tolerance during germination in tomato by selective genotyping. Mol Breed 3:269–277

Foolad MR, Chen FQ, Lin GY (1998a) RFLP mapping of QTLs conferring salt tolerance during germination in an interspecific cross of tomato. Theor Appl Genet 97:1133–1144

Foolad MR, Chen FQ, Lin GY (1998b) RFLP mapping of QTLs conferring cold tolerance during seed germination in an interspecific cross of tomato. Mol Breed 4:519–529

Foolad MR, Lin GY, Chen FQ (1999) Comparison of QTLs for seed germination under non-stress, cold stress and salt stress in tomato. Plant Breed 118: 167–173

Foolad MR, Zhang LP, Lin GY (2001) Identification and validation of QTLs for salt tolerance during vegetative growth in tomato by selective genotyping. Genome 44:444–454

Foolad MR, Zhang LP, Khan AA et al (2002) Identification of QTLs for early blight (*Alternaria solani*) resistance in tomato using backcross populations of a *Lycopersicon esculentum* × *L. hirsutum* cross. Theor Appl Genet 104:945–958

Foolad MR, Zhang LP, Subbiah P (2003) Genetics of drought tolerance during seed germination in tomato: inheritance and QTL mapping. Genome 46:536–545

Foolad MR, Subbiah P, Zhang LP (2007) Common QTL affect the rate of tomato seed germination under different stress and nonstress conditions. Int J Plant Genom. doi:10.1155/2007/97386

Frary A, Nesbitt TC, Frary A et al (2000) *fw2.2*: a quantitative trait locus key to the evolution of tomato fruit size. Science 289:85–88

Frary A, Doganlar S, Frampton A et al (2003) Fine mapping of quantitative trait loci for improved fruit characteristics from *Lycopersicon chmielewskii* chromosome 1. Genome 46:235–243

Frary A, Fulton TM, Zamir D et al (2004a) Advanced backcross QTL analysis of a *Lycopersicon esculentum* × *L. pennellii* cross and identification of possible orthologs in the Solanaceae. Theor Appl Genet 108:485–496

Frary A, Fritz LA, Tanksley SD (2004b) A comparative study of the genetic bases of natural variation in tomato leaf, sepal, and petal morphology. Theor Appl Genet 109:523–533

Frary A, Göl D, Keleş D et al (2010) Salt tolerance in *Solanum pennellii*: antioxidant response and related QTL. BMC Plant Biol 10:58

Frary A, Keles D, Pinar H et al (2011) NaCl tolerance in *Lycopersicon pennellii* introgression lines: QTL related to physiological responses. Biol Plant 55 (3):461–468

Fridman E, Pleban T, Zamir D (2000) A recombination hotspot delimits a wild-species quantitative trait locus for tomato sugar content to 484 bp within an invertase gene. Proc Natl Acad Sci USA 97:4718–4723

Fridman E, Liu YS, Carmel-Goren L et al (2002) Two tightly linked QTLs modify tomato sugar content via different physiological pathways. Mol Genet Genom 266:821–826

Fridman E, Carrari F, Liu YS et al (2004) Zooming in on a quantitative trait for tomato yield using interspecific introgressions. Science 305:1786–1789

Fulton TM, Beck-Bunn T, Emmatty D et al (1997) QTL analysis of an advanced backcross of *Lycopersicon peruvianum* to the cultivated tomato and comparisons with QTLs found in other wild species. Theor Appl Genet 95:881–894

Fulton TM, Grandillo S, Beck-Bunn T et al (2000) Advanced backcross QTL analysis of a *Lycopersicon esculentum* × *L. parviflorum* cross. Theor Appl Genet 100:1025–1042

Fulton TM, Bucheli P, Voirol E (2002) Quantitative trait loci (QTL) affecting sugars, organic acids and other biochemical properties possibly contributing to flavor, identified in four advanced backcross populations of tomato. Euphytica 127:163–177

Geldermann H (1975) Investigation on inheritance of quantitative characters in animals by gene markers. I. Methods. Theor Appl Genet 46:319–330

Georgelis N, Scott JW, Baldwin EA (2004) Relationship of tomato fruit sugar concentration with physical and chemical traits and linkage of RAPD markers. J Am Soc Hort Sci 129:839–845

Georgiady MS, Whitkus RW, Lord EM (2002) Genetic analysis of traits distinguishing outcrossing and self-pollinating forms of currant tomato, *Lycopersicon pimpinellifolium* (Jusl.) Mill. Genetics 161:333–344

Godfray HC, Beddington JR, Crute IR et al (2010) Food security: the challenge of feeding 9 billion people. Science 327:812–818

Goldman IL, Paran I, Zamir D (1995) Quantitative trait locus analysis of a recombinant inbred line population derived from a *Lycopersicon esculentum* × *L. cheesmanii* cross. Theor Appl Genet 90:925–932

Gong P, Zhang J, Li H et al (2010) Transcriptional profiles of drought-responsive genes in modulating transcription signal transduction, and biochemical pathways in tomato. J Exp Bot 61(13):3563–3575

Gonzalo MJ, van der Knaap E (2008) A comparative analysis into the genetic bases of morphology in tomato varieties exhibiting elongated fruit shape. Theor Appl Genet 116:647–656

Goodstal FJ, Kohler GR, Randall LB et al (2005) A major QTL introgressed from wild *Lycopersicon hirsutum* confers chilling tolerance to cultivated tomato (*Lycopersicon esculentum*). Theor Appl Genet 111:898–905

Gorguet B, Eggink PM, Ocaña J et al (2008) Mapping and characterization of novel parthenocarpy QTLs in tomato. Theor Appl Genet 116:755–767

Grandillo S (2013) Introgression libraries with wild relatives of crops. In: Tuberosa R, Graner A, Frison E (eds) Genomics of plant genetics resources (chapt 4). Springer, Dordrecht, pp 87–122

Grandillo S, Tanksley SD (1996a) QTL analysis of horticultural traits differentiating the cultivated tomato from the closely related species *Lycopersicon pimpinellifolium*. Theor Appl Genet 92:935–951

Grandillo S, Tanksley SD (1996b) Genetic analysis of RFLPs, GATA microsatellites and RAPDs in a cross between *L. esculentum* and *L. pimpinellifolium*. Theor Appl Genet 92:957–965

Grandillo S, Ku HM, Tanksley SD (1996) Characterization of *fs8.1*, a major QTL influencing fruit shape in tomato. Mol Breed 2:251–260

Grandillo S, Ku HM, Tanksley SD (1999) Identifying the loci responsible for natural variation in fruit size and shape in tomato. Theor Appl Genet 99: 978–987

Grandillo S, Tanksley SD, Zamir D (2008) Exploitation of natural biodiversity through genomics. In: Varshney RK, Tuberosa R (eds) Genomics assisted crop improvement, vol I: genomics approaches and platforms. Springer, Dordrecht, pp 121–150

Grandillo S, Chetelat R, Knapp S et al (2011) *Solanum* sect. *Lycopersicon*. In: Kole C (ed) Wild crop relatives: genomic and breeding resources, vol 5: vegetables. Springer, Dordrecht, pp 129–215

Grandillo S, Termolino P, van der Knaap E (2013) Molecular mapping of complex traits in tomato. In: Kole C (ed) Genetics, genomics and breeding of crop plants. Volume: Liedl BE, Labate JA, Slade AJ, Stommel JR, Kole C (vol eds) Genetics, genomics and breeding of tomato. Science Publishers, Enfield, pp 150–227

Grandillo S, Cammareri M, Palombieri S, Fei Z, Xu Y, McQuinn R, Giovannoni J (2014) RNA-seq analysis in a set of *Solanum habrochaites* LA1777 introgression lines. In: 58th Italian society of agricultural genetics annual congress, 15–18 September, Alghero, Italy, ISBN 978-88-904570-4-3

Griffiths PD, Scott JW (2001) Inheritance and linkage of *Tomato mottle virus* resistance genes derived from *Lycopersicon chilense* accession LA 1932. J Am Soc Hort Sci 126:462–467

Guo M, Simmons CR (2011) Cell number counts—the *fw2.2* and *CNR* genes and implications for controlling plant fruit and organ size. Plant Sci 181:1–7

Guo M, Rupe MA, Dieter JA et al (2010) *Cell number regulator1* affects plant and organ size in maize: implications for crop yield enhancement and heterosis. Plant Cell 22:1057–1073

Gupta PK, Rustgi S, Kulwal PL (2005) Linkage disequilibrium and association studies in higher plants: present status and future prospects. Plant Mol Biol 57(4):461–485

Gur A, Zamir D (2004) Unused natural variation can lift yield barriers in plant breeding. PLoS Biol 2(10):e245

Gur A, Osorio S, Fridman E et al (2010) *hi2-1*, a QTL which improves harvest index, earliness and alters metabolite accumulation of processing tomatoes. Theor Appl Genet 121(8):1587–1599

Gur A, Semel Y, Osorio S et al (2011) Yield quantitative trait loci from wild tomato are predominately expressed by the shoot. Theor Appl Genet 122 (2):405–420

Haggard JE, Johnson EB, St Clair DA (2013) Linkage relationships among multiple QTL for horticultural traits and late blight (*P. infestans*) resistance on chromosome 5 introgressed from wild tomato *Solanum habrochaites*. G3 (Bethesda) 3(12):2131–2146. doi:10.1534/g3.113.007195

Hanson P, Schafleitner R, Huang SM et al (2014) Characterization and mapping of a QTL derived from *Solanum habrochaites* associated with elevated rutin content (quercetin-3-rutinoside) in tomato. Euphytica 200:441–454

Holtan HE, Hake S (2003) Quantitative trait locus analysis of leaf dissection in tomato using *Lycopersicon pennellii* segmental introgression lines. Genetics 165:1541–1550

Huang Z, van der Knaap E (2011) Tomato *fruit weight 11.3* maps close to *fasciated* on the bottom of chromosome 11. Theor Appl Genet 123:465–474

Huang Z, Van Houten J, Gonzalez G et al (2013) Genome-wide identification, phylogeny and expression analysis of *SUN*, *OFP* and *YABBY* gene family in tomato. Mol Genet Genomics 288(3–4):111–129

Hutton SF, Scott JW, Yang W et al (2010) Identification of QTL associated with resistance to bacterial spot race T4 in tomato. Theor Appl Genet 121(7):1275–1287

Hutton SF, Scott JW, Vallad GE (2014) Association of the Fusarium Wilt Race 3 Resistance Gene, I-3, on Chromosome 7 with Increased Susceptibility to Bacterial Spot Race T4 in Tomato. J Am Soc Hortic Sci 139(3):282–289

Ikeda H, Hiraga M, Shirasawa K et al (2013) Analysis of a tomato introgression line, IL8-3, with increased Brix content. Sci Hort 153:103–108

Jiang N, Gao D, Xiao H et al (2009) Genome organization of the tomato *sun* locus and characterization of the unusual retrotransposon *Rider*. Plant J 60(1):181–193

Jiménez-Gómez JM, Alonso-Blanco C, Borja A et al (2007) Quantitative genetic analysis of flowering time in tomato. Genome 50:303–315

Johnson EB, Haggard JE, St Clair DA (2012) Fractionation, stability, and isolate-specificity of QTL for resistance to *Phytophthora infestans* in cultivated tomato (*Solanum lycopersicum*). G3 (Bethesda) 2 (10):1145–1159

Kabelka E, Franchino B, Francis DM (2002) Two loci from *Lycopersicon hirsutum* LA407 confer resistance to strains of *Clavibacter michiganensis* subsp. *michiganensis*. Phytopathology 92:504–510

Kabelka E, Yang WC, Francis DM (2004) Improved tomato fruit color within an inbred backcross line derived from *Lycopersicon esculentum* and *L. hirsutum* involves the interaction of loci. J Am Soc Hort Sci 129:250–257

Kadirvel P, de la Pena R, Schafleitner R et al (2013) Mapping of QTLs in tomato line FLA456 associated with resistance to a virus causing tomato yellow leaf curl disease. Euphytica 190(2):297–308

Kamenetzky L, Asís R, Bassi S et al (2010) Genomic analysis of wild tomato introgressions determining metabolism- and yield-associated traits. Plant Physiol 152:1772–1786

Kazmi RH, Khan N, Willems LAJ et al (2012) Complex genetics controls natural variation among seed quality phenotypes in a recombinant inbred population of an interspecific cross between *Solanum lycopersicum* × *Solanum pimpinellifolium*. Plant, Cell Environ 35(5):929–951

Kerem BS, Rommens JM, Buchanan JA et al (1989) Identification of the cystic fibrosis gene: genetic analysis. Science 245:1073–1080

Khan N, Kazmi RH, Willems LAJ et al (2012) Exploring the natural variation for seedling traits and their link with seed dimensions in tomato. PLoS ONE. doi:10.1371/journal.pone.0043991

Kinkade MP, Foolad MR (2013) Validation and fine mapping of lyc12.1, a QTL for increased tomato fruit lycopene content. Theor Appl Genet 126(8):2163–2175

Kochevenko A, Fernie AR (2011) The genetic architecture of branched-chain amino acid accumulation in tomato fruits. J Exp Bot 62(11):3895–3906

Korte A, Farlow A (2013) The advantages and limitations of trait analysis with GWAS: a review. Plant Methods 9:29

Kromdijk J, Bertin N, Heuvelink E et al (2014) Crop management impacts the efficiency of quantitative trait loci (QTL) detection and use: case study of fruit load QTL interactions. J Exp Bot 65(1):11–22

Ku HM, Doganlar S, Chen KY et al (1999) The genetic basis of pear-shaped tomato fruit. Theor Appl Genet 99:844–850

Ku HM, Grandillo S, Tanksley SD (2000) *fs8.1*, a major QTL, sets the pattern of tomato carpel shape well before anthesis. Theor Appl Genet 101:873–878

Labate JA, Grandillo S, Fulton T et al (2007) Tomato. In: Kole C (ed) Genome mapping and molecular breeding in plants, vol 5: vegetables. Springer, Berlin, pp 1–96

Lahaye M, Quemener B, Causse M et al (2012) Hemicellulose fine structure is affected differently during ripening of tomato lines with contrasted texture. Int J Biol Macromol 1(4):462–470

Lahaye M, Devaux MF, Poole M et al (2013) Pericarp tissue microstructure and cell wall polysaccharide chemistry are differently affected in lines of tomato with contrasted firmness. Postharvest Biol Technol 76:83–90

Larsson SJ, Lipka AE, Buckler ES (2013) Lessons from *Dwarf8* on the strengths and weaknesses of structured association mapping. PLoS Genet 9(2):e1003246

Lawson DM, Lunde CF, Mutschler MA (1997) Marker-assisted transfer of acylsugar-mediated pest resistance from the wild tomato, *Lycopersicon pennellii*, to the cultivated tomato, *Lycopersicon esculentum*. Mol Breed 3:307–317

Leckie BM, De Jong DM, Mutschler MA (2012) Quantitative trait loci increasing acylsugars in tomato breeding lines and their impacts on silverleaf whiteflies. Mol Breed 30(4):1621–1634

Leckie BM, De Jong DM, Mutschler MA (2013) Quantitative trait loci regulating sugar moiety of acylsugars in tomato. Mol Breed 31(4):957–970

Lecomte L, Duffé P, Buret M et al (2004a) Marker-assisted introgression of five QTLs controlling fruit quality traits into three tomato lines revealed interactions between QTLs and genetic backgrounds. Theor Appl Genet 109:658–668

Lecomte L, Saliba-Colombani V, Gautier A et al (2004b) Fine mapping of QTLs of chromosome 2 affecting the fruit architecture and composition of tomato. Mol Breed 13:1–14

Lee JM, Joung JG, McQuinn R et al (2012) Combined transcriptome, genetic diversity and metabolite profiling in tomato fruit reveals that the ethylene response factor SlERF6 plays an important role in ripening and carotenoid accumulation. Plant J 70:191–204

Li J, Liu L, Bai Y et al (2011a) Seedling salt tolerance in tomato. Euphytica 178(3):403–414

Li J, Liu L, Bai Y et al (2011b) Identification and mapping of quantitative resistance to late blight (*Phytophthora infestans*) in *Solanum habrochaites* LA1777. Euphytica 179(3):427–438

Libault M, Zhang XC, Govindarajulu M (2010) A member of the highly conserved *FWL* (tomato *FW2.2-like*) gene family is essential for soybean nodule organogenesis. Plant J. 62:852–864

Lin KH, Yeh WL, Chen HM et al (2010) Quantitative trait loci influencing fruit-related characteristics of tomato grown in high-temperature conditions. Euphytica 174 (1):119–135

Lindhout P, Heusden S, Pet G et al (1994) Perspectives of molecular marker assisted breeding for earliness in tomato. Euphytica 79:279–286

Lippman Z, Tanksley SD (2001) Dissecting the genetic pathway to extreme fruit size in tomato using a cross between the small-fruited wild species *Lycopersicon pimpinellifolium* and *L. esculentum* var. Giant Heirloom. Genetics 158:413–422

Lippman ZB, Semel Y, Zamir D (2007) An integrated view of quantitative trait variation using tomato interspecific introgression lines. Curr Opin Genet Dev 17:545–552

Liu H, Ouyang B, Zhang J et al (2012) Differential modulation of photosynthesis, signaling, and transcriptional regulation between tolerant and sensitive tomato genotypes under cold stress. PLoS ONE 7(11): e50785

Liu J, Van Eck J, Cong B et al (2002) A new class of regulatory genes underlying the cause of pear-shaped tomato fruit. Proc Natl Acad Sci USA 99:13302–13306

Liu J, Cong B, Tanksley SD (2003a) Generation and analysis of an artificial gene dosage series in tomato to study the mechanisms by which the cloned quantitative trait locus *fw2.2* controls fruit size. Plant Physiol 132:292–299

Liu J, Gur A, Ronen G et al (2003b) There is more to tomato fruit colour than candidate carotenoid genes. Plant Biotech J 1:195–207

Mackay TFC, Stone EA, Ayroles JF (2009) The genetics of quantitative traits: challenges and prospects. Nat Rev Genet 10:565–577

Mageroy MH, Tieman DM, Floystad A et al (2012) A *Solanum lycopersicum* catechol-O-methyltransferase involved in synthesis of the flavor molecule guaiacol. Plant J 69(6):1043–1051

Maliepaard C, Bas N, van Heusden S et al (1995) Mapping of QTLs for glandular trichome densities and *Trialeurodes vaporariorum* (greenhouse whitefly) resistance in an F2 from *Lycopersicon esculentum* × *Lycopersicon hirsutum f. glabratum*. Heredity 75:425–433

Mangin B, Thoquet P, Olivier J et al (1999) Temporal and multiple quantitative trait loci analyses of resistance to bacterial wilt in tomato permit the resolution of linked loci. Genetics 151:1165–1172

Martin B, Nienhuis J, King G et al (1989) Restriction fragment length polymorphisms associated with water use efficiency in tomato. Science 243:1725–1728

Mather K (1941) Variation and selection of polygenic characters. J Genet 41:159–193

Mather K (1949) Biometrical genetics, the study of continuous variation. Methuen & Co/Dover Publications, London

Mathieu S, Dal Cin V, Fei Z et al (2009) Flavour compounds in tomato fruits: identification of loci and potential pathways affecting volatile composition. J Exp Bot 60:325–337

Mazzucato A, Papa R, Bitocchi E et al (2008) Genetic diversity, structure and marker-trait associations in a collection of Italian tomato (*Solanum lycopersicum* L.) landraces. Theor Appl Genet 116:657–669

Miller JC, Tanksley SD (1990) RFLP analysis of phylogenetic relationships and genetic variation in the genus *Lycopersicon*. Theor Appl Genet 80:437–448

Minutolo M, Amalfitano C, Evidente A et al (2013) Polyphenol distribution in plant organs of tomato introgression lines. Nat Prod Res 27(9):787–795

Mitchell-Olds T (2010) Complex-trait analysis in plants. Genome Biol 11(4):113

Momotaz AS, Scott JV, Schuster DJ (2010) Identification of quantitative trait loci conferring resistance to Bemisia tabaci in an F2 population of *Solanum lycopersicum* × *Solanum habrochaites* accession LA1777. J Am Soc Hortic Sci 135(2):134–142

Monforte AJ, Tanksley SD (2000a) Development of a set of near isogenic and backcross recombinant inbred lines containing most of the *Lycopersicon hirsutum* genome in a *L. esculentum* genetic background: a tool for gene mapping and gene discovery. Genome 43:803–813

Monforte AJ, Tanksley SD (2000b) Fine mapping of a quantitative trait locus (QTL) from *Lycopersicon hirsutum* chromosome 1 affecting fruit characteristics and agronomic traits: breaking linkage among QTLs affecting different traits and dissection of heterosis for yield. Theor Appl Genet 100:471–479

Monforte AJ, Asìns MJ, Carbonell EA (1996) Salt tolerance in *Lycopersicon* species. IV. High efficiency of marker-assisted selection to obtain salt-tolerant breeding lines. Theor Appl Genet 93:765–772

Monforte AJ, Asìns MJ, Carbonell EA (1997a) Salt tolerance in *Lycopersicon* species. V. Does genetic variability at quantitative trait loci affect their analysis? Theor Appl Genet 95:284–293

Monforte AJ, Asìns MJ, Carbonell EA (1997b) Salt tolerance in *Lycopersicon* species. VI. Genotype by salinity interaction in quantitative trait loci detection: constitutive and response QTLs. Theor Appl Genet 95:706–713

Monforte AJ, Asìns MJ, Carbonell EA (1999) Salt tolerance in *Lycopersicon* spp. VII. Pleiotropic action of genes controlling earliness on fruit yield. Theor Appl Genet 98:593–601

Monforte AJ, Friedman E, Zamir D et al (2001) Comparison of a set of allelic QTL-NILs for chromosome 4 of tomato: deductions about natural variation and implications for germplasm utilization. Theor Appl Genet 102:572–590

Monforte AJ, Diaz AI, Caño-Delgado A et al (2014) The genetic basis of fruit morphology in horticultural crops: lessons from tomato and melon. J Exp Bot 65 (16):4625–4637

Morgan MJ, Osorio S, Gehl B et al (2013) Metabolic engineering of tomato fruit organic acid content guided by biochemical analysis of an introgression line. Plant Physiol 161(1):397–407

Moyle LC, Graham EB (2005) Genetics of hybrid incompatibility between *Lycopersicon esculentum* and *L. hirsutum*. Genetics 169:355–373

Moyle LC, Nakazato T (2008) Comparative genetics of hybrid incompatibility: sterility in two Solanum species crosses. Genetics 179:1437–1453

Muños S, Ranc N, Botton E et al (2011) Increase in tomato locule number is controlled by two single-nucleotide polymorphisms located near *WUSCHEL*. Plant Physiol 156(4):2244–2254

Mutschler MA, Doerge RW, Liu SC et al (1996) QTL analysis of pest resistance in the wild tomato *Lycopersicon pennellii*: QTLs controlling acylsugar level and composition. Theor Appl Genet 92:709–718

Myles S, Peiffer J, Patrick J (2009) Brown Association mapping: critical considerations shift from genotyping to experimental design. Plant Cell 21:2194–2202

Nesbitt TC, Tanksley SD (2002) Comparative sequencing in the genus *Lycopersicon*: implications for the evolution of fruit size in the domestication of cultivated tomatoes. Genetics 162:365–379

Nienhuis J, Helentjaris T, Slocum M et al (1987) Restriction fragment length polymorphism analysis of loci associated with insect resistance in tomato. Crop Sci 27:797–803

Orsi CH, Tanksley SD (2009) Natural variation in an ABC transporter gene associated with seed size evolution in tomato species. PLoS Genet 5:e1000347

Osborn TC, Alexander DC, Fobes JF (1987) Identification of restriction fragment length polymorhisms linked to genes controlling soluble solids content in tomato. Theor Appl Genet 73:350–356

Overy SA, Walker HJ, Malone S et al (2005) Application of metabolite profiling to the identification of traits in a population of tomato introgression lines. J Expt Bot 56:287–296

Pan Q, Liu YS, Budai-Hadrian O et al (2000) Comparative genetics of nucleotide binding site-leucine rich repeat resistance gene homologues in the genomes of two dicotyledons: tomato and Arabidopsis. Genetics 155:309–322

Paran I, Zamir D (2003) Quantitative traits in plants: beyond the QTL. Trends Genet 19(6):303–306

Paran I, Goldman I, Tanksley SD et al (1995) Recombinant inbred lines for genetic mapping in tomato. Theor Appl Genet 90:542–548

Paran I, Goldman I, Zamir D (1997) QTL analysis of morphological traits in a tomato recombinant inbred line population. Genome 40:242–248

Pascual L, Xu J, Biais B et al (2013) Deciphering genetic diversity and inheritance of tomato fruit weight and composition through a systems biology approach. J Exp Bot 64:5737–5752

Pascual L, Desplat N, Huang BE et al (2015) Potential of a tomato MAGIC population to decipher the genetic control of quantitative traits and detect causal variants in the resequencing era. Plant Biotechnol J 13(4):565–577

Paterson AH, Lander ES, Hewitt JD et al (1988) Resolution of quantitative traits into Mendelian factors by using a complete linkage map of restriction fragment length polymorphisms. Nature 335:721–726

Paterson AH, DeVerna JW, Lanini B et al (1990) Fine mapping of quantitative trait loci using selected overlapping recombinant chromosomes, in an interspecies cross of tomato. Genetics 124:735–742

Paterson AH, Damon S, Hewitt JD et al (1991) Mendelian factors underlying quantitative traits in tomato: comparison across species, generations, and environments. Genetics 127:181–197

Peralta IE, Spooner DM, Knapp S (2008) Taxonomy of wild tomatoes and their relatives (*Solanum* sections *Lycopersicoides*, *Juglandifolia*, *Lycopersicon*; Solanaceae). Syst Bot Monogr 84:1–186

Pereira da Costa JH, Rodríguez GR, Pratta GR et al (2013) QTL detection for fruit shelf life and quality traits across segregating populations of tomato. Sci Hort 156:47–53

Perez-Fons L, Wells T, Corol DI et al (2014) A genome-wide metabolomic resource for tomato fruit from *Solanum pennellii*. Sci Rep. doi:10.1038/srep03859

Pratta GR, Rodriguez GR, Zorzoli R et al (2011) Phenotypic and molecular characterization of selected tomato recombinant inbred lines derived from the cross *Solanum lycopersicum* × *S. pimpinellifolium*. J Genet 90(2):229–237

Prudent M, Causse M, Génard M et al (2009) Genetic and physiological analysis of tomato fruit weight and composition: influence of carbon availability on QTL detection. J Exp Bot 60:923–937

Prudent M, Bertin N, Génard M et al (2010) Genotype-dependent response to carbon availability in growing tomato fruit. Plant, Cell Environ 33 (7):1186–1204

Prudent M, Lecomte A, Bouchet JP et al (2011) Combining ecophysiological modelling and quantitative trait locus analysis to identify key elementary processes underlying tomato fruit sugar concentration. J Exp Bot 3:907–919

Quadrana L, Almeida J, Asis R et al (2014) Natural occurring epialleles determine vitamin E accumulation in tomato fruits. Nat Commun 5:3027

Rafalski JA (2010) Association genetics in crop improvement. Curr Opin Plant Biol 13:174–180

Ranc N, Munos S, Xu J et al (2012) Genome-wide association mapping in tomato (*Solanum lycopersicum*) is possible using genome admixture of *Solanum lycopersicum* var. cerasiforme. G3 (Bethesda) 2(8):853–864

Rick CM (1982) The potential of exotic germplasm for tomato improvement. Vasil I K, Scowcroft WR, Frey KJ (eds) Plant improvement and somatic cell genetics. Academic Press, New York, pp 1–28

Robert VJM, West MAL, Inai S et al (2001) Marker-assisted introgression of blackmold resistance QTL alleles from wild *Lycopersicon cheesmanii* to cultivated tomato (*L. esculentum*) and evaluation of QTL phenotypic effects. Mol Breed 8:217–233

Robbins MD, Sim SC, Yang W et al (2011) Mapping and linkage disequilibrium analysis with a genome-wide collection of SNPs that detect polymorphism in cultivated tomato. J Exp Bot 62(6):1831–1845

Rodriguez GR, Moyseenko JB, Robbins MD et al (2010) Tomato analyzer: a useful software application to collect accurate and detailed morphological and colorimetric data from two-dimensional objects. J Vis Exp 37:e1856

Rodriguez GR, Muños S, Anderson C et al (2011) Distribution of *SUN*, *OVATE*, *LC*, and *FAS* in the tomato germplasm and the relationship to fruit shape diversity. Plant Physiol 156:275–285

Rodríguez GR, Kim HJ, van der Knaap E (2013) Mapping of two suppressors of *OVATE* (*sov*) loci in tomato. Heredity 111(3):256–264

Ron M, Dorrity MW, de Lucas M et al (2013) Identification of novel loci regulating inter-specific variation in root morphology and cellular development in tomato. Plant Physiol 162(2):755–768

Rousseaux MC, Jones CM, Adams D et al (2005) QTL analysis of fruit antioxidants in tomato using *Lycopersicon pennellii* introgression lines. Theor Appl Genet 111:1396–1408

Sacco A, Di Matteo A, Lombardi N et al (2013) Quantitative trait loci pyramiding for fruit quality traits in tomato. Mol Breed 31(1):217–222

Saito K, Matsuda F (2010) Metabolomics for functional genomics, systems biology, and biotechnology. Annu Rev Plant Biol 61:463–489

Saliba-Colombani V, Causse M, Langlois D et al (2001) Genetic analysis of organoleptic quality in fresh market tomato. 1. Mapping QTLs for physical and chemical traits. Theor Appl Genet 102:259–272

Salinas M, Capel C, Alba JM et al (2013) Genetic mapping of two QTL from the wild tomato *Solanum pimpinellifolium* L. controlling resistance against two-spotted spider mite (*Tetranychus urticae* Koch). Theor Appl Genet 26(1):83–92

Salvi S, Tuberosa R (2005) To clone or not to clone plant QTLs: present and future challenges. Trends Plant Sci 10(6):297–304

Sandbrink JM, van Ooijen J, Purimahua CC et al (1995) Localization of genes for bacterial canker resistance in *Lycopersicon peruvianum* using RFLPs. Theor Appl Genet 90:444–450

Sauvage C, Segura V, Bauchet G et al (2014) Genome-wide association in tomato reveals 44 candidate loci for fruit metabolic traits. Plant Physiol 165 (3):1120–1132

Sax K (1923) Association of size differences with seed-coat pattern and pigmentation in *Phaseolus vulgaris*. Genetics 8:552–560

Schauer N, Semel Y, Roessner U et al (2006) Comprehensive metabolic profiling and phenotyping of interspecific introgression lines for tomato improvement. Nat Biotechnol 24:447–454

Schauer N, Semel Y, Balbo I et al (2008) Mode of inheritance of primary metabolic traits in tomato. Plant Cell 20:509–523

Schilmiller A, Shi F, Kim J et al (2010) Mass spectrometry screening reveals widespread diversity in trichome specialized metabolites of tomato chromosomal substitution lines. Plant J 62(3):391–403

Schilmiller AL, Charbonneau AL, Last RL (2012) Identification of a BAHD acetyltransferase that produces protective acyl sugars in tomato trichomes. Proc Natl Acad Sci USA 109(40):16377–16382

Segura V, Vilhjálmsson BJ, Platt A et al (2012) An efficient multi-locus mixed-model approach for genome-wide association studies in structured populations. Nat Genet 44:825–830

Semel Y, Nissenbaum J, Menda N et al (2006) Overdominant quantitative trait loci for yield and fitness in tomato. Proc Natl Acad Sci USA 103:12981–12986

Severin AJ, Peiffer GA, Xu WW et al (2010) An integrative approach to genomic introgression mapping. Plant Physiol 154:3–12

Shirasawa K, Fukuoka H, Matsunaga H et al (2013) DNA marker applications to molecular genetics and genomics in tomato. Breed Sci 63(1):21–30

Shivaprasad PV, Dunn RM, Santos BA et al (2012) Extraordinary transgressive phenotypes of hybrid tomato are influenced by epigenetics and small silencing RNAs. EMBO J 31(2):257–266

Schmalenbach I, March TJ, Bringezu T et al (2011) High-resolution genotyping of wild barley introgression lines and fine-mapping of the threshability locus thresh-1 using the Illumina GoldenGate assay. G3 (Bethesda) 1:187–196

Sim SC, Van Deynze A, Stoffel K et al (2012) High-density SNP genotyping of tomato (*Solanum lycopersicum* L.) reveals patterns of genetic variation due to breeding. PLoS ONE 7(9):e45520

Smart CD, Tanksley SD, Mayton H, Fry WE (2007) Resistance to *Phytophthora infestans* in *Lycopersicon pennellii*. Plant Dis 91(8):1045–1049

Steinhauser MC, Steinhauser D, Gibon Y et al (2011) Identification of enzyme activity quantitative trait loci in a *Solanum lycopersicum* x *Solanum pennellii* introgression line population. Plant Physiol 157 (3):998–1014

Stevens R, Buret M, Duffé P et al (2007) Candidate genes and quantitative trait loci affecting fruit ascorbic acid content in three tomato populations. Plant Physiol 143:1943–1953

Stevens R, Page D, Gouble B et al (2008) Tomato fruit ascorbic acid content is linked with monodehydroascorbate reductase activity and tolerance to chilling stress. Plant, Cell Environ 31:1086–1096

Stommel JR, Zhang Y (2001) Inheritance and QTL analysis of anthracnose resistance in the cultivated tomato (*Lycopersicon esculentum*). Acta Hort 542:303–310

Sumugat MR, Sugiyama N (2010) Quantitative trait loci analysis of flowering time and vegetative traits in tomato plants grown using different seedling raising methods. Hortic Environ Biotechnol 51(4):326–334

Sumugat MR, Lee ON, Nemoto K et al (2010) Quantitative trait loci analysis of flowering-time-related traits in tomato. Sci Hort 123(3):343–349

Sumugat MR, Lee ON, Mine Y et al (2011) Quantitative trait analysis of transplanting time and other root-growth-related traits in tomato. Sci Hort 129 (4):622–628

Sun YD, Liang Y, Wu JM et al (2012) Dynamic QTL analysis for fruit lycopene content and total soluble solid content in a Solanum lycopersicum & #x00D7. S. pimpinellifolium cross. Genet Mol Res 11(4): 3696–3710

Tadmor Y, Fridman E, Gur A et al (2002) Identification of *malodorous*, a wild species allele affecting tomato aroma that was selected against during domestication. J Agri Food Chem 50:2005–2009

Takagi H, Abe A, Yoshida K et al (2013) QTL-seq: rapid mapping of quantitative trait loci in rice by whole genome resequencing of DNA from two bulked populations. Plant J 74(1):174–183

Tanksley SD (1993) Mapping polygenes. Annu Rev Genet 27:205–233

Tanksley SD (2004) The genetic, developmental, and molecular bases of fruit size and shape variation in tomato. Plant Cell 16:S181–S189

Tanksley SD, McCouch SR (1997) Seed banks and molecular maps: unlocking genetic potential from the wild. Science 277:1063–1066

Tanksley SD, Nelson JC (1996) Advanced backcross QTL analysis: a method for the simultaneous discovery and transfer of valuable QTLs from unadapted germplasm into elite breeding lines. Theor Appl Genet 92:191–203

Tanksley SD, Medina-Filho H, Rick CM (1982) Use of naturally-occuring enzyme variation to detect and map genes controlling quantitative traits in an interspecific cross of tomato. Heredity 49:11–25

Tanksley SD, Ganal MW, Prince JP et al (1992) High density molecular linkage maps of the tomato and potato genomes. Genetics 132:1141–1160

Tanksley SD, Grandillo S, Fulton TM et al (1996) Advanced backcross QTL analysis in a cross between an elite processing line of tomato and its wild relative *L. pimpinellifolium*. Theor Appl Genet 92:213–224

Tecle IY, Menda N, Buels RM et al (2010) solQTL: a tool for QTL analysis, visualization and linking to genomes at SGN database. BMC Bioinformatics 11:525

Termolino P, Fulton T, Perez O et al (2010) Advanced backcross QTL analysis of a *Solanum lycopersicum* × *Solanum chilense* cross. In: Proceedings of the SOL2010 7th solanaceae conference, Dundee (Scotland), 5–9 September, p 56

Thoday JM (1961) Location of polygenes. Nature 191:368–370

Thoquet P, Olivier J, Sperisen C et al (1996a) Quantitative trait loci determining resistance to bacterial wilt in tomato cultivar Hawaii 7996. Mol Plant Microbe Interact 9:826–836

Thoquet P, Olivier J, Sperisen C et al (1996b) Polygenic resistance of tomato plants to bacterial wilt in the French West Indies. Mol Plant Microbe Interact 9:837–842

Tieman DM, Zeigler M, Schmelz EA et al (2006) Identification of loci affecting flavour volatile emissions in tomato fruits. J Exp Bot 57:887–896

Tomato Genome Consortium (2012) The tomato genome sequence provides insights into fleshy fruit evolution. Nature 485(7400):635–641

Toubiana D, Semel Y, Tohge T et al (2012) Metabolic profiling of a mapping population exposes new insights in the regulation of seed metabolism and seed, fruit, and plant relations. PLoS Genet. doi:10.1371/journal.pgen.1002612

Tripodi P, Di Dato F, Maurer S et al (2010) A genetic platform of tomato multi-species introgression lines: new tools for QTL analysis, gene cloning and molecular breeding. 54° Convegno della Società di Genetica Agraria. Matera, 27–30 Settembre, ISBN 978-88-904570-0-5

Truco MJ, Randall LB, Bloom AJ et al (2000) Detection of QTLs associated with shoot wilting and root ammonium uptake under chilling temperatures in an interspecific backcross population from *Lycopersicon esculentum* × *L. hirsutum*. Theor Appl Genet 101:1082–1092

Trujillo-Moya C, Gisbert C, Vilanova S et al (2011) Localization of QTLs for *in vitro* plant regeneration in tomato. BMC Plant Biol 11:140

Uozumi A, Ikeda H, Hiraga M et al (2012) Tolerance to salt stress and blossom-end rot in an introgression line, IL8-3, of tomato. Sci Hort 138:1–6

Vallejos CE, Tanksley SD (1983) Segregation of isozyme markers and cold tolerance in an interspecific backcross of tomato. Theor Appl Genet 66:241–247

Van Schalkwyk A, Wenzl P, Smit S et al (2012) Bin mapping of tomato diversity array (DArT) markers to genomic regions of *Solanum lycopersicum* x *Solanum pennellii* introgression lines. Theor Appl Genet 124(5):947–956

Van der Hoeven RS, Monforte AJ, Breeden D et al (2000) Genetic control and evolution of sesquiterpene biosynthesis in *Lycopersicon esculentum* and *L. hirsutum*. Plant Cell 12:2283–2294

van der Knaap E, Tanksley SD (2001) Identification and characterization of a novel locus controlling early fruit development in tomato. Theor Appl Genet 103:353–358

van der Knaap E, Tanksley SD (2003) The making of a bell pepper-shaped tomato fruit: identification of loci controlling fruit morphology in Yellow Stuffer tomato. Theor Appl Genet 107:139–147

van der Knaap E, Lippman ZB, Tanksley SD (2002) Extremely elongated tomato fruit controlled by four quantitative trait loci with epistatic interactions. Theor Appl Genet 104:241–247

van der Knaap E, Sanyal A, Jackson SA et al (2004) High-resolution fine mapping and fluorescence in situ hybridization analysis of *sun*, a locus controlling tomato fruit shape, reveals a region of the tomato genome prone to DNA rearrangements. Genetics 168:2127–2140

van der Knaap E, Chakrabarti M, Chu YH et al (2014) What lies beyond the eye: the molecular mechanisms regulating tomato fruit weight and shape. Front Plant Sci 5(227):1–13

van Heusden AW, Koornneef M, Voorrips RE et al (1999) Three QTLs from *Lycopersicon peruvianum* confer a high level of resistance to *Clavibacter michiganensis* ssp. *michiganensis*. Theor Appl Genet 99:1068–1074

Villalta I, Bernet GP, Carbonell EA et al (2007) Comparative QTL analysis of salinity tolerance in terms of fruit yield using two Solanum populations of F7 lines. Theor Appl Genet 114:1001–1017

Villalta I, Reina-Sánchez A, Bolarín MC et al (2008) Genetic analysis of Na(+) and K(+) concentrations in leaf and stem as physiological components of salt tolerance in tomato. Theor Appl Genet 116:869–880

Víquez-Zamora M, Vosman B, van de Geest H et al (2013) Tomato breeding in the genomics era: insights from a SNP array. BMC Genom. doi:10.1186/1471-2164-14-354

Visscher PM, Brown MA, McCarthy MI et al (2012) Five years of GWAS discovery. Am J Hum Genet 90:7–24

Wang JF, Olivier J, Thoquet P et al (2000) Resistance of tomato line Hawaii7996 to *Ralstonia solanacearum* Pss4 in Taiwan is controlled mainly by a major strain-specific locus. Mol Plant Microbe Interact 13:6–13

Wang JF, Ho FI, Hai THT et al (2013) Identification of major QTLs associated with stable resistance of tomato cultivar 'Hawaii 7996' to *Ralstonia solanacearum*. Euphytica 190(2):241–252

Wayne ML, McIntyre LM (2002) Combining mapping and arraying: an approach to candidate gene identification. Proc Natl Acad Sci USA 99(23):14903–14906

Weller JI, Soller M, Brody T (1988) Linkage analysis of quantitative traits in an interspecific cross of tomato

(*Lycopersicon esculentum* × *Lycopersicon pimpinellifolium*) by means of genetic markers. Genetics 118:329–339

Wu F, Mueller LA, Crouzillat D et al (2006) Combining bioinformatics and phylogenetics to identify large sets of single-copy orthologous genes (COSII) for comparative, evolutionary and systematic studies: a test case in the euasterid plant clade. Genetics 174:1407–1420

Wu S, Xiao H, Cabrera A et al (2011) *SUN* regulates vegetative and reproductive organ shape by changing cell division patterns. Plant Physiol 157(3):1175–1186

Xiao H, Jiang N, Schaffner EK et al (2008) A retrotransposon-mediated gene duplication underlies morphological variation of tomato fruit. Science 319:1527–1530

Xiao H, Radovich C, Welty N et al (2009) Integration of tomato reproductive developmental landmarks and expression profiles, and the effect of *SUN* on fruit shape. BMC Plant Biol 9:49

Xu J, Zhao Q, Du P et al (2010) Developing high throughput genotyped chromosome segment substitution lines based on population whole-genome re-sequencing in rice (*Oryza sativa* L.). BMC Genom 11:656

Xu J, Ranc N, Muños S et al (2013) Phenotypic diversity and association mapping for fruit quality traits in cultivated tomato and related species. Theor Appl Genet 126(3):567–581

Xu X, Martin B, Comstock JP et al (2008) Fine mapping a QTL for carbon isotope composition in tomato. Theor Appl Genet 117:221–233

Yang W, Sacks EJ, Lewis Ivey ML et al (2005) Resistance in *Lycopersicon esculentum* intraspecific crosses to race T1 strains of *Xanthomonas campestris* pv. *vesicatoria* causing bacterial spot of tomato. Phytopathology 95:519–527

Yates HE, Frary A, Doganlar S et al (2004) Comparative fine mapping of fruit quality QTLs on chromosome 4 introgressions derived from two wild species. Euphytica 135:283–296

Yogendra KN, Ramanjini Gowda PH (2013) Phenotypic and molecular characterization of a tomato (*Solanum lycopersicum* L.) F2 population segregation for improving shelf life. Genet Mol Res 12(1):506–518

Zamir D (2001) Improving plant breeding with exotic genetic libraries. Nat Rev Genet 2:983–989

Zamir D (2013) Where have all the crop phenotypes gone? PLoS Biol 11(6):e1001595

Zamir D, Tal M (1987) Genetic analysis of sodium, potassium and chloride ion content in *Lycopersicon*. Euphytica 36:187–191

Zamir D, Selia Ben-David T, Rudich J et al (1984) Frequency distributions and linkage relationships of 2-tridecanone in interspecific segregating generation of tomato. Euphytica 33:481–488

Zanor MI, Rambla JL, Chaïb J et al (2009) Metabolic characterization of loci affecting sensory attributes in tomato allows an assessment of the influence of the levels of primary metabolites and volatile organic contents. J Exp Bot 60(7):2139–2154

Zhang L, Lin GY, Nino-Liu D et al (2003a) Mapping QTLs conferring early blight (*Alternaria solani*) resistance in a *Lycopersicon esculentum* × *L. hirsutum* cross by selective genotyping. Mol Breed 12:3–19

Zhang LP, Lin GY, Foolad MR (2003b) QTL comparison of salt tolerance during seed germination and vegetative growth in a *Lycopersicon esculentum* × *L. pimpinellifolium* RIL population. Acta Hort 618: 59–67

Zhang N, Brewer MT, van der Knaap E (2012) Fine mapping of *fw3.2* controlling fruit weight in tomato. Theor Appl Genet 125(2):273–284

Zhu C, Gore M, Buckler ES et al (2008) Status and prospects of association mapping in plants. Plant Genome 1(1):1–19

Tomato Resources for Functional Genomics

5

Christophe Rothan, Cécile Bres, Virginie Garcia and Daniel Just

Abstract

Tomato is currently the model species for fleshy fruit development and for *Solanaceae* species. The recent completion of a high-quality genome sequence of the inbred tomato (*Solanum lycopersicum*) cultivar 'Heinz 1706' allowed the prediction and in silico annotation of ca 35,000 genes. Assigning a biological function to these genes is among the priorities of the tomato community, especially for genes contributing to fleshy fruit development and quality, and to other major agronomical traits in tomato and *Solanaceae*. More than a decade of research using genomic tools, mostly transcriptome and metabolome, combined with genetic mapping approaches, provided first cues on the possible function of tomato genes by describing where, when, and with which other gene/metabolite these genes are expressed. Current advances in sequencing technologies now allow the exhaustive inventory of tomato transcripts in various plant organs, tissues and even cell types. To cope with the need to assign biological functions to a large number of genes, tomato mutant resources based on several technologies [T-DNA and transposon insertional mutants, fast-neutron, γ-ray and ethyl methanesulfonate (EMS) mutants] have been developed in the recent years. Among them, the Targeting Induced Local Lesions In Genomes (TILLING) technology, based on the generation by EMS of high density point mutations evenly distributed in the genome and on the subsequent detection of mutations in target genes is presently the most established. The present chapter will describe the main

C. Rothan (✉) · C. Bres · V. Garcia · D. Just
INRA, UMR 1332 Biologie du Fruit et Pathologie, 71 Av Edouard Bourleaux, 33140 Villenave d'Ornon, France
e-mail: christophe.rothan@bordeaux.inra.fr

C. Rothan · C. Bres · V. Garcia · D. Just
UMR 1332 Biologie du Fruit et Pathologie, Université de Bordeaux, 33140 Villenave d'Ornon, France

M. Causse et al. (eds.), *The Tomato Genome*, Compendium of Plant Genomes,
DOI 10.1007/978-3-662-53389-5_5

resources, strategies and tools currently available for linking genes to phenotype in tomato.

Keywords
Tomato · Mutants · TILLING · Reverse genetics · EMS

Introduction

The recent completion of a high-quality genome sequence of the inbred tomato (*Solanum lycopersicum*) cultivar 'Heinz 1706' and of a draft sequence of the *S. pimpinellifolium* LA1589 genotype allowed deciphering tomato genome organisation and features. In the ca 900 megabase (Mb) genome size of cultivated tomato, computational annotation supported by RNA sequencing (RNAseq) data predicted the existence of 34,727 protein coding genes and of 96 miRNAs (Tomato Genome Consortium 2012). Basic gene structure (*cis*-regulating regions, untranslated regions, introns, exons) of tomato genes, location (chromosomal region) and polymorphism with the *S. pimpinellifolium* wild ancestor can be predicted using web databases and tools (Fig. 5.1), among which the central tomato genomics data repository at the Solanaceae Genome Network database (SGN). Additional data helping to refine the tomato genome annotation are continuously updated thanks to the improvement of the current next generation sequencing (NGS) tools allowing precise inventory of transcripts in plant organs, tissues and cells (Matas et al. 2011; Park et al. 2012) and to the availability of whole genome sequences of an increasing number of cultivated tomato genotypes (Kobayashi et al. 2014) and wild relatives (Sato et al. 2013). Despite these progresses, in silico gene predictions cannot be considered as absolutely accurate; moreover, alternative gene models can be found for one single locus. In the post-genome era, there is a strong need to assign a biological function to tomato genes and this is one of the current challenges facing the Solanaceae community.

Tomato is unique in being a model for both the fleshy fruits and the *Solanaceae*. Though fruit-expressed genes are clearly among the main priorities (Gady et al. 2012; Jones et al. 2012; Baldet et al. 2013; Di Matteo et al. 2013), other targets are also of utmost importance, e.g. those regulating the fitness of tomato plants challenged with new environmental conditions or pathogen attacks (Leide et al. 2007; Piron et al. 2010; Kimbara et al. 2013; Petit et al. 2014; Shi et al. 2013), their adaptation to evolving cultural practices (Martín-Trillo et al. 2011) or the fruit yield (Krieger et al. 2010). Existing genomics resources in the plant models rice and Arabidopsis can be extremely useful for analysing the function of orthologous genes sharing similar functions in tomato and in other species. However, each plant species has its own specific features, which sometimes makes difficult the functional study of a tomato gene in a different species. For example, unlike Arabidopsis, tomato fruit is fleshy and its genome includes 727 gene groups confined to fleshy fruit species (tomato, grape and potato; Tomato Genome Consortium 2012). In addition, for a number of traits, for example cuticle composition and properties (Yeats et al. 2012; Petit et al. 2014), Arabidopsis is not representative of many plant families including tomato.

What tools do we have and what strategies are currently being developed to study the relationships between gene function and plant phenotype in tomato?

In addition to tomato genome annotation and prediction of gene function based on DNA sequence homology, the information on where and when a gene is expressed provides the first cues on the possible function of a gene *in planta*. With the development of high throughput

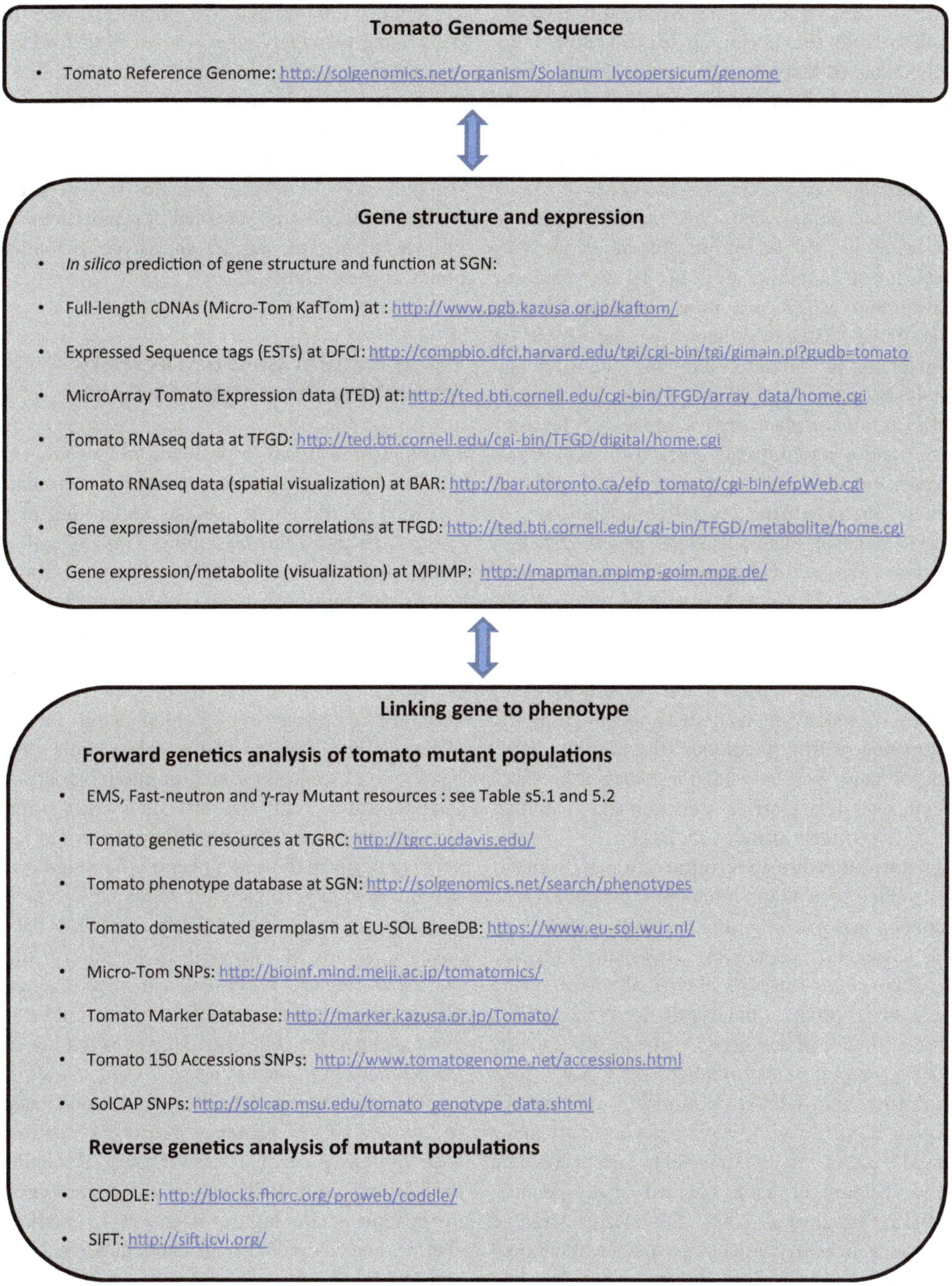

Fig. 5.1 Linking gene to phenotype in tomato. Information on tomato genome sequence and gene structure and expression can be combined with the available tomato genetic resources and tools for gene discovery and assigning biological functions to tomato proteins

technologies, large efforts have been devoted in tomato from the beginning to constitute large collections of Expressed Sequence Tags (ESTs) from various plant tissues and of full-length cDNAs (Fig. 5.1). This information allowed the construction of gene expression arrays used to monitor the expression of tomato genes in a large variety of organs and conditions. Exhaustive inventory of gene transcripts obtained from NGS experiments (Matas et al. 2011; Tomato Genome Consortium 2012) are now available through web-based databases such as the BAR at University of Toronto, allowing the in silico analysis of the expression pattern of a gene in various tomato plant organs, stages of development and even fruit cell types (Fig. 5.1). Gene expression data can be further combined with other genomics data, typically metabolome, to assign putative functions to the genes. Data can be first examined using visualisation tools such as MapMan (Urbanczyk-Wochniak et al. 2005) and further analysed using various statistical means such as correlation network analysis. This strategy recently enabled the identification of genes of unknown function implicated in the regulation of fruit flavonoids (Ozaki et al. 2010) and of major developmental and metabolic shifts occurring during fruit development (Mounet et al. 2009; Rohrmann et al. 2011).

However, correlative information is by itself not sufficient to assign a function to a gene. Gene function and role *in planta* is usually inferred by the analysis of phenotypic alterations triggered by changes in transcript level or alteration of the gene under study. Analysis of the function of a single gene or of few genes is classically done by stable genetic transformation of tomato with Agrobacterium (RNAi or amiRNA or chimeric repressor silencing, overexpression; Fernandez et al. 2009) or by transient expression via agro-injection or Virus-Induced-Gene-Silencing (VIGS; Orzaez et al. 2009). On a larger scale, in tomato as in other model plant species, functional genomics typically rely on the generation and analysis of mutant collections (T-DNA or transposon-tagged lines, fast-neutron and EMS mutants). Using tomato germplasm or mutant collections displaying artificially induced genetic variability, linking gene to phenotype can be done using two approaches known as (i) forward (classical) genetics, in which tomato genetic resources are first screen for phenotypes-of-interest and the underlying genes are further identified by map-based cloning or association mapping and (ii) reverse genetics in which the mutant collection is screened for mutations in known target gene and the phenotype of mutant plants is then analysed.

Using Natural Genetic Diversity for Linking Phenotype to Gene

Germplasm resources represent a large source of wild and cultivated genetic variability for tomato in which natural allelic variants underlying phenotypic changes can be found. Tomato collections may include related species, various accessions with high genetic diversity often collected near the centre of origin of the species, and heirloom and cultivated lines obtained by breeders worldwide. These collections provide very useful resources for identifying natural alleles, mine the available phenotypic and genotypic diversity in search of allelic variations linked with a trait and test their association. Considerable effort has been devoted in the last years to perform thorough phenotypic analysis of natural diversity, mostly in cultivated tomato, and to store these data in web accessible databases that can be browsed in search of trait variations (Fig. 5.1). These can include not only classical descriptors, e.g. fruit shape or colour or plant architecture, but also other information on fruit quality traits including fruit composition in sugars, secondary metabolites or aroma. Thanks to the parallel generation of genetic populations such as Introgression Lines (IL) and Recombinant Inbred Lines (RIL) and to the development of genetic maps highly saturated in markers (Fernie and Klee 2011; Tomato genome Consortium 2012), the chromosomal regions harbouring many of the traits of interest can be further located in silico.

Natural variations found in cultivated tomato or in wild relatives have been instrumental in the last 15 years to discover the function of several key proteins controlling fruit weight and fruit shape variations in domesticated tomato (Frary et al. 2000; Chakrabarti et al. 2013), to decipher the ripening regulatory complex (e.g. the RIN gene; Klee and Giovannoni 2011) and to identify allelic variants underlying variations in fruit sugar content and aroma (Fridman et al. 2000; Fernie and Klee 2011). This has been done essentially through map-based cloning or positional cloning of Mendelian mutations and of Quantitative Trait Loci (QTL), a process by which the genetic basis of a phenotypic variation is identified by looking for linkage of phenotype to markers with known physical location. Hundreds or even thousands of Mendelian mutations and QTLs have now been mapped. This is an ongoing process since the refinement of analytical methods and gene mapping approaches now allow for example the decomposition of previously identified complex fruit composition traits into multiple single quantitative traits (Schauer et al. 2006; Fernie and Klee 2011).

Breakthrough advances in the last few years including the tomato genome sequencing (Tomato Genome Consortium 2012), the availability of tens of thousands of genetic markers distributed over the whole genome (Shirasawa et al. 2010; Kobayashi et al. 2014; Ranc et al. 2012) and the development and availability of high throughput methods for detecting DNA polymorphism such as the SolCAP SNP genotyping array (Sim et al. 2012) has greatly facilitated the identification of causal alleles responsible for a particular phenotype. As a consequence, the map-based cloning process which could take several years in tomato not long ago has been considerably reduced.

In addition, complementary or parallel strategies can be undertaken to identify the source of phenotypic variation. In the candidate gene approach, the location of target genes with functions related to the traits studied are compared with the map location of the mutation/QTL for that trait (Causse et al. 2004) and the candidate gene is screened for genetic/epigenetic variations possibly responsible for the phenotypic alteration. Recently, Causse and coworkers also demonstrated that the powerful Genome Wide Association (GWA) mapping approach, by which the tomato genome is screened for significant associations between SNPs and specific phenotypic alterations, was possible in tomato using genome admixture of *Solanum lycopersicum* var. *cerasiforme* (Ranc et al. 2012). Association mapping was further shown to be an effective tool for assessing the molecular basis of fruit developmental and quality traits in tomato and for the discovery of causal SNPs (Chakrabarti et al. 2013; Xu et al. 2013). With the current availability of NGS technologies at low cost, non-targeted whole genome sequencing (WGS) of tomato germplasm and its comparison with reference tomato genome may also help detecting polymorphism responsible for modifications in protein function (e.g. the splice junction, nonsense and missense mutations) and link them with possible downstream phenotypic variations (Causse et al. 2013; Hirakawa et al. 2013). Once the causal polymorphism has been identified, the linkage between polymorphism/protein function changes and the particular trait studied has to be confirmed using the range of tools available for *in planta* functional analysis of genes in tomato, which include the mutant resources described below.

Using Artificially Induced Genetic Diversity for Assigning a Function to Tomato Genes

Using natural genetic diversity for assigning a function to a gene has several limitations. The most interesting sources of natural genetic variation are often found in wild type tomato species (Ichihashi and Sinha 2014). Studying the relationship between a gene polymorphism and the corresponding phenotypic variation can be difficult for some complex traits, due to the large genetic and phenotypic variation brought by the wild parental line. Even in Nearly Isogenic Lines (NILs), the introgressed fragment may carry tenths of genes susceptible to affect the trait studied. Conversely, the genetic variability of cultivated tomato has been much reduced by domestication (Frary et al. 2000; Tomato Genome

Consortium 2012). The new variability generated by spontaneous mutations and retained in heirloom varieties or during subsequent breeding process is often far from that found in artificially induced mutant collections. In mutant collections, the genetic background is identical for all mutants generated in a given genotype, except for the genetic variability induced by the mutagenesis (gene disruption or footprints with T-DNA or transposon, deletions with fast-neutron mutagenesis and γ-rays, point mutations with EMS). Mutant analysis therefore provides a widely used tool for defining the function of a gene in model plant species.

T-DNA and Transposon Tagging

Efficient and routine methods for homologous recombination are not yet available in plants. Hence, insertional mutagenesis using transferred DNA (T-DNA) or transposons has been the method of choice for the generation of genetic diversity in plants. It is commonly used for gene discovery and functional analysis of genes-of-interest in the model plants *Arabidopsis* and rice. In tomato, T-DNA from the Ti plasmid from *Agrobacterium tumefasciens* and the nonautonomous mobile elements Activator(Ac)/ Dissociation(Ds) from maize have been used to generate mutant collections (Meissner et al. 1997, 2000; Emmanuel and Levy 2002; Gidoni et al. 2003; Mathews et al. 2003). The T-DNA and Ac/Ds elements are transferred to tomato through *Agrobacterium*-mediated genetic transformation using somatic embryogenesis. Their insertion into tomato mostly causes loss-of-function mutations which can be easily detected. In addition, the Ac/Ds system provides some interesting features. Ds elements are Ac elements that have undergone deletions and lost all or part of the transposase activity necessary to excise Ac or Ds mobile elements. Following genetic transformation, plants carrying stable transposition events can be obtained by selecting against plants carrying the Ac element (Meissner et al. 2000). Crossing them with transposase plants allows the mobile element to be excised and reinserted close by in the tomato genome, thereby creating a new insertion site. Excision of the mobile element will leave a footprint (minor sequence changes) at the donor site. Depending on the DNA context of the transposition site, revertants can be obtained or new alleles with possibly new properties can be created.

Knockout mutants can therefore be generated in tomato by insertional mutagenesis (Fig. 5.2) and used as reverse genetics tools for analysing the function *in planta* of tomato target genes (Vogg et al. 2004). Combination of T-DNA or transposon with reporter genes or enhancers can be further used for discovering genes and/or transcriptional regulators (Meissner et al. 2000), as shown for the discovery of a transcriptional activator of the anthocyanin pathway using tomato T-DNA activation tagging lines (Mathews et al. 2003). However, though these strategies are very interesting, several limitations inherent to tomato have hindered their development in this species. Because the probability to hit the gene-of-interest is lower for small genes than for large genes, loss-of-function mutants for the target gene are not always identified and very large numbers of mutagenized plants are needed to reach near saturation of the collection. In Arabidopsis, it is estimated that the number of mutations should exceed by five to tenfold the number of genes in the genome (Alonso and Ecker 2006). Given the size of tomato genome, which is almost sixfold larger than that of Arabidopsis, some estimates (Emmanuel and Levy 2002) consider than near to 200,000–300,000 transposon-tagged lines are necessary to obtain 95 % saturation of the genome. In contrast to Arabidopsis that is easily transformed by the floral dipping method, tomato genetic transformation is based on in vitro somatic embryogenesis, which remains a low throughput technology. This limitation can be partially overcome using transposon tagging rather than T-DNA tagging. While T-DNA insertion number per plant typically varies from one to three, a low number of initial primary transformants carrying non-autonomous mobile elements can generate hundreds or thousands of transposon-tagged lines. Model tomato varieties such as the

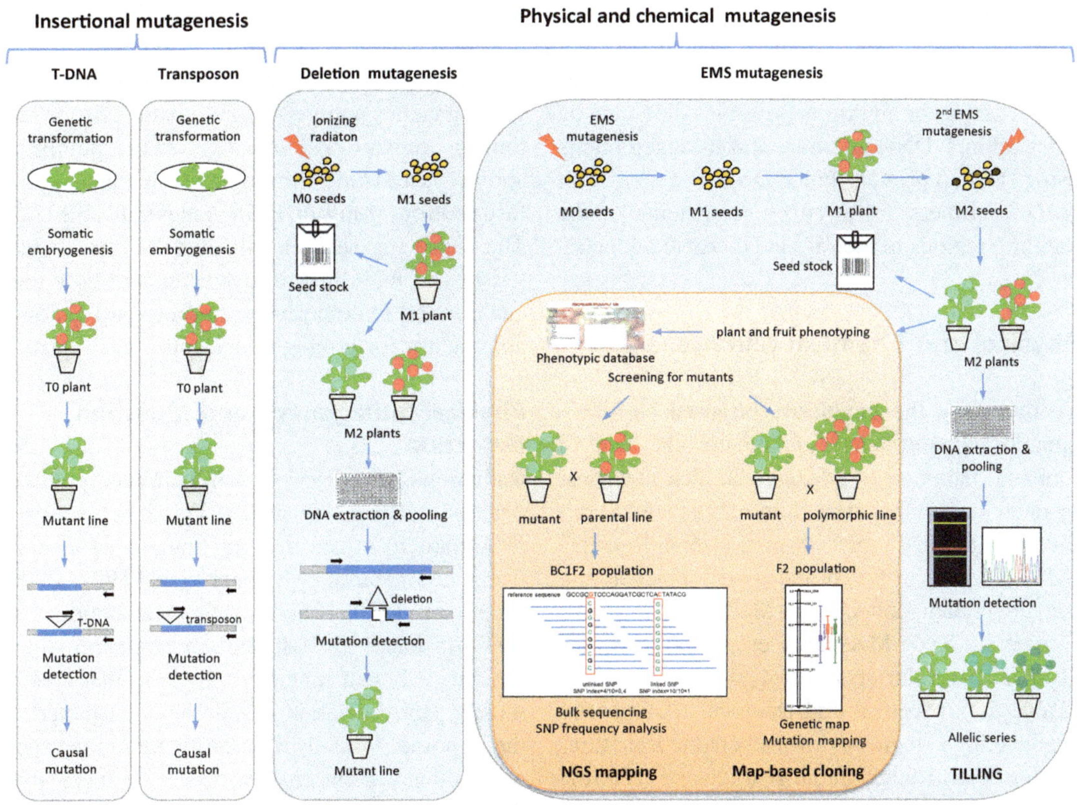

Fig. 5.2 Generation of tomato mutant resources and use through forward and reverse genetic approaches. **a** T-DNA insertional mutants can be generated by *Agrobacterium*-mediated genetic transformation and regeneration via somatic embryogenesis. Homozygous mutant lines displaying a phenotype-of-interest can be PCR-screened to identify the T-DNA flanking regions using T-DNA specific primer and random primer. **b** Transposon insertional mutants can be generated by *Agrobacterium*-mediated genetic transformation and regeneration via somatic embryogenesis. Homozygous mutant lines displaying a phenotype-of-interest can be PCR-screened to identify the transposon flanking regions. **c** Deletion tomato mutants are generated by submitting seeds (M0) to ionising radiations (fast neutron or γ-ray). The mutagenized seeds (M1) produce M1 plants, from which M2 seeds are collected for seed stock or sown to produce M2 plants. DNA is collected from individual M2 plants or from M2 families, pooled and used for PCR-detection of deletions in genes-of-interest using gene-specific primers. **d** Tomato mutants carrying point mutations are generated by submitting tomato seeds (M0) to EMS treatment. The mutagenized seeds (M1) produce M1 plants, from which M2 seeds are collected for seed stock or sown to produce M2 plants. For identification by TILLING of unknown mutations in genes-of-interest, the DNA is collected from individual M2 plants or from M2 families, pooled and used for the detection of point mutations e.g. by electrophoresis-based techniques or NGS sequencing. For forward genetic approach, phenotypic data are collected on the M2 family or individual plants, stored in a database and mined for identifying mutants carrying the traits of interest. Homozygous mutant line can be crossed with polymorphic tomato line (e.g. *S. pimpinellifolium*). The causal mutation is identified by traditional map-based cloning through genetic mapping of the mutant trait using the segregating F2 population. Alternatively, mutant line can be crossed with the wild type parental line to identify the causal mutation by NGS mapping. Bulks displaying or not the mutant trait are constituted from the F2 population issued from the back-cross (BC1F2) and submitted to deep sequencing. Chromosomal region harbouring the causal SNP is identified by comparison of SNP frequencies in the two bulks

miniature tomato cultivar Micro-Tom suitable for high throughput reverse genetics approaches can further be used (Meissner et al. 1997). However, until now, the existing limitations have prevented the development of large insertional mutant collections in tomato.

Detection of T-DNA or transposon insertional mutations is straightforward (Fig. 5.2). The DNA regions flanking the insertion in a gene can be PCR amplified using primers specific to the gene-of-interest and to the T-DNA. Alternatively, single or multiple insertion sites can be detected in T-DNA and transposon-tagged plants using random primers and T-DNA or Ds element specific primers, followed by sequencing of the flanking regions to identify the disrupted gene(s).

Physical and Chemical Mutagenesis

To overcome the limitations imposed by stable genetic transformation of tomato, the use of ionising radiations (γ-radiations, fast-neutrons) or chemical mutagenesis with ethyl methanesulfonate (EMS) or *N*-methyl-*N*-nitrosourea (MNU) has been tested and used for generating tomato mutant collections (Menda et al. 2004; Dan et al. 2007; Matsukura et al. 2007; Minoia et al. 2010; Saito et al. 2011; Just et al. 2013) (Table 5.1). Physical and chemical mutagenesis produce high frequency of irreversible mutations randomly distributed in the genome and therefore independent from genome size. In addition, each type of mutagenesis has its own specificities conditioning the method used for detecting the mutations and the possible utilisation of the mutants.

Artificially induced genetic variability has been used for decades to modify existing traits or to create new valuable traits in cultivated varieties. Crop improvement through mutation breeding has produced a set of commercial varieties in a wide range of species including tomato (Kharkwal and Shu 2009). Recently, the technological developments allowing the detection of unknown mutations in mutant collections through PCR-based methods such as DeleteA-Gene (Li et al. 2001) for detecting fast-neutron or γ-ray mutations and Targeting induced Local Lesions IN Genomes (TILLING; Colbert et al. 2001) for EMS mutations has renewed the interest of the tomato community for creating highly mutagenized tomato mutant collections for assigning gene function in tomato. More recently, the continuous technological improvement and cost reduction in NGS technologies has led to the successful use of these technologies for mutation detection in both reverse (from gene to phenotype) and forward (from phenotype to gene) genetics approaches in tomato. The current tomato mutant resources already published and/or accessible through web are either fast-neutron, γ-ray or EMS mutants (Table 5.1). The following section will describe the characteristics of these mutant resources and how they can be used in both reverse and forward genetics approaches for linking genes and phenotypes.

Physical Mutagenesis and Mutation Detection

Radiations have been shown to induce physical deletions of genes in plants. This has been used in tomato to create deletion mutant collections using fast neutron (Meissner et al. 1997, 2000; Menda et al. 2004) and γ-rays (Matsukura et al. 2007) (Table 5.1). Fast-neutron bombardment is a highly efficient mutagenic method that creates mostly DNA deletions randomly distributed in the genome. Size distribution of mutations typically ranges between few bases to more than 30 kb. Gamma-ray irradiation also causes deletion and chromosomal rearrangements whose severity will depend on the dose used. With both methods of mutagenesis, loss-of-function mutants are mostly generated, like for T-DNA and transposon mutagenesis. However, much higher mutation frequencies, which are independent of genome size (Li and Zhang 2002), can be obtained. This considerably reduces the size of the mutant collection necessary to be screened for mutations in gene-of-interest. Few thousands mutants are required with physical deletion instead of tens or hundreds of thousand mutants with insertional mutagenesis. It is therefore well adapted to tomato. Since the large deletions generated may overlap with several genes, physical deletion can be useful when duplicated genes, which often show functional redundancy, are arranged in tandem repeats (Li and Zhang 2002). In these cases, tight genetic linkage between the genes often prevents the generation of double mutants by crossing. Conversely, large deletions may be problematic. Several genes may be deleted at the same locus,

Table 5.1 Tomato EMS and ionising irradiation mutant resources

Cultivar	Mutagenesis	Mutant population	Mutant population screened by TILLING	TILLed DNA sequence (kb)	Mutation frequency	References	Web links
Micro-Tom INRA (France)	EMS 1 % EMS 1 % + EMS 1 %	4500 M2 families 3500 M2 families	7296 M2 families (12 plants/family)	49.9	1/663 to 1/130 kb	Dan et al. (2007), Just et al. (2013), Petit et al. (2014)	
Micro-Tom NBRP (Japan)	0.5 % EMS 1 % EMS	2180 M2 families 872 M2 families	3052 M2 families (10 plants/family)	15.3?	1/1710 kb 1/737 kb	Okabe et al. (2011), Saito et al. (2011)	http://www.tomatoma.nbrp.jp/index.jsp
TPAADASU	EMS 1 %	8225 M2 families 7030 M3 families	8025 M2 families 6692 M3 families	0.85	1/737 kb	Gady et al. (2009)	
M82	EMS 0.5 %	6000 M2 families	4759 M3 families	30.9	1/574 kb	Menda et al. (2004), Piron et al. (2010)	http://zamir.sgn.cornell.edu/mutants/ http://www-urgv.versailles.inra.fr/tilling/tomato.htm http://urgv.evry.inra.fr/cgi-bin/projects/Tilling/index.pl
Red Setter	EMS 0.7 % EMS 1 %	4156 M3 families 1352 M3 families	3924 M3 families (8 plants/family) 1297 M3 families (8 plants/family)	9.5	1/574 kb 1/322 kb	Minoia et al. (2010)	http://www.agrobios.it/tilling/
Heinz 1706	EMS 1 %	4500 M2 families	512 M2 families		1/450 kb	–	http://www.tilling.ucdavis.edu/index.php/Tomato_Tilling
Best of all	EM 0.75 % EMS 1 %	5000 M2 families	–	–	–	–	http://www-urgv.versailles.inra.fr/tilling/tomato.htm

(continued)

Table 5.1 (continued)

Cultivar	Mutagenesis	Mutant population	Mutant population screened by TILLING	TILLed DNA sequence (kb)	Mutation frequency	References	Web links
Money maker	EM 0.75 % EMS 1 %	5000 M2 families	–	–	–	–	http://www-urgv.versailles.inra.fr/tilling/tomato.htm
Micro-Tom NBRP (Japan)	γ-ray	6422 M2 families	–	–	nd	Matsukura et al. (2007), Saito et al. (2011)	http://www.tomatoma.nbrp.jp/index.jsp
M82	Fast neutron 12–15 Gy	7000 M2 families	–	–	nd	Menda et al. (2004)	http://zamir.sgn.cornell.edu/mutants/

which may alter the subsequent genetic analyses and the functional analysis of target genes.

Tomato Fast Neutron and γ-Ray Mutagenesis Construction of fast-neutron or γ-ray deletion mutant collections is straightforward, once pilot studies have been performed for determining the optimal dose/rate of mutations (Li and Zhang 2002; Sikder et al. 2013; Yuan et al. 2014). Typically, half of the mutagenized M1 plants should be fertile enough to give M2 seeds. For creating the deletion mutant resource, a large number of M0 seeds (wild type seeds) are mutagenized to give M1 seeds (seeds carrying heterozygous mutations) that are sown (Fig. 5.2). The M2 seeds are then collected from M1 plants and sown for collecting DNA and/or phenotyping M2 plants, or stored. Early studies based on experiments performed in Arabidopsis suggest that *ca* 50,000 mutagenized lines would be necessary to obtain deletion mutants for 80 % of the targeted loci in various plant species (Li and Zhang 2002). In tomato, growing 50,000 mutant lines and harvesting seeds from them remains a considerable task, especially if cultivars of normal plant size and indeterminate growth are chosen. For these reasons, tomato mutant collections issued from physical mutagenesis have been produced until now in determinate processing tomato varieties carrying the *sp* mutation (M82, fast neutron mutagenesis) and in the determinate miniature cultivar Micro-Tom (fast-neutron and γ-ray mutagenesis). Available tomato deletion mutant populations range from 6400 M2 families for γ-ray-treated Micro-Tom (Matsukura et al. 2007) to 7000 M2 families in fast neutron mutagenized M82 (Menda et al. 2004) (Table 5.1).

Detection of Deletion Mutants
Detection of unknown mutations in target genes can be done on pooled DNA from M2 plants using a simple PCR-based technique (Li et al. 2001) (Fig. 5.2). Now that high-quality tomato genome sequence is available, specific primers can be designed for any locus targeted. The PCR extension time is adjusted so that deletions can be detected by PCR using DNA pools of up to 2500 lines. Individual mutants in the pools are further PCR-identified by deconvolution of the pools, i.e. reducing the complexity of the pools to subpools with fewer lines and then to individual plants. The mutations are finally confirmed by DNA sequencing. However, until now, fast neutron mutagenesis has been used in tomato in forward genetic screens for identifying mutants displaying various phenotypic traits, e.g. resistance to infection by fungal spores (David-Schwartz et al. 2001) or to parasitic weed (Dor et al. 2011). To date, no examples of successful screening of

fast-neutron or γ-ray tomato for detecting mutations in genes-of-interest have been provided until now. In the next future, mutant phenotyping combined with increased availability of tomato gene sequences of high quality and cost-effective NGS should trigger the identification by forward genetics approaches of genes underlying phenotypic variations in tomato.

Chemical Mutagenesis and Mutation Detection

Chemical mutagenesis with EMS or MNU generates a greater diversity of mutations than insertional mutagenesis. In tomato, attempts to perform mutagenesis with MNU have been largely unsuccessful and the chemical mutant collections published to date have been generated using EMS. Like MNU, the EMS induces single nucleotide changes by alkylation of specific nucleotides and produces mostly G/C to A/T transitions (Greene et al. 2003; Henikoff and Comai 2003). These point mutations, often termed SNP for Single Nucleotide Polymorphism, are randomly distributed in the genome at high density (Greene et al. 2003). Hundreds to thousands of mutations can be found in each individual plant and it is therefore possible to find a mutation in any given gene by screening few thousands of tomato plants (Gady et al. 2009; Minoia et al. 2010; Piron et al. 2010; Okabe et al. 2011; Baldet et al. 2013).

In addition, chemical mutagenesis offers specific advantages over the insertional or irradiation mutations, which tend to produce knockouts (complete loss-of-function mutations) by disrupting or deleting the genes-of-interest. A gene knockout can be lethal when the target gene is essential to the plant. In contrast, EMS produces allelic series including truncation mutations, e.g. splicing site mutations or nonsense mutations resulting in gene knockout, but also missense mutations due to a single base change in a given codon. Both truncation and missense mutations may affect the function of the protein; they represent ~5 and ~45 % of mutations respectively (Greene et al. 2003). Amino acid substitutions due to missense mutations can be conservative (similar function is expected) or nonconservative (modification of the function). A large range of alleles can therefore be obtained by screening an EMS mutant population, including not only strong alleles but also hypomorphic alleles that are highly informative for functional studies of target genes. As indicated above, hypomorphic alleles may be preferable to complete knockouts that can be lethal. The EMS-induced point mutations may also produce dominant-negative mutants, which are very useful for assessing the biological function of proteins, for example enzymes undergoing feedback regulation by metabolites or transcription factors or kinases involved in regulatory networks (Diévart and Clark 2003; Ostergaard and Yanofsky 2004).

Tomato EMS Mutagenesis

As for fast-neutron or γ-ray deletion mutants, creating tomato EMS mutant collections is straightforward (Fig. 5.2), once pilot experiments have been performed (Meissner et al. 1997; Menda et al. 2004; Minoia et al. 2010). EMS effect and mutation frequencies will depend on many conditions such as tomato genotype, seed physiological state, EMS concentration, etc. … EMS doses used for EMS mutagenesis in tomato typically range from 0.5 to 1 %, with most of the mutant populations being obtained with 0.7 to 1 % EMS (Table 5.2). The M0 seeds are mutagenized to give M1 mutated seeds which are sown. After selfing, M2 seeds are harvested from M1 plants. The DNA is collected for mutation detection from M2 or M3 plants since M1 plants carry somatic mutations. M2 and M3 plants can be further analysed for their phenotype, and phenotypic data can be stored in a database. Because only two to three cells are at the origin of the gametes in tomato, (A. Levy, personal communication), mutation segregation patterns different from the expected ones can be observed in M2 families when M2 seeds are collected in bulk from M1 plants. To increase mutation frequency, the M2 mutagenized seeds can also be subjected to a new round of EMS mutagenesis before reentering the same process (Fig. 5.2).

To date, the largest EMS mutant collections available in tomato have been generated in the

Table 5.2 TILLING identification of tomato allelic variants with possible effect on protein function

Cultivar	Gene	Biological process	Screened size (kb)	Number of plants screened	Deleterious mutations	Mutation detection technology	References (mutant collection and TILLING)
Micro-Tom INRA France	*GDP galactose phosphorylase* (*SlGGP2*)	Vitamin C	1.5	7296 M2 families ~88,000 plants	1 Splice junction	Endo1/Li-Cor	Baldet et al. (2013)
Micro-Tom NBRJ Japan	*GDP mannose pyrophosphorylase* (*SlGMP2*)	Vitamin C	1.4	3052 M2 families ~30,000 plants	1 Nonsense 1Missense	Endo1/Li-Cor	Baldet et al. (2013)
	GDP mannose epimerase (*SlGME1*)	Vitamin C	1.0	3052 M2 families ~30,000 plants	1 Missense	Endo1/Li-Cor	Baldet et al. (2013)
	GDP galactose phosphorylase (*SlGGP2*)	Vitamin C	1.5	3052 M2 families ~30,000 plants	1 Nonsense 1 Missense	Endo1/Li-Cor	Baldet et al. (2013)
	Ethylene Resistant (*SlETR1*)	Ethylene perception	1.4	3052 M2 families ~30,000 plants	6 Missense	Endo1/Li-Cor	Okabe et al. (2011)
	Ethylene Resistant (*SlETR5*)	Ethylene perception	1.5	3052 M2 families ~30,000 plants	1 Nonsense	Endo1/Li-Cor	Okabe et al. (2011)
TPAADASU	*Proline dehydrogenase* (*ProDH*)	Proline degradation	0.8	8025 M2 families 6703 M3 families		HRM	Gady et al. (2009)
	Auxin response factor 7	Auxin signalling	0.9	5000 M2 families		CSCE	Gady et al. (2009)
	Phytoene synthase (*PSY1*)	Carotenoid biosynthesis	0.8	8025 M2 families	1 Nonsense 3 Missenses	HRM/CSCE	Gady et al. (2012)

(continued)

Table 5.2 (continued)

Cultivar	Gene	Biological process	Screened size (kb)	Number of plants screened	Deleterious mutations	Mutation detection technology	References (mutant collection and TILLING)
M82	*Elongation factor (SleIF4e)*	Development virus resistance	0.9	3008 M2 families (15,040 plants)	2 Missenses	454 (GS)FLX sequencer	Rigola et al. (2009)
	eIF4E1, eIF4E2, eIF(iso)4E, eIF4G, eIF(iso)4G, DET1, COP1like, DDB1a, COP10, NAM, ACO1, E8, DHS, RAB11a, PG, MET1, Exp1, CRTISO, CUL4	Fruit ripening, Development Signalling virus resistance	30.9		1 Splice junction 6 Nonsense 85 Missenses	Endo1/Li-Cor	Piron et al. (2010)
	SlDET1, SlCOP1, SlCOP10, SlCOP1-like, SlCUL4, SlDDB1	Light signalling	9.7	4759 M3 families	3 Nonsense 13 Missenses	Endo1/Li-Cor	Jones et al. (2012)
	Branched (SlBRC1a)	Development	–		3 Missenses	Endo1/Li-Cor	Martín-Trillo et al. (2011)
	Blind (SlBli2)	Development	–		1 Nonsense 1 Missense	Endo1/Li-Cor	Busch et al. (2011)
	Terminating flower (Tmf)	Development	–		1 Missense	Endo1/Li-Cor	MacAlister et al. (2012)
Red Setter	*RIN, Gr, Rab11a, Exp1, PG, Lcy-b, Lcy-e*	Fruit ripening, Softening, Carotenoid synthesis,	9.5	5221 M3 families	41 Missenses	Endo1/Li-Cor	Minoia et al. (2010)
	Ethylene Response factor (ERF1)	Development/Stress	1	5221 M3 families	1 Missense	Endo1/Li-Cor	Di Matteo et al. (2013)
	Branched (SlBRC1a)	Development	–		3 Missenses	Endo1/Li-Cor	Martín-Trillo et al. (2011)

Point mutations in target genes were identified by screening EMS-mutagenized tomato populations. Only mutations reported to have possible deleterious effect on protein function are indicated miniature determinate cultivar Micro-Tom, well fitted for functional studies of target genes since it can be grown at high density year-round in greenhouse and has a short life cycle (four generations/year) (Meissner et al. 1997, 2000). More than 13,000 M2 EMS mutant families have been generated in the last 10 years in this cultivar, of which more than 10,000 have been further used for TILLING (Okabe et al. 2011; Baldet et al. 2013; Just et al. 2013). Most other cultivars are determinate processing tomatoes easy to grow in open fields for harvesting seeds: M82 cultivar, which had been used for generating the *S. pennellii* introgression lines in cultivated tomato (Eshed and Zamir 1995), Heinz 1706 used for the reference tomato sequence (Tomato genome Consortium 2012) and Red Setter. In addition, the semi-determinate variety Arka Vikas has been used (Sreelakshmi et al. 2010) and, more recently, the indeterminate cultivars Money Maker and Best of All (Table 5.2).

Similar EMS mutation frequencies are expected whatever the genome size of the plant species though genome redundancy confers tolerance to EMS mutations as recently shown in Arabidopsis (Tsai et al. 2013). Mutation frequencies are therefore higher in polyploid species (1 mutation/25 kb in hexaploid wheat; Slade et al. 2005). Mutation frequencies observed in tomato EMS mutant populations are function of the gene sequences analysed and, above all, of the total number of genes TILLed. Despite these limitations, the mean mutation frequencies observed in the various tomato mutant populations were quite similar and ranged between 1 mutation/320 kb to 1 mutation/730 kb for the most highly mutagenized populations. Decreasing EMS concentration leads to decreased mutation frequencies. Conversely, increasing mutagenesis pressure by performing two rounds of mutagenesis as performed for Micro-Tom mutant population in our lab (Fig. 5.2) further increases the mutation load. Mutation frequencies of up to ~1 mutation/130 kb were observed by TILLING for some genes (Table 5.2). More recently, whole genome sequencing of mutants from the same population led to the same conclusions. In contrast to Arabidopsis, in which the vast majority of EMS mutations observed were G/C to A/T transitions (Greene et al. 2003), spectrum of EMS-induced mutations observed in tomato is much wider and also includes transversions (Minoia et al. 2010; Piron et al. 2010). Our own observations on Micro-Tom EMS populations are in agreement with these data. Since Micro-Tom plants are grown in insect-proof greenhouses, possibility of cross-contamination with other tomato genotypes is much reduced. In addition, any pollen contamination from normal-sized tomatoes would result in the appearance of large tomato plants in the progeny since the miniature size of Micro-Tom is controlled by recessive alleles (Dan et al. 2007). This never happened to date, indicating that a wide spectrum of mutations can be induced by EMS in tomato.

Mutation Detection by TILLING

TILLING (Targeting Induced Local Lesions IN Genomes) combines random chemical mutagenesis by EMS with PCR-based methods for detecting unknown point mutations in regions of interest in target genes (Colbert et al. 2001). Since the early description of TILLING using heteroduplex analysis with denaturing HPLC, the detection of unknown mutations in mutant collections has been done using a large array of techniques including direct sequencing, capillary electrophoresis, conformation sensitive capillary electrophoresis (CSCE), capillary electrophoresis single strand conformation polymorphism (CE-SSCP), high resolution melt (HRM), MALDI-TOF, and infrared- or fluorescence-based sequencing (Julio et al. 2008; Gady et al. 2009; Rigola et al. 2009; Sikora et al. 2011; Okabe et al. 2011; Gady et al. 2012). Most of these technologies can be automated and are therefore suitable for high throughput screening of mutant collections. However, with the exception of the infrared-based LI-COR system, most of them display best results when DNA fragment sizes range from 300 to 600 bp. This can considerably increase the time and cost of TILLING when screening a large number of genes.

Therefore, enzymatic mismatch cleavage using endonuclease enzymes members of the S1 nuclease family, followed by electrophoresis separation of the cleaved fragments, a strategy originally described by Colbert et al. (2001), has been widely used in the recent years.

Screening tomato mutant collection with the endonuclease/electrophoresis detection system is very simple (Fig. 5.2). CEL1, the first mismatch cleavage enzyme used for TILLING, was originally extracted from celery and later from other plant species and is produced as a recombinant enzyme (Colbert et al. 2001). ENDO1, an additional S1 type endonuclease performing similar functions has first been identified from Arabidopsis and cloned to produce recombinant protein (Triques et al. 2007). It was later identified from tomato in which genetically transformed plants overproducing the enzyme have been obtained (Okabe et al. 2011). Using these enzymes, mutant detection can be done in most labs having robust and sensitive sequencing equipment, for example the LI-COR system. A DNA fragment of 0.5–2 kb of the target gene is first PCR amplified from DNA pools (four to eightfold pools usually) with differentially labelled primers. Primer labelling will depend on the electrophoresis equipment used: infrared-based sequencers such as LI-COR or fluorescence-based sequencers. The choice of the target gene region to be amplified depends on a number of factors. Among these is the presence of conserved domains in the protein, of substrate binding or catalytic sites, of protein–protein interaction domains, of DNA binding sites etc.… It also depends on intron/exon gene structure and on gene composition since these will affect the probability to find splice junction, nonsense and missense mutations. Tools such as CODDLE (Codons Optimized to detect Deleterious Lesions) (Fig. 5.1) have been developed to scan the gene sequence in search of the most favourable region to find deleterious mutations. Following amplification, high temperature-denaturation of the amplified fragment followed by low temperature re-annealing creates DNA homoduplexes and heteroduplexes. Heteroduplexes are then cleaved next to the mismatch by CEL1/ENDO1 endonuclease while homoduplexes are left intact by the enzyme (Fig. 5.2). Electrophoresis on denaturing gel will further separate the cleaved end-labelled DNA fragments from the non-cleaved ones. The use of differentially labelled primers allows the precise location on the gel of the two cleaved fragments and hence of the position of the point mutation in the DNA sequence. Once a mutant is detected in a pool of families or of individual plants, the deconvolution of the pool, identification of the plant mutant and confirmation of the mutation by Sanger sequencing is done as described for deletion mutants.

To date, detection of mutated alleles in tomato EMS mutant collections has been mostly done using the Endo1/LI-COR technology, though HRM, CSCE and next generation (GS)FLX sequencer have also been successfully used (Gady et al. 2009; Rigola et al. 2009; Gady et al. 2012) (Table 5.2). The current technical progresses in NGS technologies and concomitant cost reductions now trigger the development of TILLING by sequencing in tomato, using technologies such as Illumina or Ion Torrent sequencing, as already done in Arabidopsis (Tsai et al. 2011, 2013).

Besides the identification of mutations in target genes, mutant collections can be screened for mutations in proteins when high throughput technologies allowing the discrimination between wild type proteins and mutant proteins are available. This has recently been done by screening a Micro-Tom mutant collection for alterations in kinetic parameters of enzymes from central metabolism, allowing the discovery of two mutants in triose-phosphate isomerase (Ménard et al. 2013).

Reverse Genetic Approach: Linking Mutation to Phenotype Using Induced Genetic Variability

In tomato, more than 40 TILLed genes, for which deleterious mutations susceptible to affect the biological function of the encoded protein were identified (Table 5.2), have been published in the last four years. Biological processes investigated were largely focused on development and sensorial and nutritional quality of fruit (Minoia et al. 2010; Okabe et al. 2011; Gady et al. 2012; Jones et al. 2012; Baldet et al. 2013; Di Matteo et al. 2013), on major agronomical traits such as plant branching and yield control (Busch et al. 2011; Martín-Trillo et al. 2011; MacAlister et al. 2012) and on stress and virus resistance (Gady et al. 2009; Rigola et al. 2009; Piron et al. 2010). When a non-synonymous SNP has been detected within an exon, the amino acid changes in the protein can be further analysed using PARSESNP (Project Aligned Related Sequences and Evaluate SNPs) and the possible effect of amino acid changes on the function of the encoded protein can be checked using Sorting Tolerant From Intolerant (SIFT) (Fig. 5.1). These predictions can be used to select the allelic series that will be studied further. This is usually done by selecting plants carrying homozygous mutation in the progeny (when mutation is non-lethal) and by characterising their phenotype. Since hundreds of unrelated EMS mutations introduced by EMS mutagenesis are present in each plant, it is necessary to use plants from the same family that do not carry the mutated allele as controls. Mutant plant carrying the mutated allele of interest can be further backcrossed with wild type plants, to reduce the mutation load, as was done in wheat in which four backcrosses were estimated sufficient to derive lines similar to the wild type parent (Slade et al. 2005). However, purifying the mutation by several backcrosses can take a long time. For linking a plant phenotypic change to a mutation, the association between the mutation and the phenotype can be studied (i) using a large segregating population; (ii) several independent alleles displaying the same phenotypic effect that do not complement each other can be obtained (Henikoff and Comai 2003; MacAlister et al. 2012); (iii) and/or other strategies such as RNAi silencing (Busch et al. 2011) or VIGS can be used.

Forward Genetic Approach: Linking Phenotype to Mutation Using Induced Genetic Variability

Besides TILLING, the tomato EMS mutant collections can be further exploited through forward genetic approach aiming at identifying the mutation underlying the mutant trait. One of the advantages of this approach is that it does not require prior assumptions of the function of a gene. It relies first on the phenotypic characterization of the mutant trait, which can now be done using portable data acquisition devices (Vankudavath et al. 2012). Several EMS or irradiation tomato mutant populations designed for TILLING have already been thoroughly screened for phenotypic alterations. These include M82 (Menda et al. 2004), Red Setter (Minoia et al. 2010) and Micro-Tom (Saito et al. 2011; Just et al. 2013; Petit et al. 2014) cultivars. Thousands to tens of thousands of phenotypic traits were identified, classified into up to 48 (Menda et al. 2004; Saito et al. 2011) and 150 (Just et al. 2013) categories and subcategories and used to build in silico databases allowing the association between mutant line and phenotypic categories.

Mining phenotypic databases allows the identification of mutant lines for a given trait. In most populations described, several mutant alleles were found per locus indicating the nearly saturation of the mutant collections. Allelism tests can be performed to determine whether one or several mutated loci are responsible for the mutant trait. Next step is the identification of the gene mutation or other allelic variation underlying the phenotypic alteration observed. This can be done as for natural genetic variation through map-based cloning, which can be combined or not with candidate gene approach (Fig. 5.2). This strategy has been considerably eased by the availability of reference tomato genome sequence (Tomato Genome Consortium 2012), the development of genetic markers (Shirasawa et al. 2010; Kobayashi et al. 2014; Ranc et al. 2012) and of SNP genotyping arrays in tomato (Sim et al. 2012), and the availability of gene expression atlas in plant organs, tissues and cells (Matas et al. 2011; Park et al. 2012). Map-based cloning of mutations has been very successful in the recent years to identify allelic variants responsible for fruit cuticle alterations in M82 and Micro-Tom mutant populations (Isaacson et al. 2009; Yeats et al. 2012; Kimbara et al. 2013; Shi et al. 2013; Petit et al. 2014). However, this strategy still requires crossing the mutant line with polymorphic tomato genotype, such as the *S. pimpinellifolium* wild ancestor of cultivated tomato. The F2 progeny may therefore present for the trait under scrutiny a large phenotypic diversity independent from the mutation studied.

Thanks to the large amount of sequence data produced by current sequencing technologies, a very promising approach is to use the EMS mutations as genetic markers in order to identify the mutation by NGS mapping, as described in Arabidopsis (Hartwig et al. 2012) and rice (Abe et al. 2012). Since this strategy involves a cross between the mutant and its wild type parent (Fig. 5.2), no undesirable variations other than those due to mutagenesis are observed in the F2 progeny segregating for the mutation. Whole genome sequencing is performed on two bulks of F2 segregants displaying or not the mutant phenotype. SNP frequencies are then compared between the two bulks. In case of recessive mutation, a 100 % SNP frequency is expected in the bulk showing mutant phenotype whereas a 33 % SNP frequency is expected in the bulk without mutant phenotype. Using NGS mapping, we recently identified the mutation responsible for a fruit colour variation in our Micro-Tom EMS mutant population. The 600 plants from the F2 segregating population were cultivated in a greenhouse on only 4 m^2. These results indicate that NGS mapping combined with the use of Micro-Tom EMS mutants can be successfully used for the identification of EMS mutants in tomato, opening the way to the plant biology community for using tomato as a model for gene discovery and functional studies in plants.

References

Abe A, Kosugi S, Yoshida K, Natsume S, Takagi H, Kanzaki H, Matsumura H, Mitsuoka C, Tamiru M, Innan H, Cano L, Kamoun S, Terauchi R (2012) Genome sequencing reveals agronomically important loci in rice using MutMap. Nat Biotechnol 30:174–178

Alonso JM, Ecker JR (2006) Moving forward in reverse: genetic technologies to enable genome-wide phenomic screens in Arabidopsis. Nat Rev Genet 7:524–536

Baldet P, Bres C, Mauxion J-P, Just D, Bournonville C, Ferrand C, Mori K, Okabe Y, Ezura H, Rothan C (2013) TILLING identification of ascorbate biosynthesis tomato mutants for investigating vitamin C in tomato. Plant Biotechnol 30:309–314

Busch BL, Schmitz G, Rossmann S, Piron F, Ding J, Bendahmane A, Theres K (2011) Shoot branching and leaf dissection in tomato are regulated by homologous gene modules. Plant Cell 23:3595–3609

Causse M, Desplat N, Pascual L, Le Paslier MC, Sauvage C, Bauchet G, Bérard A, Bounon R, Tchoumakov M, Brunel D, Bouchet JP (2013) Whole genome resequencing in tomato reveals variation associated with introgression and breeding events. BMC Genom 14:791

Causse M, Duffe P, Gomez MC, Buret M, Damidaux R, Zamir D, Gur A, Chevalier C, Lemaire-Chamley M, Rothan C (2004) A Genetic map of candidate genes and QTLs involved in tomato fruit size and composition. J Exp Bot 55:1671–1685

Chakrabarti M, Zhang N, Sauvage C, Muños S, Blanca J, Cañizares J, Diez MJ, Schneider R, Mazourek M, McClead J, Causse M, van der Knaap E (2013) A cytochrome P450 regulates a domestication trait in cultivated tomato. Proc Natl Acad Sci USA 110:17125–17130

Colbert T, Till BJ, Tompa R, Reynolds S, Steine MN, Yeung AT, McCallum CM, Comai L, Henikoff S (2001) High-throughput screening for induced point mutations. Plant Physiol 126:480–484

Dan Y, Fei Z, Rothan C (2007) Micro-Tom—a new model plant for genomics. Genes Genomes Genomics 1:167–179

David-Schwartz R, Badani H, Smadar W, Levy AA, Galili G, Kapulnik Y (2001) Identification of a novel genetically controlled step in mycorrhizal colonization: plant resistance to infection by fungal spores but not extra-radical hyphae. Plant J 27:561–569

Diévart A, Clark SE (2003) Using mutant alleles to determine the structure and function of leucine-rich repeat receptor-like kinases. Curr Opin Plant Biol 6:507–516

Di Matteo A, Ruggieri V, Sacco A, Rigano MM, Carriero F, Bolger A, Fernie AR, Frusciante L, Barone A (2013) Identification of candidate genes for phenolics accumulation in tomato fruit. Plant Sci 205–206:87–96

Dor E, Yoneyama K, Wininger S, Kapulnik Y, Yoneyama K, Koltai H, Xie X, Hershenhorn J (2011) Strigolactone deficiency confers resistance in tomato line SL-ORT1 to the parasitic weeds Phelipanche and Orobanche spp. Phytopathology 101: 213–222

Emmanuel E, Levy AA (2002) Tomato mutants as tools for functional genomics. Curr Opin Plant Biol 5: 112–117

Eshed Y, Zamir D (1995) An introgression line population of Lycopersicon pennellii in the cultivated tomato enables the identification and fine mapping of yield-associated QTL. Genetics 141:1147–1162

Fernandez AI, Viron N, Alhagdow M, Karimi M, Jones M, Amsellem Z, Sicard A, Czerednik A, Angenent G, Grierson D, May S, Seymour G, Eshed Y, Lemaire-Chamley M, Rothan C, Hilson P (2009) Flexible tools for gene expression and silencing in tomato. Plant Physiol 151:1729–1740

Fernie AR, Klee HJ (2011) The use of natural genetic diversity in the understanding of metabolic organization and regulation. Front Plant Sci 2:59

Frary A, Nesbitt TC, Grandillo S, Knaap E, Cong B, Liu J, Meller J, Elber R, Alpert KB, Tanksley SD (2000) fw2.2: a quantitative trait locus key to the evolution of tomato fruit size. Science 289:85–88

Fridman E, Pleban T, Zamir D (2000) A recombination hotspot delimits a wild-species quantitative trait locus for tomato sugar content to 484 bp within an invertase gene. Proc Natl Acad Sci USA 97:4718–4723

Gady AL, Hermans FW, Van de Wal MH, van Loo EN, Visser RG, Bachem CW (2009) Implementation of two high through-put techniques in a novel application: detecting point mutations in large EMS mutated plant populations. Plant Methods 5:13

Gady AL, Vriezen WH, Van de Wal MH, Huang P, Bovy AG, Visser RG, Bachem CW (2012) Induced point mutations in the phytoene synthase 1 gene cause differences in carotenoid content during tomato fruit ripening. Mol Breed 29:801–812

Gidoni D, Fuss E, Burbidge A, Speckmann GJ, James S, Nijkamp D, Mett A, Feiler J, Smoker M, de Vroomen MJ et al (2003) Multi-functional T-DNA/Ds tomato lines designed for gene cloning and molecular and physical dissection of the tomato genome. Plant Mol Biol 51:83–98

Greene EA, Codomo CA, Taylor NE, Henikoff JG, Till BJ, Reynolds SH, Enns LC, Burtner C, Johnson JE, Odden AR, Comai L, Henikoff S (2003) Spectrum of chemically induced mutations from a large-scale reverse-genetic screen in Arabidopsis. Genetics 164:731–740

Hartwig B, James GV, Konrad K, Schneeberger K, Turck F (2012) Fast isogenic mapping-by-sequencing of ethyl methanesulfonate-induced mutant bulks. Plant Physiol 160:591–600

Henikoff S, Comai L (2003) Single-nucleotide mutations for plant functional genomics. Annu Rev Plant Biol 54:375–401

Hirakawa H, Shirasawa K, Ohyama A, Fukuoka H, Aoki K, Rothan C, Sato S, Isobe S, Tabata S (2013) Genome-wide SNP genotyping to infer the effects on gene functions in tomato. DNA Res 20:221–233

Ichihashi Y, Sinha NR (2014) From genome to phenome and back in tomato. Curr Opin Plant Biol 18C:9–15

Isaacson T, Kosma DK, Matas AJ, Buda GJ, He Y, Yu B, Pravitasari A, Batteas JD, Stark RE, Jenks MA, Rose JKC (2009) Cutin deficiency in the tomato fruit cuticle consistently affects resistance to microbial infection and biomechanical properties, but not transpirational water loss. Plant J 60:363–377

Jones MO, Piron-Prunier F, Marcel F, Piednoir-Barbeau E, Alsadon AA, Wahb-Allah MA, Al-Doss AA, Bowler C, Bramley PM, Fraser PD, Bendahmane A (2012) Characterisation of alleles of tomato light signalling genes generated by TILLING. Phytochemistry 79:78–86

Julio E, Laporte F, Reis S, Rothan C, Dorlhac de Borne F (2008) Targeted mutation breeding as a tool for tobacco crop improvement. Mol Breed 21:369–381

Just D, Garcia V, Fernandez L, Bres C, Mauxion J, Petit J, Jorly J, Assali J, Bournonville C, Ferrand C, Baldet P, Lemaire-Chamley M, Mori K, Okabe Y, Ariizumi T, Asamizu E, Ezura H, Rothan C (2013) Micro-Tom mutants for functional analysis of target genes and discovery of new alleles in tomato. Plant Biotechnol 30:225–231

Kharkwal MC, Shu QY (2009) The role of induced mutations in world food security. In: Shu QY (ed) induced plant mutations in the genomics era. Food and Agricultural Organization of the United nations, Rome, pp 33–38

Kimbara J, Yoshida M, Ito H, Kitagawa M, Takada W, Hayashi K, Shibutani Y, Kusano M, Okazaki Y, Nakabayashi R, Mori T, Saito K, Ariizumi T, Ezura H (2013) Inhibition of CUTIN DEFICIENT 2 causes defects in cuticle function and structure and metabolite changes in tomato fruit. Plant Cell Physiol 54:1535–1548

Klee HJ, Giovannoni JJ (2011) Genetics and control of tomato fruit ripening and quality attributes. Annu Rev Genet 45:41–59

Kobayashi M, Nagasaki H, Garcia V, Just D, Bres C, Mauxion JP, Le Paslier MC, Brunel D, Suda K, Minakuchi Y, Toyoda A, Fujiyama A, Toyoshima H, Suzuki T, Igarashi K, Rothan C, Kaminuma E, Nakamura Y, Yano K, Aoki K (2014) Genome-wide analysis of intraspecific DNA polymorphism in 'Micro-Tom', a model cultivar of tomato (*Solanum lycopersicum*). Plant Cell Physiol 55:445–454

Krieger U, Lippman ZB, Zamir D (2010) The flowering gene SINGLE FLOWER TRUSS drives heterosis for yield in tomato. Nat Genet 42:459–463

Leide J, Hildebrandt U, Reussing K, Riederer M, Vogg G (2007) The developmental pattern of tomato fruit wax accumulation and its impact on cuticular transpiration barrier properties: effects of a deficiency in a beta-ketoacyl-coenzyme A synthase (LeCER6). Plant Physiol 144:1667–1679

Li X, Song Y, Century K, Straight S, Ronald P, Dong X, Lassner M, Zhang Y (2001) A fast neutron deletion mutagenesis-based reverse genetics system for plants. Plant J 27:235–242

Li X, Zhang Y (2002) Reverse genetics by fast neutron mutagenesis in higher plants. Funct Integr Genom 2:254–258

MacAlister CA, Park SJ, Jiang K, Marcel F, Bendahmane A, Izkovich Y, Eshed Y, Lippman ZB (2012) Synchronization of the flowering transition by the tomato TERMINATING FLOWER gene. Nat Genet 44:1393–1398

Martín-Trillo M, Grandío EG, Serra F, Marcel F, Rodríguez-Buey ML, Schmitz G, Theres K, Bendahmane A, Dopazo H, Cubas P (2011) Role of tomato BRANCHED1-like genes in the control of shoot branching. Plant J 67:701–714

Mathews H, Clendennen SK, Caldwell CG, Liu XL, Connors K, Matheis N, Schuster DK, Menasco DJ, Wagoner W, Lightner J, Wagner DR (2003) Activation tagging in tomato identifies a transcriptional regulator of anthocyanin biosynthesis, modification, and transport. Plant Cell 15:1689–1703

Matas AJ, Yeats TH, Buda GJ, Zheng Y, Chatterjee S, Tohge T, Ponnala L, Adato A, Aharoni A, Stark R, Fernie AR, Fei Z, Giovannoni JJ, Rose JK (2011) Tissue- and cell-type specific transcriptome profiling of expanding tomato fruit provides insights into metabolic and regulatory specialization and cuticle formation. Plant Cell 23:3893–3910

Matsukura C, Yamaguchi I, Inamura M, Ban Y, Kobayashi Y, Yin YG, Saito T, Kuwata C, Imanishi S, Nishimura S (2007) Generation of gamma irradiation-induced mutant lines of the miniature tomato (*Solanum lycopersicum* L.) cultivar 'Micro-Tom'. Plant Biotechnol 24:39–44

Ménard G, Biais B, Prodhomme D, Ballias P, Petit J, Just D, Rothan C, Rolin D, Gibon Y (2013) High throughput biochemical phenotyping for plants. Adv Bot Res 67:407–439

Menda N, Semel Y, Peled D, Eshed Y, Zamir D (2004) In silico screening of a saturated mutation library of tomato. Plant J 38:861–872

Meissner R, Jacobson Y, Melamed S, Levyatov S, Shalev G, Ashri A, Elkind Y, Levy AA (1997) A new model system for Tomato Genetics. Plant J 12:1465–1472

Meissner R, Chague V, Zhu Q, Emmanuel E, Elkind Y, Levy AA (2000) A high throughput system for transposon tagging and promoter trapping in tomato. Plant J 38:861–872

Minoia S, Petrozza A, D'Onofrio O, Piron F, Mosca G, Sozio G, Cellini F, Bendahmane A, Carriero F (2010) A new mutant genetic resource for tomato crop improvement by TILLING technology. BMC Res Notes 3:69

Mounet F, Moing A, Garcia V, Petit J, Maucourt M, Deborde C, Bernillon S, Le Gall G, Colquhoun I, Defernez M, Giraudel J-L, Rolin D, Rothan C, Lemaire-Chamley M (2009) Gene and metabolite regulatory network analysis of early developing fruit tissues highlights new candidate genes for the control of tomato fruit composition and development. Plant Physiol 149:1505–1528

Okabe Y, Asamizu E, Saito T, Matsukura C, Ariizumi T, Brès C, Rothan C, Mizoguchi T, Ezura H (2011) Tomato TILLING technology: development of a reverse genetics tool for the efficient isolation of mutants from Micro-Tom mutant libraries. Plant Cell Physiol 52:1994–2005

Orzaez D, Medina A, Torre S, Fernández-Moreno JP, Rambla JL, Fernández-Del-Carmen A, Butelli E, Martin C, Granell A (2009) A visual reporter system for virus-induced gene silencing in tomato fruit based on anthocyanin accumulation. Plant Physiol 150:1122–1134

Ostergaard L, Yanofsky MF (2004) Establishing gene function by mutagenesis in *Arabidopsis thaliana*. Plant J 39:682–696

Ozaki S, Ogata Y, Suda K, Kurabayashi A, Suzuki T, Yamamoto N, Iijima Y, Tsugane T, Fujii T, Konishi C, Inai S, Bunsupa S, Yamazaki M, Shibata D, Aoki K (2010) Coexpression analysis of tomato genes and experimental verification of coordinated expression of genes found in a functionally enriched coexpression module. DNA Res 17:105–116

Park SJ, Jiang K, Schatz MC, Lippman ZB (2012) Rate of meristem maturation determines inflorescence architecture in tomato. Proc Natl Acad Sci USA 109:639–644

Petit J, Bres C, Just D, Garcia V, Marion D, Bakan B, Joubes J, Domergue F, Rothan C (2014) Analyses of tomato fruit brightness mutants uncover strong cutin-deficient mutants among which a new allele of GDSL lipase. Plant Physiol 164:888–906

Piron F, Nicolaï M, Minoïa S, Piednoir E, Moretti A, Salgues A, Zamir D, Caranta C, Bendahmane A (2010) An induced mutation in tomato eIF4E leads to immunity to two potyviruses. PLoS ONE 5:e11313

Ranc N, Muños S, Xu J, Le Paslier MC, Chauveau A, Bounon R, Rolland S, Bouchet JP, Brunel D, Causse M (2012) Genome-wide association mapping in tomato (*Solanum lycopersicum*) is possible using genome admixture of *Solanum lycopersicum* var. cerasiforme. Genes Genom Genet 2:853–864

Rigola D, van Oeveren J, Janssen A, Bonné A, Schneiders H, van der Poel HJ, van Orsouw NJ, Hogers RC, de Both MT, van Eijk MJ (2009) High-throughput detection of induced mutations and natural variation using KeyPoint technology. PLoS ONE 4:e4761

Rohrmann J, Tohge T, Alba R, Osorio S, Caldana C, McQuinn R, Arvidsson S, Van der Merwe MJ, Riaño-Pachón DM, Mueller-Roeber B, Fei Z, Nesi AN, Giovannoni JJ, Fernie AR (2011) Combined transcription factor profiling, microarray analysis and metabolite profiling reveals the transcriptional control of metabolic shifts occurring during tomato fruit development. Plant J 68:999–1013

Saito T, Ariizumi T, Okabe Y, Asamizu E, Hiwasa-Tanase K, Fukuda N, Mizoguchi T, Yamazaki Y, Aoki K, Ezura H (2011) TOMATOMA: a novel tomato mutant database distributing Micro-Tom mutant collections. Plant Cell Physiol 52:283–296

Sato S, Shirasawa K, Tabata S (2013) Structural analyses of the tomato genome. Plant Biotechnol 30:257–263

Schauer N, Semel Y, Roessner U, Gur A, Balbo I, Carrari F, Pleban T, Perez-Melis A, Brudigam C, Kopka J, Willmitzer L, Zamir D, Fernie AR (2006) Genetics of metabolite in fruits of interspecific introgressions of tomato. Nat Biotechnol 24:447–454

Shi JX, Adato A, Alkan N, He Y, Lashbrooke J, Matas AJ, Meir S, Malitsky S, Isaacson T, Prusky D, Leshkowitz D, Schreiber L, Granell AR, Widemann E, Grausem B, Pinot F, Rose JK, Rogachev I, Rothan C, Aharoni A (2013) The tomato SlSHINE3 transcription factor regulates fruit cuticle formation and epidermal patterning. New Phytol 197:468–480

Shirasawa K, Isobe S, Hirakawa H, Asamizu E, Fukuoka H, Just D, Rothan C, Sasamoto S, Fujishiro T, Kishida Y, Kohara M, Tsuruoka H, Wada T, Nakamura Y, Sato S, Tabata S (2010) SNP discovery and linkage map construction in cultivated tomato. DNA Res 17:381–391

Sikder S, Biswas P, Hazra P, Akhtar S, Chattopadhyay A, Badigannavar AM, D'Souza SF (2013) Induction of mutation in tomato (*Solanum lycopersicum* L.) by gamma irradiation and EMS. Indian J Genet Plant Breeding 73:392–399

Sikora P, Chawade A, Larsson M, Olsson J, Olsson O (2011) Mutagenesis as a tool in plant genetics, functional genomics, and breeding. Int J Plant Genom. doi:10.1155/2011/314829

Sim SC, Durstewitz G, Plieske J, Wieseke R, Ganal MW, Van Deynze A, Hamilton JP, Buell CR, Causse M, Wijeratne S, Francis DM (2012) Development of a large SNP genotyping array and generation of high-density genetic maps in tomato. PLoS ONE 7: e40563

Slade AJ, Fuerstenberg SI, Loeffler D, Steine MN, Facciotti D (2005) A reverse genetic, nontransgenic approach to wheat crop improvement by TILLING. Nat Biotechnol 23:75–81

Sreelakshmi Y, Gupta S, Bodanapu R, Chauhan VS, Hanjabam M, Thomas S, Mohan V, Sharma S, Srinivasan R, Sharma R (2010) NEATTILL: a simplified procedure for nucleic acid extraction from arrayed tissue for TILLING and other high-throughput reverse genetic applications. Plant Methods 6:3

Tomato Genome Consortium TG (2012) The tomato genome sequence provides insights into fleshy fruit evolution. Nature 485:635–641

Triques K, Sturbois B, Gallais S, Dalmais M, Chauvin S, Clepet C, Aubourg S, Rameau C, Caboche M, Bendahmane A (2007) Characterization of *Arabidopsis thaliana* mismatch specific endonucleases: application to mutation discovery by TILLING in pea. Plant J 51:1116–1125

Tsai H, Howell T, Nitcher R, Missirian V, Watson B, Ngo KJ, Lieberman M, Fass J, Uauy C, Tran RK, Khan AA, Filkov V, Tai TH, Dubcovsky J, Comai L (2011) Discovery of rare mutations in populations: TILLING by sequencing. Plant Physiol 156:1257–1268

Tsai H, Missirian V, Ngo KJ, Tran RK, Chan SR, Sundaresan V, Comai L (2013) Production of a high-efficiency TILLING population through polyploidization. Plant Physiol 161:1604–1614

Urbanczyk-Wochniak E, Usadel B, Thimm O, Nunes-Nesi A, Carrari F, Davy M, Blasing O, Kowalczyk M, Weicht D, Polinceusz A, Meyer S, Stitt M, Fernie AR (2005) Conversion of MapMan to allow the analysis of transcript data from *Solanaceous species*: effects of genetic and environmental alterations in energy metabolism in the leaf. Plant Mol Biol 60:773–792

Vankudavath RN, Bodanapu R, Sreelakshmi Y, Sharma R (2012) High-throughput phenotyping of plant populations using a personal digital assistant. Methods Mol Biol 918:97–116

Vogg G, Fischer S, Leide J, Emmanuel E, Jetter R, Levy AA, Riederer M (2004) Tomato fruit cuticular waxes and their effects on transpiration barrier properties: functional characterization of a mutant deficient in a very-long-chain fatty acid beta-ketoacyl-CoA synthase. J Exp Bot 55:1401–1410

Xu J, Ranc N, Muños S, Rolland S, Bouchet JP, Desplat N, Le Paslier MC, Liang Y, Brunel D, Causse M (2013) Phenotypic diversity and association mapping for fruit quality traits in cultivated tomato and related species. Theor Appl Genet 126:567–581

Yeats TH, Martin LB, Viart HM, Isaacson T, He Y, Zhao L, Matas AJ, Buda GJ, Domozych DS, Clausen MH, Rose JK (2012) The identification of cutin synthase: formation of the plant polyester cutin. Nat Chem Biol 8:609–611

Yuan L, Dou Y, Kianian SF, Zhang C, Holding DR (2014) Deletion mutagenesis identifies a haploinsufficient role for γ-zein in opaque2 endosperm modification. Plant Physiol 164:119–130

The Sequencing: How it was Done and What it Produced

6

Marco Pietrella and Giovanni Giuliano

Abstract

The tomato genome sequencing was part of a larger international project whose final aim was to develop a network of resources focusing on the biology of the plant and to address key questions about adaptation and diversification in the Solanaceae family. *Solanum lycopersicum* was chosen as a model system by virtue of the wealth of its genetic resources and was sequenced by the International SOL Consortium including 10 different countries. Initially the project started with a BAC-by-BAC strategy with the support of dense genetic and physical maps. With the advent of Next Generation Sequencing (NGS) techniques, strategies were revised to a primarily Whole Genome Shotgun (WGS) approach. The published genome is the result of the combination of the data obtained from both methodologies and represents a golden standard among Solanaceae that will serve different aspects of basic and applied research.

Keywords

Tomato • BAC-by-BAC • Whole genome sequencing • Physical mapping • FISH

Introduction

The Tomato Genome Sequencing Project was launched on November 3, 2003, at a workshop held in Washington, DC, where a large international group of scientists discussed the feasibility, utility, strategy, and level of international interest for sequencing the tomato genome as a reference for the family Solanaceae and other closely related plant families. In 2004, tomato genome sequencing, as part of the larger "International Solanaceae Genome Project (SOL): Systems Approach to Diversity and Adaptation" initiative (Mueller et al. 2005) was finalized and a white paper was drafted. The project involved 10 different countries, including the USA, South Korea, China, UK, India, The Netherlands, France, Japan, Spain, and Italy.

M. Pietrella · G. Giuliano (✉)
Italian National Agency for New Technologies, Energy and Sustainable Development (ENEA), Casaccia Research Center, Via Anguillarese 301, 00123 Rome, Italy
e-mail: giovanni.giuliano@enea.it

M. Causse et al. (eds.), *The Tomato Genome*, Compendium of Plant Genomes,
DOI 10.1007/978-3-662-53389-5_6

The aim of the project was to provide a high-quality genome assembly that could serve as a golden reference for other species in the Solanaceae family. In particular, tomato was chosen because it represents the *Solanum* genus, containing close to 50 % of the total number of *Solanaceae* species and several crop plants, such as tomato, potato (*S. tuberosum*), eggplant (*S. melongena*), and pepino (*S. muricatum*). Large populations of Recombinant Inbred and Backcross Inbred (Introgression) lines derived from interspecific crosses with wild tomato species, mutant collections, inbred lines, and BAC libraries were available (solgenomics.net) and made tomato the ideal species for a sequencing effort in the Solanaceae genus. The Heinz 1706 cultivar (Ozminkowski 2004), a progenitor of many modern canning varieties, from which deep coverage Bacterial Artificial Chromosome (BAC) libraries were available, was chosen for the sequencing.

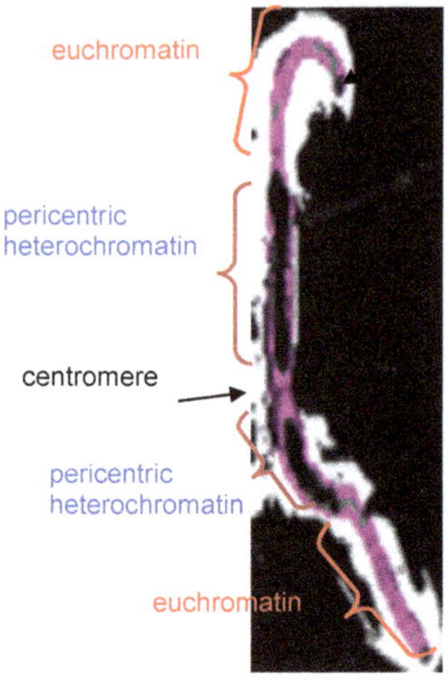

Fig. 6.1 Eu- and hetero- chromatin distribution on a tomato chromosome

BAC-by-BAC Sequencing

BAC-by-BAC genome walking was initially the proposed strategy, starting from "seed" BACs anchored on the genetic map. This approach had been successfully used for the sequencing of the 125-Mb Arabidopsis genome (The Arabidopsis Genome 2000), while the first sequences of the 450-Mb rice genome had been obtained by Whole Genome Shotgun (WGS) approaches (Goff et al. 2002; Yu et al. 2002).

Because of the relatively large size of the tomato genome (950 Mbases), the initial goal of the project was to sequence the euchromatic regions of all 12 chromosomes. The main motivations were the high cost of Sanger sequencing, and the availability of over 2500 anchored markers on an F2 *Solanum lycopersicum* × *Solanum pennellii* mapping population (Fulton et al. 2002; Frary et al. 2005), of which the majority located on gene-rich euchromatin. Furthermore, observations showed that the tomato genome was structured into gene-poor pericentromeric and telomeric heterochromatin and distal, gene-rich euchromatin (Fig. 6.1; Peterson et al. 1996; Wang et al. 2006; Chang et al. 2008). Despite the latter encompassed only about 25 % of total genome sequence, approximately 90 % of all non-transposon genes were calculated to reside there (Van der Hoeven 2002).

The sum of euchromatic portions of the genome was estimated to be about 220 Mb, with a projected sequencing cost less than twice that required to sequence the Arabidopsis genome. The proposed BAC-by-BAC sequencing strategy was based on the anchoring of BACs to a reference genetic map (called "seed" BACs). Three BAC libraries were available: a HindIII library, consisting of 129,024 clones (Budiman et al. 2000), an EcoRI one (75,264 clones) and an MboI (52,992 clones) library, resulting in more than 25 × coverage of the tomato genome. In addition to these libraries, in order to strengthen the genomic coverage and to accelerate the finishing, a BAC library (80,256 clones) and a fosmid library (>100,000 clones) were prepared from random sheared DNA (Table 6.1).

All libraries were end-sequenced (the BAC libraries by US partners and the fosmid libraries by the Wellcome Trust Sanger Institute and the University of Padua), yielding >340,000 high-quality reads (0.2 × genome coverage) and >180,000 reads (0.15 × genome coverage), respectively (Table 6.2).

Using the sequence-ends, a minimal tiling path of BAC clones mapping on the euchromatic

Table 6.1 BAC and fosmid libraries used for the sequencing project

Library name	Enzyme used	Clones (*n*)	Length span (mean)		Genome equivalents[b]
			Theorical (kb)	Calculated[a] (kb)	
SL_Hind (LE_Hba)	HindIII	129,024	117	105.2	14.3×
SL_Eco	EcoRI	75,264	100	103.2	8.2×
SL_Mbo	MboI	52,992	135	121.4	6.7×
SL_Fos	–	153,600	38	37.2	6.0×
Random sheared	–	80,256			

[a]Calculated on genome (v2.40) by remapping ends
[b]Calculated on a 950 Mb genome size

Table 6.2 Sanger clone end (SCE) sequence data used for the *S. lycopersicum* genome sequencing project

Library type	Fragment length (kb)	Read length (bp)	Raw		Filtered single		Filtered paired	
			Reads (*n*°)	Bases (Mb)	Reads (*n*°)	Bases (Mb)	Reads (*n*°)	Bases (Mb)
HindIII BAC	105.2	621	143,602	89,203	17,507	9682	125,538	79,134
EcoRI BAC	103.2	601	76,975	46,292	10,031	5489	66,944	40,803
MboI BAC	121.4	504	88,728	44,789	10,410	4777	78,060	39,943
Fosmid	37.2	549	151,301	83,159	21,855	10,973	129,444	72,185
Total			460,606	263,444	59,803	30,922	399,986	232,065

"arms" for each chromosome was calculated (Mueller et al. 2005).

The seed BACs, once sequenced, were further extended by identifying overlapping BACs (extension BACs) on the minimal tiling path. This process was iterated until exhausting the tiling path around a seed BAC.

The random sheared libraries were of particular use to fill the voids left due to nonrandom distribution of restriction sites on the genome. The usefulness of such resources had been already demonstrated during the finishing process of the rice genome (Ammiraju et al. 2005). Furthermore, the defined insert length of the fosmids could be used as an analytical tool to detect potential misassemblies. The shorter insert length from the fosmid clones was ideal for filling smaller gaps, minimizing redundant sequencing.

The final goal of the Tomato Genome Sequencing Project was to provide researchers with a high-quality, "golden standard" assembly representing a reference genome for the other Solanaceae. The standards of quality and completion agreed on by the consortium were comparable to those of the international rice genome sequencing project (The International Rice Genome Sequencing 2005) including:

- an error rate of less than 1:10,000 bases and continuous sequence across the entire BAC (HTGS phase 3)
- average of eightfold redundancy in sequencing coverage with a minimum of one high-quality read in both directions at any given location
- being as gap-free as possible, given all reasonable state-of-the-art gap-filling approaches available at the time of sequencing

The 12 chromosomes were assigned to the different participating countries (Fig. 6.2) and at Cornell University, seed BACs were anchored on the genetic map and shipped to the respective

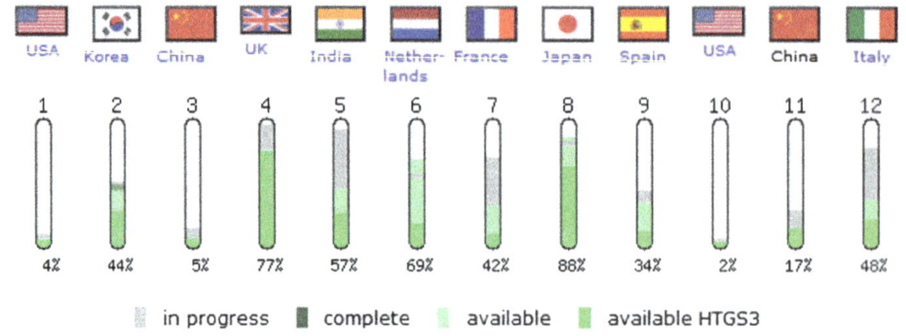

BACs													Total
Chr Total	391	268	274	193	111	213	277	175	164	186	135	113	2,500
In progress	9	2	20	39	54	5	106	5	14	1	18	48	321
Complete	19	171	0	133	27	152	57	155	67	4	6	24	815
Available	19	119	15	150	64	147	119	155	57	4	23	55	927
HTGS 1	5	0	0	0	0	118	69	0	13	0	1	21	227
HTGS 2	0	0	0	0	44	0	61	2	31	0	10	11	159
HTGS 3	14	71	15	148	28	37	23	112	18	4	22	23	515
% Done	5%	50%	4%	78%	46%	55%	32%	82%	35%	2%	13%	37%	

Overall Stats

27% of sequencing is complete
33% of BACs are reported finished
37% of BACs have downloadable sequence

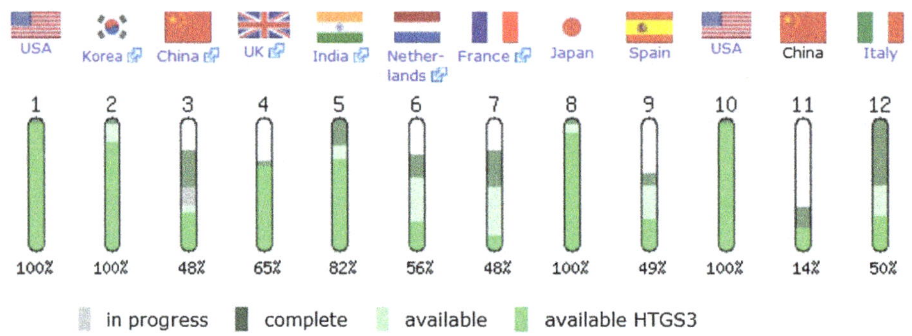

BACs													Total
Chr Total	391	268	274	193	111	213	277	175	164	186	135	113	2,500
In progress	28	4	79	2	28	1	52	5	0	14	18	62	293
Complete	604	370	127	126	83	153	141	203	94	625	7	32	2,565
Available	632	268	91	129	91	155	162	203	95	638	24	80	2,568
HTGS 1	47	22	2	0	1	114	29	0	16	72	0	25	328
HTGS 2	97	14	17	0	11	0	107	2	44	62	0	28	382
HTGS 3	488	227	72	127	78	41	23	160	35	504	24	27	1,806
% Done	100%	100%	48%	65%	82%	56%	48%	100%	49%	100%	14%	50%	

Overall Stats

100% of BAC sequencing is complete
106% of BACs are reported finished
106% of BACs have downloadable sequence

Fig. 6.2 Picture of the 12 tomato chromosomes showing, for each chromosome, the responsible country. The sequencing status for each chromosome (Chr) is shown as of October 2008, when the WGS strategy was adopted (*top*), and as of November 2013 (*bottom*)

sequencing centers. The actual sequencing effort started in the fall of 2004.

The high-density reference genetic map used to select the seed BACs contained DNA markers of different origin (SSR, AFLP, EST, etc.) obtained from the Tomato-EXPEN 2000 mapping population. The map, visible on the SGN website (http://solgenomics.net/cview/map.pl?map_id=9), was derived from an F_2 population of 83 individuals developed from a cross of the cultivated tomato (*S. lycopersicum*) line "LA925" with a line (LA716) of the wild tomato relative *S. pennellii* (Fulton et al. 2002; Frary et al. 2005). The resulting linkage map accounted for a total of 2604 markers, including a smaller subset of restriction fragment length polymorphism (RFLP) markers from the old Tomato-EXPEN 1992 map (Tanksley et al. 1992) and a larger subset of Conserved Ortholog Set (COS) markers (Fulton et al. 2002; Wu et al. 2006) derived from a comparison of a tomato EST database against the entire Arabidopsis genome. Only single/low copy COS markers with significant matches to putative orthologous *loci* in Arabidopsis were selected, useful for identifying chromosomal inversions, duplications, and other large-scale genome rearrangements. More recently, a similar map, based on the same material as the Tomato-EXPEN 2000 but containing new SSR markers, has been produced by Shirasawa et al. (2010) and can be viewed at http://www.kazusa.or.jp/tomato/. This new marker set contains 2116 loci, covering 1503 cM and was also used to further anchor BACs onto the genetic map (Fig. 6.3).

To anchor BACs to the genetic map, libraries were screened with "overgo" probes, designed on sequenced markers of the EXPEN 2000

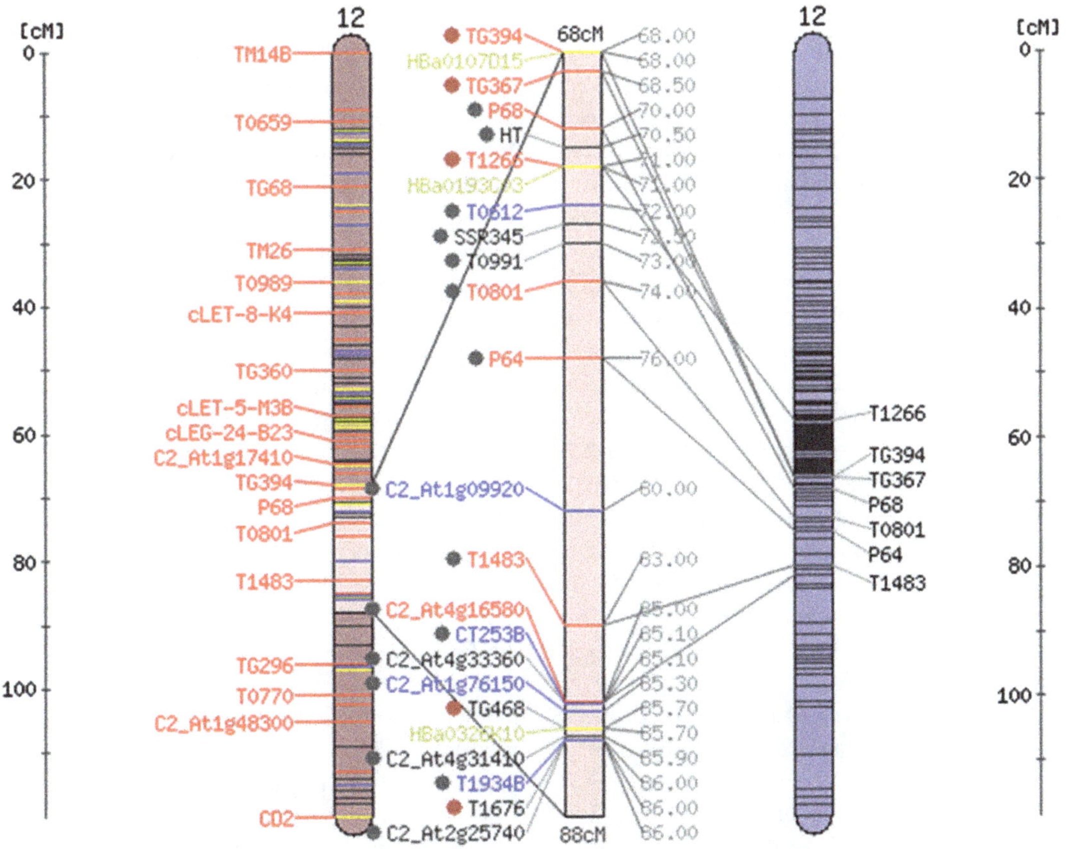

Fig. 6.3 Comparison between Tomato-EXPEN 2000 map (*left*) and Kazusa F2-2000 genetic maps (*right*) for tomato chromosome 12 (from: http://solgenomics.net/)

map. A total of 1536 probes (i.e., one every 143 Kbases of euchromatin on average) were used to screen a total of 128,560 BACs resulting in 7972 high-quality probe-BAC associations. A summary of these results can be found on the SGN website (http://solgenomics.net/maps/physical/overgo_stats.pl). Although overgo screening is simple and efficient, spurious hybridization may cause both false-positive and false-negative BAC associations (Han et al. 2000; Romanov et al. 2003; Peters et al. 2009). As an example, the RFLP marker cLET-5-M3 was initially mapped onto chromosome 12 with high confidence (LOD = 3) and, with lower confidence, on chromosome 7 (http://solgenomics.net/marker/SGN-M2981/details) while FISH mapping (see below) localized this marker on chromosome 6 (Peters et al. 2009).

Knowing the shortcomings of the map and the necessity to univocally identify and place the seed BACs, markers needed to be confirmed by PCR and re-sequencing of the loci. Each seed BAC was then used to extend out into the minimum tiling path. The "BAC walking" approach, where the next BAC in the euchromatin was identified by BLASTing against a BAC/fosmid end sequence database (http://solgenomics.net/tools/blast/index.pl) was used without a priori knowledge of the clone position in the genome, instead of using the physical map to sort clones. To avoid too much redundant sequencing, candidate extension BACs were selected, having an overlap of 5–10 kb with seed BACs. After a couple of instances in which the BLAST approach alone resulted in "jumps" on extension BACs located on non-related chromosomes, an additional verification step was introduced, based on PCR mapping of extension BACs on an Introgression Line mapping population (Eshed and Zamir 1995) (see below). To help identifying suitable extension BACs, two dedicated software tools were also developed, namely TOPAAS (Peters et al. 2006) and PABS (Todesco et al. 2008).

An additional, albeit laborious, step to validate the mapping of seed BACs was FISH (Fluorescence In Situ Hybridization) with BAC clones. The exact map locations of genetic markers and the relative positions between markers is sometimes difficult to determine, especially in genomic regions in which recombination is suppressed (Sherman and Stack 1995). Therefore, verification of the positions of seed BACs by FISH proved to be an important tool. FISH was performed in two laboratories using two slightly different techniques: stirred spreads (de Jong lab, (Szinay et al. 2008) and synaptonemal complex (SC) spreads (Stack lab, (Stack et al. 2009), with good levels of inter-laboratory reproducibility (The Tomato Genome Sequencing 2012). Using these data, a cytological map consisting of tomato pachytene chromosomes has been developed which can be visualized at http://solgenomics.net/cview/map.pl?map_version_id=25. Besides validating the chromosomal localization of seed BACs on the euchromatic parts of chromosome, such FISH map was used to assist and guide the extension of the euchromatic tiling path and to determine when the heterochromatin and telomeric regions had been reached on each arm, thus preventing the sequencing of undesired chromosomal parts. While repeated sequences can interfere with both BAC walking and FISH, this problem often can be minimized for FISH by chromosomal in situ suppression (CISS) hybridization with unlabeled tomato Cot 100 DNA (Szinay et al. 2008).

A valuable alternative (or better a prerequisite of the BAC prior to BAC/FISH mapping) consisted in the genetic mapping of the BAC on tomato chromosomes with the use of introgression line (IL) populations (Eshed and Zamir 1995 http://solgenomics.net/cview/map.pl?map_id=il6). This approach was based on identification of polymorphisms between *S. lycopersicum* and *S. pennellii* on BAC end sequences that were amplified by PCR. SNP polymorphisms were at a 2.5 % level, based on sample sequencing of the two genotypes. The workflow used for the mapping is shown in Fig. 6.4. The utility and the robustness of this process were demonstrated by the concordant results obtained from FISH analyses run in parallel (http://solgenomics.net/search/genomic/clones and data not shown). Assessments of chromosomal positions of the BACs with these techniques indicated that seed

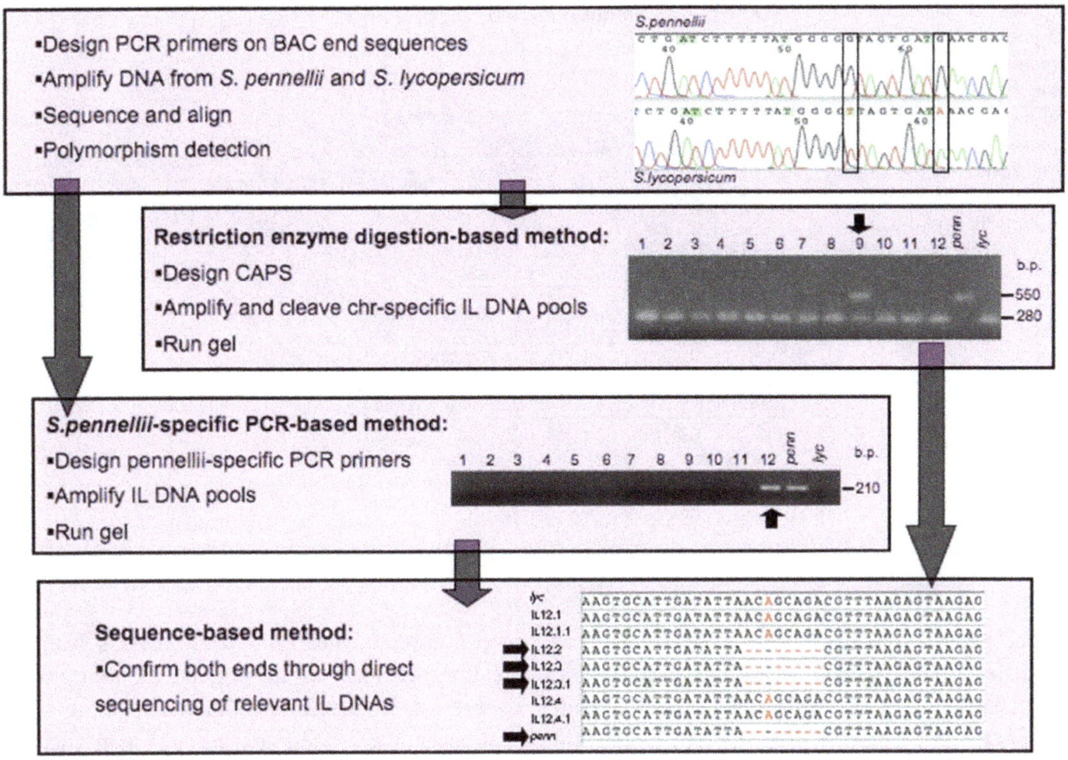

Fig. 6.4 Workflow of introgression lines-based BAC mapping

BACs were generally in agreement with the genetically mapped marker order from the EXPEN 2000 map, although inconsistencies have been found, mainly regarding erroneous mapping or misplaced order. The latter could also be due to rearrangements that may exist between the genotype of the EXPEN 2000 map and that of the species used for IL/FISH mapping [(Peters et al. 2009), data not shown]: in fact several inversions have been identified between the cultivated tomato and *S. pennellii*, parents used for the reference map (van der Knaap 2004).

Due to an uneven distribution of markers or to inherent problems with the overgo mapping process, early in the project a number of regions were identified lacking seed BACs. To overcome this problem, additional screenings of BAC libraries were performed with the already identified overgo sequences or with new markers that were not yet included in the overgo process. A helpful tool to easily screen large BAC populations was developed by INRA-CNRGV (http://cnrgv.toulouse.inra.fr/) that relies on the use of 3D BAC pools to screen an entire genomic library by PCR searching for marker sequences (Fig. 6.5).

To this date (July 2016) 2500 BACs have been sequenced and anchored to the 12 chromosomes (Fig. 6.2): of these, 328 were sequenced up to HTGS (High-Throughput Genome Sequence) 1 phase, 382 to HTGS2 and 1806 to HTGS3 phase, representing 290 Mb, including overlaps. Seventy-five additional BAC clones (four in HTGS1, two in HTGS2 and 69 in HTGS3 phase), accounting for a total of 8,716,369 sequenced bases have been sequenced but not anchored to a specific chromosome. These represent problematic (chimeric, etc.) clones whose localization could not be confirmed. The sequenced BACs are available for download at SGN (http://solgenomics.net/organism/Solanum_lycopersicum/clone_sequencing) and GenBank (http://www.ncbi.nlm.nih.gov/). Although a first high-quality draft of the

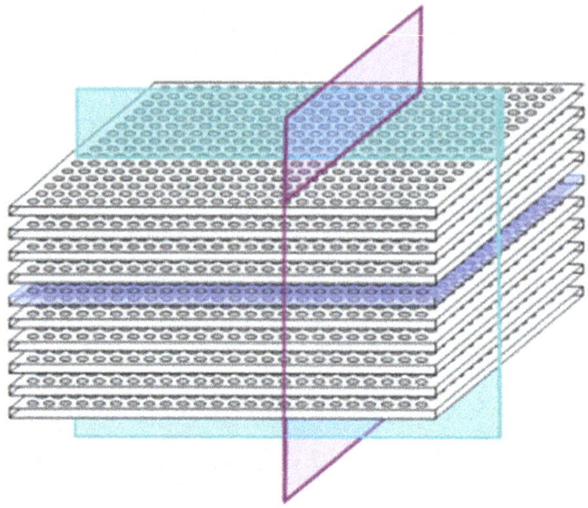

Fig. 6.5 Organization of a 3D pool for BAC library screenings. Each well represents a single BAC and the colored planes represent the 3D pools; the identification of the single BAC is possible by crossing the coordinates (*line*, *column* and *plate*)

tomato genome has already been assembled and published (The Tomato Genome Sequencing 2012) both the BAC sequencing and FISHing has progressed, and now 544 FISHed clones are available, a number of which has been found to localize in gaps between sequenced scaffolds and can thus be included in the assembly (S. Stack, pers. comm.).

Physical Mapping

Physical mapping is an integral part of the reconstruction of the tiling path as it provides the backbone for ordering and joining sequence data. Initially, physical maps were built by fingerprinting BACs from the HindIII library, and contigs of overlapping BACs were generated using the fingerprinted contigs (FPC) tool (Soderlund et al. 2000). This yielded 644 markers, and resulted in 4385 contigs (http://www.genome.arizona.edu/fpc/tomato/). Similar results were obtained using the MboI library. However, the technique used to produce the maps has been found to introduce gaps and false overlaps (Meyers et al. 2004), so that a new FPC map was built, based on the more precise capillary-based method (high-information-content fingerprinting, HICF (Luo et al. 2003). The derived SNaPshot map comprised nearly 340,000 BACs from all the libraries used in the project, including 4123 overgo and electronic markers. BAC end sequences were used to link physical mapped clones to tomato unigenes in the SNG database (ftp://ftp.solgenomics.net/unigene_builds), individually sequenced BACs, and sequenced markers.

Contextually, a Whole Genome Profiling (WGP) physical map was constructed using KeyGene proprietary technology (http://www.keygene.com/products-tech/wgp2-0/). For this, 92,160 BACs, representing an approximate 11X genome coverage, were used. A sequence-based physical BAC map was assembled using an improved version of the FPC software (Keygene N.V.) that is capable of processing sequence-based BAC fingerprint (WGP) data instead of fragment mobility information as used in the original FPC (Soderlund et al. 1997; van Oeveren et al. 2011). WGP data were used as input in the FPC map assembly yielding a physical map consisting of 2521 contigs in size and containing including over 52,000 BACs (Table 6.3).

Together with the high-density genetic map and the genome-wide BAC FISH, the two BAC-based physical maps constituted the framework for anchoring single BACs, contigs

Table 6.3 WGP and SNaPshot mapping results

	WGP map	SNaPshot map
Total # of BACs in FPC	66,084	82,784
# of contigs	2521	1835
# BACs in contigs	52,617	66,810
# Singleton BACs	13,467	15,974
Coverage (Mbp)	953	859
Average # BACs/contig	21	36
N50 contig size (# BACs)	26	67
Average contig size (Mbp)	0.378	0.39
N50 contig size (Mbp)	0.563	0.697

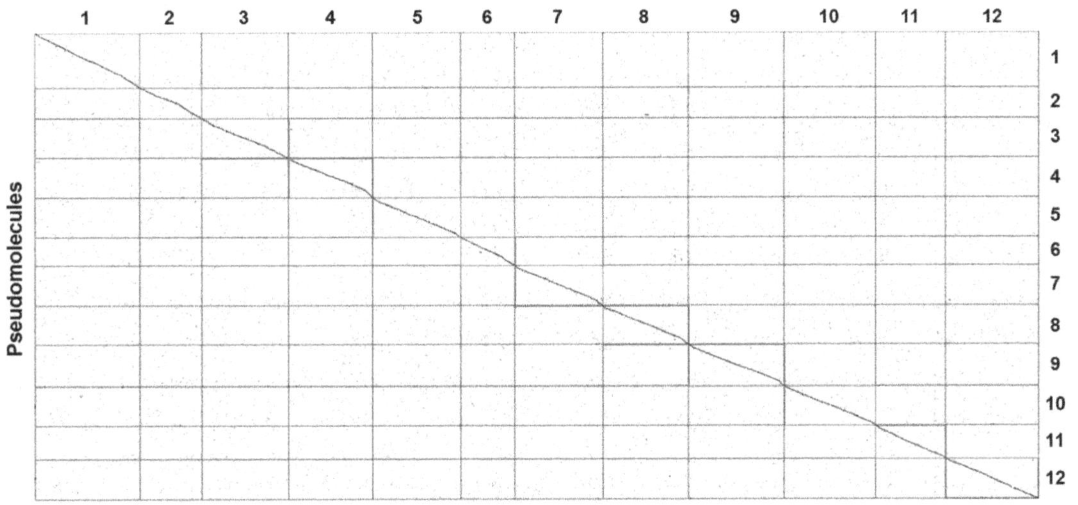

Fig. 6.6 Dotplot showing whole genome alignment of the physical contigs to pseudomolecules

and scaffolds originating from all sequencing techniques used and helped to the reconstitute the correct sequence and orientation of the genomic sequenced fragments into chromosome-like structures (Fig. 6.6).

Whole Genome Shotgun (WGS) Approach

By the fall of 2008, despite the great efforts made by all participating countries, it became evident that the overall progress of the BAC-by-BAC sequencing project was problematic. Large regions of the euchromatic genome were lacking seed BACs and extension of the majority of BAC contigs became increasingly difficult or had come to an end. Furthermore, the conviction that the heterochromatic part of the genome, initially excluded from sequencing, would be virtually devoid of genes (Wang et al. 2006) appeared not to be true. This assumption could be confirmed when FISH was used to identify BACs from euchromatin–heterochromatin boundaries or heterochromatic regions (Szinay et al. 2008). Sequencing a number these BACs demonstrated that they were actually more gene-rich than expected (Peters et al. 2009). This observation raised the possibility that a reasonably high amount of genic sequences would be missed in a euchromatin-only approach.

Meanwhile, from 2005 to 2006 on, a series of "Next Generation Sequencing" (NGS) technologies, including Roche/454, Applied Biosystems/SOLiD, and Illumina/Solexa became commercially available. These technologies offered much higher throughput and much lower costs than traditional Sanger sequencing (Mardis 2008). In particular, in late 2008 the 454 Titanium technology, with read lengths of up to 400 bp and mate-pairs spanning up to 20 kb became available. Already in 2007, 454 sequencing of tomato BACs had been initiated, significantly accelerating the release of sequenced BACs. However, the caveats of the BAC-by-BAC approach still represented a significant bottleneck for a timely completion of the project.

In parallel the Japanese team developed in 2007 an additional WGS approach, based on BAC pools denoted as SBM (Selected BAC Mixtures). These pools comprised of 30,800 BAC clones with only one or neither end containing repetitive sequences. These BAC mixtures, thought to represent mainly the gene-rich fraction of the genome, were shotgun-sequenced through Sanger technology to obtain >4 million reads (ca 3.5 × genome coverage).

At the SOL meeting in Cologne, in October 2008, three countries (Netherlands, Japan, and Italy) proposed to adopt a WGS approach based on the use of the SBM sequence and complemented by a substantial amount of Roche/454 sequencing and WGP mapping. This was an innovative proposal, since at that time no genome of the size of tomato had been completed or published through the predominant use of NGS technologies. The approach was broadly adopted by the whole consortium and the actual sequencing started in spring 2009. The first assembly was obtained in October 2009 and the first annotation (iTAG 1.0, very similar to the 2.3 version actually in use) appeared in December 2009. Through this approach, no distinctions were made between the different chromosomes or regions (hetero- or euchromatin). As such, the process contributed to all existing national chromosome projects and included also those regions that initially were excluded in the BAC-by-BAC approach.

In total 28.4 Gb of sequence data was generated using Roche/454 Titanium technology. This consisted in 14.4 Gb of shotgun reads (corresponding to a 15 × coverage of the tomato genome, reflected by more than 40 million of reads with a mean length of 350 bp), 7.1 Gb of 3 kb mate-pair reads (>7 × coverage), 3.9 Gb of 8-kb mate-pair reads (>4 × coverage), and 3.0 Gb of 20-kb mate-pair reads (>3 × coverage). Together with the SBM Sanger reads, these reads contributed to the backbone for the tomato genome assembly. Additionally, 133 Gb SOLiD reads (140 × coverage) were generated for both shotgun and mate-pairs with different insert sizes (1, 4, and 8 kb). Finally, two Illumina paired-end libraries with insert sizes of ~450 and 500 bp and four mate-pair libraries of 2, 3, 4, and 5 kb, were sequenced, representing an 86 × coverage of the tomato genome.

Mate-pairs were particularly important for a de novo genome assembly: because they are pairs of reads spanning a known distance span that ranged from 1 to 20 kb, they helped joining contigs that were separated by problematic regions (difficult to sequence or repeated sequences). Mate-pairs are thus particularly useful when trying to assemble eukaryotic genomes, containing large amounts of repeated sequences. In the case those repeats or low complexity genomic fragments remain shorter than the mate-pair span, the genomic region can be assembled. In this respect, BAC and fosmid paired ends can be considered as mate-pairs with a larger span. Moreover, mate-pairs can be used to assess the structural integrity of an assembled genome, because they are oriented and can help evaluating if the contigs have been assembled in the correct orientation and direction.

The 29 × 454 coverage represented a high level of redundancy, when compared with other sequenced genomes that used WGS approaches: *Vitis vinifera* had a Sanger coverage of 12×, (Jaillon et al. 2007), and an even lower coverage was sufficient to assemble *Sorghum bicolor* (Paterson et al. 2009).

Data Preprocessing

The total amount of data, the high coverage, the presence of redundant information due to duplicated materials, contaminants, and low-quality reads, was a heavy challenge for the computational resources needed to assemble the reference genome. Furthermore, these could result in chimerisms or other artefactual results. Because of the large amounts of 454 reads, and the lack of efficient software to generate hybrid assemblies, Illumina and SOLiD data were not used in the actual assembly but for base error correction of the typical 454-related errors (indels in homopolymers).

Preprocessing involved screening and removal of bacterial sequence contaminations from all the HTGS2 and HTGS3 BACs. These were then assembled into a nonredundant set of BAC contigs. The same was done for the SBM reads, where additionally, low-quality data were removed with Phrap. The data were further screened for the presence of remnants of cloning vector sequence with Cross_match and NCBI Blast. Similarly, duplicated reads were removed. 454 read filtering included the removal of duplicate reads; additionally, reads shorter than 50 bases or longer than 450 bases but containing more than one ambiguous base call (N) were likewise discarded.

SOLiD reads were quality-trimmed according to their quality scores; afterwards, those reads with trimmed lengths below 35 bp for the 50 bp libraries, or lengths below 20 bp for the 35 bp libraries, or an average quality below 15 were discarded. Similarly to the treatment of 454 reads, duplicated reads were also removed. The adjusted reads were aligned against the assembly using PASS (Campagna et al. 2009) and coupled using the pairing option of PASS. This step was essential to produce the data required for the evaluation of the structural correctness of the de novo assembly. The resulting polished reads are summarized in Table 6.2 for BAC and Fosmid ends and in Table 6.4 for the other materials. This material has been used for the final tomato genome assembly.

Table 6.4 Sequence data for the *S. lycopersicum* genome

Reads class	Library type	Fragment length	Read length (bp)	Raw		Filtered	
				Reads (*n*)	Total bases (Gb)	Reads (*n*)	Total bases (Gb)
Selected BAC mixture	Mate-pair	2.5 kb	881	4,039,383	3.558	3,797,957	3.137
454	Shotgun	700 bp	353	40,113,556	14.390	28,741,862	10.881
	Mate-pair	3 kb	336	20,055,779	7.101	14,908,129	5.581
	Mate-pair	8 kb	335	11,690,684	3.928	8,583,068	3.011
	Mate-pair	20 kb	342	8,639,567	2.945	3,880,727	1.399
	Total			80,499,585	28.364	56,113,785	20.872
SOLiD	Mate-pair	1 kb	2 × 25	816,569,620	20.414	518,915,022	11.416
	Mate-pair	4 kb	2 × 25	1,168,816,240	29.22	932,988,223	20.526
	Fragment	7 kb	50	408,291,426	20.414	128,650,283	5.146
	Mate-pair	8 kb	2 × 50	1,259,868,973	62.993	687,471,776	27.499
	Total			3,653,546,259	133.041	2,268,025,304	64.587
Illumina	Paired-end	312 bp	2 × 90	774,073,174	69.667		
	Mate-pair	2 kb	2 × 54	59,383,914	3.207		
	Mate-pair	3 kb	2 × 54	61,880,468	3.342		
	Mate-pair	4 kb	2 × 54	56,466,436	3.049		
	Mate-pair	5 kb	2 × 54	57,196,140	3.089		
	Total			1,009,000,132	82.354		

Conclusion and Outlook

The importance of tomato as a genetic model for Asterids, a model for fleshy fruit ripening and an important horticultural crop can hardly be overstated. This was the main reason for launching the genome sequencing effort as early as 2003, when genome sequencing was in its infancy and just two plant genomes (Arabidopsis and rice) had been completed and published. The initial goal was to sequence only the 220 Mb of euchromatin, predicted to contain the majority of tomato genes, through a BAC-by-BAC approach. In spite of the switch to a WGS approach in 2009, to date more than 2500 BACs have been sequenced and made available to the community through the SOL portal (http://solgenomics.net/; Fig. 6.2).

However, despite the large efforts devoted to BAC sequencing, the turning point of the project was the switch to a WGS approach based on a mix of Next Generation Sequencing, Sanger sequencing, and physical mapping. This approach produced, in less than a year, one of the best quality assemblies available in Asterids, with an error rate of less than 1 base in 7000 (falling below 1 in 15,000 in coding regions) (The Tomato Genome Sequencing 2012) and 742 Mb (i.e., 83 % of the 900-Mb genome) assembled in just 91 chromosome-anchored, oriented scaffolds. These metrics are far better than any other dicot genome published to date, with the exception of Arabidopsis, which is sevenfold smaller than tomato. Other chapters of this book are devoted to the details of the assembly effort and to the annotation by the international Tomato Annotation Group (iTAG), which is of comparably high quality.

The effort to improve the assembly is still ongoing with funding from the US and Dutch governments. Extensive FISH mapping, optical mapping, pooled gap spanning, 454 BAC sequencing and PacBio (Pacific Biosciences) long sequences incorporated into the assembly using PBJelly (English et al. 2012) are major players in this effort and have resulted to date in extensive gap filling (Table 6.5) and the reorientation and rearrangement of 45 of the 91 scaffolds (Shearer et al. 2014), located mostly in heterochromatin and comprising 34 % of the sequenced DNA.

The final goal is to reduce the scaffold number to 12, corresponding to the 12 tomato

Table 6.5 Gap filling using 454-sequenced BACs and PacBio sequences

Chromosome	Ch01	Ch02	Ch03	Ch04	Ch05	Ch06	Ch07	Ch08	Ch09	Ch10	Ch11	Ch12
≥ 25 bp gaps in tomato assembly before merging 454 sequenced US BACs	2431	1398	2121	1752	2114	1326	1862	1623	1745	2185	1746	1703
≥ 25 bp gaps in tomato assembly after merging 454 sequenced US BACs	1204	989	1670	1565	1650	1001	1423	1221	1317	946	1269	1431
Gaps remaining after PacBio sequencing and merging with PBJelly	352	345	589	552	553	362	489	470	426	327	448	563
≥ 25 bp gaps remaining after 2 PBJelly runs	305	310	532	485	495	321	423	420	375	284	400	501
Gaps remaining (%)	13	22	25	28	23	24	23	26	21	13	23	29

chromosomes, covering at least 85 % of the projected genome size and at least 98 % of the genes, with an error rate of less than 1 in 10,000. That is when the genome sequencing effort will be considered "reasonably" complete.

Acknowledgments We are grateful to the Tomato Genome Consortium for producing the data used in this chapter; to Lukas Mueller and the Solanaceae Genomics Network for providing Figs. 6.1, 6.2 and 6.3; to Helene Berges for granting us the use of Fig. 6.5, to Jose Luis Goicoechea and Bruce Roe for Fig. 6.6; Tables 6.3 and 6.5. We also thank Helene Berges, Jose Luis Goicoechea, Bruce Roe, Stephane Rombauts, and Steve Stack for carefully revising the manuscript.

References

Ammiraju JSS, Yu Y, Luo M, Kudrna D, Kim H, Goicoechea JL, Katayose Y, Matsumoto T, Wu J, Sasaki T, Wing RA (2005) Random sheared fosmid library as a new genomic tool to accelerate complete finishing of rice (Oryza sativa spp. Nipponbare) genome sequence: sequencing of gap-specific fosmid clones uncovers new euchromatic portions of the genome. Theor Appl Genet 111(8):1596–1607

Budiman MA, Mao L, Wood TC, Wing RA (2000) A deep-coverage tomato BAC library and prospects toward development of an STC framework for genome sequencing. Genome Res 10(1):129–136

Campagna D, Albiero A, Bilardi A, Caniato E, Forcato C, Manavski S, Vitulo N, Valle G (2009) PASS: a program to align short sequences. Bioinformatics 25(7):967–968

Chang SB, Yang TJ, Datema E, van Vugt J, Vosman B, Kuipers A, Meznikova M, Szinay D, Lankhorst RK, Jacobsen E, de Jong H (2008) FISH mapping and molecular organization of the major repetitive sequences of tomato. Chromosome Res 16(7):919–933

English AC, Richards S, Han Y, Wang M, Vee V, Qu J, Qin X, Muzny DM, Reid JG, Worley KC, Gibbs RA (2012) Mind the gap: upgrading genomes with Pacific Biosciences RS long-read sequencing technology. PLoS ONE 7(11):e47768

Eshed Y, Zamir D (1995) An introgression line population of Lycopersicon pennellii in the cultivated tomato enables the identification and fine mapping of yield-associated QTL. Genetics 141(3):1147–1162

Frary A, Xu Y, Liu J, Mitchell S, Tedeschi E, Tanksley S (2005) Development of a set of PCR-based anchor markers encompassing the tomato genome and evaluation of their usefulness for genetics and breeding experiments. Theor Appl Genet 111(2):291–312

Fulton TM, Van der Hoeven R, Eannetta NT, Tanksley SD (2002) Identification, analysis, and utilization of conserved ortholog set markers for comparative genomics in higher plants. Plant Cell 14(7):1457–1467

Goff SA, Ricke D, Lan TH, Presting G, Wang R, Dunn M, Glazebrook J, Sessions A, Oeller P, Varma H, Hadley D, Hutchison D, Martin C, Katagiri F, Lange BM, Moughamer T, Xia Y, Budworth P, Zhong J, Miguel T, Paszkowski U, Zhang S, Colbert M, Sun WL, Chen L, Cooper B, Park S, Wood TC, Mao L, Quail P, Wing R, Dean R, Yu Y, Zharkikh A, Shen R, Sahasrabudhe S, Thomas A, Cannings R, Gutin A, Pruss D, Reid J, Tavtigian S, Mitchell J, Eldredge G, Scholl T, Miller RM, Bhatnagar S, Adey N, Rubano T, Tusneem N, Robinson R, Feldhaus J, Macalma T, Oliphant A, Briggs S (2002) A draft sequence of the rice genome (Oryza sativa L. ssp. japonica). Science 296(5565):92–100

Han CS, Sutherland RD, Jewett PB, Campbell ML, Meincke LJ, Tesmer JG, Mundt MO, Fawcett JJ, Kim UJ, Deaven LL, Doggett NA (2000) Construction of a BAC contig map of chromosome 16q by two-dimensional overgo hybridization. Genome Res 10(5):714–721

Jaillon O, Aury J-M, Noel B, Policriti A, Clepet C, Casagrande A, Choisne N, Aubourg S, Vitulo N, Jubin C, Vezzi A, Legeai F, Hugueney P, Dasilva C, Horner D, Mica E, Jublot D, Poulain J, Bruyère C, Billault A, Segurens B, Gouyvenoux M, Ugarte E, Cattonaro F, Anthouard V, Vico V, Del Fabbro C, Alaux M, Di Gaspero G, Dumas V, Felice N, Paillard S, Juman I, Moroldo M, Scalabrin S, Canaguier A, Le Clainche I, Malacrida G, Durand E, Pesole G, Laucou V, Chatelet P, Merdinoglu D, Delledonne M, Pezzotti M, Lecharny A, Scarpelli C, Artiguenave F, Pè ME, Valle G, Morgante M, Caboche M, Adam-Blondon A-F, Weissenbach J, Quétier F, Wincker P (2007) The grapevine genome sequence suggests ancestral hexaploidization in major angiosperm phyla. Nature 449(7161):463–467

Luo MC, Thomas C, You FM, Hsiao J, Ouyang S, Buell CR, Malandro M, McGuire PE, Anderson OD, Dvorak J (2003) High-throughput fingerprinting of bacterial artificial chromosomes using the snapshot labeling kit and sizing of restriction fragments by capillary electrophoresis. Genomics 82(3):378–389

Mardis ER (2008) Next-generation DNA sequencing methods. Annu Rev Genomics Hum Genet 9(1):387–402

Meyers BC, Scalabrin S, Morgante M (2004) Mapping and sequencing complex genomes: let's get physical! Nat Rev Genet 5(8):578–588

Mueller LA, Tanksley SD, Giovannoni JJ, van Eck J, Stack S, Choi D, Kim BD, Chen M, Cheng Z, Li C, Ling H, Xue Y, Seymour G, Bishop G, Bryan G, Sharma R, Khurana J, Tyagi A, Chattopadhyay D, Singh NK, Stiekema W, Lindhout P, Jesse T,

Lankhorst RK, Bouzayen M, Shibata D, Tabata S, Granell A, Botella MA, Giuliano G, Frusciante L, Causse M, Zamir D (2005) The tomato sequencing project, the first cornerstone of the international solanaceae project (SOL). Comp Funct Genomics 6 (3):153–158

Ozminkowski R (2004) Pedigree of variety Heinz 1706. Rep Tomato Genet Coop 54:26

Paterson AH, Bowers JE, Bruggmann R, Dubchak I, Grimwood J, Gundlach H, Haberer G, Hellsten U, Mitros T, Poliakov A, Schmutz J, Spannagl M, Tang H, Wang X, Wicker T, Bharti AK, Chapman J, Feltus FA, Gowik U, Grigoriev IV, Lyons E, Maher CA, Martis M, Narechania A, Otillar RP, Penning BW, Salamov AA, Wang Y, Zhang L, Carpita NC, Freeling M, Gingle AR, Hash CT, Keller B, Klein P, Kresovich S, McCann MC, Ming R, Peterson DG, Mehboob ur R, Ware D, Westhoff P, Mayer KFX, Messing J, Rokhsar DS (2009) The Sorghum bicolor genome and the diversification of grasses. Nature 457(7229):551–556

Peters SA, Datema E, Szinay D, van Staveren MJ, Schijlen EG, van Haarst JC, Hesselink T, Abma-Henkens MH, Bai Y, de Jong H, Stiekema WJ, Klein Lankhorst RM, van Ham RC (2009) Solanum lycopersicum cv. Heinz 1706 chromosome 6: distribution and abundance of genes and retrotransposable elements. Plant J 58(5):857–869

Peters SA, van Haarst JC, Jesse TP, Woltinge D, Jansen K, Hesselink T, van Staveren MJ, Abma-Henkens MH, Klein-Lankhorst RM (2006) TOPAAS, a tomato and potato assembly assistance system for selection and finishing of bacterial artificial chromosomes. Plant Physiol 140(3):805–817

Peterson DG, Stack SM, Price HJ, Johnston JS (1996) DNA content of heterochromatin and euchromatin in tomato (Lycopersicon esculentum) pachytene chromosomes. Genome 39(1):77–82

Romanov MN, Price JA, Dodgson JB (2003) Integration of animal linkage and BAC contig maps using overgo hybridization. Cytogenet Genome Res 102(1–4):277–281

Shearer LA, Anderson LK, de Jong H, Smit S, Goicoechea JL, Roe BA, Hua A, Giovannoni JJ, Stack SM (2014) Fluorescence in situ hybridization and optical mapping to correct scaffold arrangement in the tomato genome. Genes Genomes Genetics 4 (8):1395–1405

Sherman JD, Stack SM (1995) Two-dimensional spreads of synaptonemal complexes from solanaceous plants. VI. High-resolution recombination nodule map for tomato (Lycopersicon esculentum). Genetics 141 (2):683–708

Shirasawa K, Asamizu E, Fukuoka H, Ohyama A, Sato S, Nakamura Y, Tabata S, Sasamoto S, Wada T, Kishida Y, Tsuruoka H, Fujishiro T, Yamada M, Isobe S (2010) An interspecific linkage map of SSR and intronic polymorphism markers in tomato. Theor Appl Genet 121(4):731–739

Soderlund C, Humphray S, Dunham A, French L (2000) Contigs built with fingerprints, markers, and FPC V4.7. Genome Res 10(11):1772–1787

Soderlund C, Longden I, Mott R (1997) FPC: a system for building contigs from restriction fingerprinted clones. Comput Appl Biosci 13(5):523–535

Stack SM, Royer SM, Shearer LA, Chang SB, Giovannoni JJ, Westfall DH, White RA, Anderson LK (2009) Role of fluorescence in situ hybridization in sequencing the tomato genome. Cytogenet Genome Res 124 (3–4):339–350

Szinay D, Chang SB, Khrustaleva L, Peters S, Schijlen E, Bai Y, Stiekema WJ, van Ham RC, de Jong H, Klein Lankhorst RM (2008) High-resolution chromosome mapping of BACs using multi-colour FISH and pooled-BAC FISH as a backbone for sequencing tomato chromosome 6. Plant J 56(4):627–637

Tanksley SD, Ganal MW, Prince JP, de Vicente MC, Bonierbale MW, Broun P, Fulton TM, Giovannoni JJ, Grandillo S, Martin GB et al (1992) High density molecular linkage maps of the tomato and potato genomes. Genetics 132(4):1141–1160

The Arabidopsis Genome I (2000) Analysis of the genome sequence of the flowering plant Arabidopsis thaliana. Nature 408(6814):796–815

The International Rice Genome Sequencing P (2005) The map-based sequence of the rice genome. Nature 436 (7052):793–800

The Tomato Genome Sequencing C (2012) The tomato genome sequence provides insights into fleshy fruit evolution. Nature 485(7400):635–641

Todesco S, Campagna D, Levorin F, D'Angelo M, Schiavon R, Valle G, Vezzi A (2008) PABS: an online platform to assist BAC-by-BAC sequencing projects. Biotechniques 44(1):60–64

Van der Hoeven R (2002) Deductions about the number, organization, and evolution of genes in the tomato genome based on analysis of a large expressed sequence tag collection and selective genomic sequencing. Plant Cell Online 14(7):1441–1456

van der Knaap E (2004) High-resolution fine mapping and fluorescence in situ hybridization analysis of sun, a locus controlling tomato fruit shape, reveals a region of the tomato genome prone to DNA rearrangements. Genetics 168(4):2127–2140

van Oeveren J, de Ruiter M, Jesse T, van der Poel H, Tang J, Yalcin F, Janssen A, Volpin H, Stormo KE, Bogden R, van Eijk MJT, Prins M (2011) Sequence-based physical mapping of complex genomes by whole genome profiling. Genome Res 21 (4):618–625

Wang Y, Tang X, Cheng Z, Mueller L, Giovannoni J, Tanksley SD (2006) Euchromatin and pericentromeric heterochromatin: comparative composition in the tomato genome. Genetics 172(4):2529–2540

Wu F, Mueller LA, Crouzillat D, Petiard V, Tanksley SD (2006) Combining bioinformatics and phylogenetics to identify large sets of single-copy orthologous genes (COSII) for comparative, evolutionary and systematic studies: a test case in the euasterid plant clade. Genetics 174(3):1407–1420

Yu J, Hu S, Wang J, Wong GK, Li S, Liu B, Deng Y, Dai L, Zhou Y, Zhang X, Cao M, Liu J, Sun J, Tang J, Chen Y, Huang X, Lin W, Ye C, Tong W, Cong L, Geng J, Han Y, Li L, Li W, Hu G, Li J, Liu Z, Qi Q, Li T, Wang X, Lu H, Wu T, Zhu M, Ni P, Han H, Dong W, Ren X, Feng X, Cui P, Li X, Wang H, Xu X, Zhai W, Xu Z, Zhang J, He S, Xu J, Zhang K, Zheng X, Dong J, Zeng W, Tao L, Ye J, Tan J, Chen X, He J, Liu D, Tian W, Tian C, Xia H, Bao Q, Li G, Gao H, Cao T, Zhao W, Li P, Chen W, Zhang Y, Hu J, Liu S, Yang J, Zhang G, Xiong Y, Li Z, Mao L, Zhou C, Zhu Z, Chen R, Hao B, Zheng W, Chen S, Guo W, Tao M, Zhu L, Yuan L, Yang H (2002) A draft sequence of the rice genome (Oryza sativa L. ssp. indica). Science 296(5565):79–92

Chloroplast and Mitochondrial Genomes of Tomato

7

Gabriel Lichtenstein, Mariana Conte, Ramon Asis and Fernando Carrari

Abstract

This chapter summarizes the main features of the tomato plastid and mitochondrial genomes in the context of the current knowledge about "orthologue" genomes from other higher plants species in a historical perspective. We have focused on the application of this knowledge to aid in deciphering the functional roles of these organelles in growth and developmental processes of the tomato plants, especially on those related to fruit ripening. It also presents an assessment of the phylogenetic position of tomato, based on the available information of plastid and chondrome sequences from other land plants; which adds to the understanding of the evolutionary history of plants.

Keywords

Tomato · Chloroplast · Mitochondria · Genome · Phylogeny

Introduction

Higher photosynthetic organisms possess many cell types and display extensive compartmentation. These characteristics make the study of the different metabolic pathways that take place throughout the life of plant cells highly complex.

Specifically, mitochondria (derived from the Greek *mitos*—a thread—and *chondros*—a grain) and chloroplasts (or plastids) (from the Greek *chloros*—green—and *plastós*—formed) are the intracellular organelles which contain the entire machinery necessary for cell respiration and photosynthesis processes, respectively. These organelles also participate in the biosynthesis of

G. Lichtenstein · M. Conte · F. Carrari (✉)
Instituto de Biotecnología, Instituto Nacional de Tecnología Agropecuaria, Buenos Aires, Argentina
e-mail: carrari.fernando@inta.gob.ar

G. Lichtenstein · R. Asis · F. Carrari
Consejo Nacional de Investigaciones Científicas y Técnicas, Buenos Aires, Argentina

R. Asis
CIBICI, Facultad de Ciencias Químicas, Universidad Nacional de Córdoba, Córdoba, Argentina

F. Carrari
Facultad de Agronomía, Universidad de Buenos Aires, Cátedra de Genética, Buenos Aires, Argentina

M. Causse et al. (eds.), *The Tomato Genome*, Compendium of Plant Genomes,
DOI 10.1007/978-3-662-53389-5_7

essential metabolites, such as amino acids, nucleotides, lipids, and starch.

Both, chloroplasts and mitochondria, are the two types of cellular power stations. The first harnesses light energy from the sun and the other "unpacks" the captured energy into smaller packets of adenosine triphosphate (ATP) which are then used as a source of chemical energy for powering the cellular work. Thus, a clear understanding of the physiological processes at the whole plant level, necessarily requires a complete comprehension of the interactions occurring between the power-station organelles with the rest of the cellular compartments. In mammals, these interactions involve transference of mainly proteins and metabolites. However, the transfer of genes from plant mitochondria and chloroplasts to the nuclei is another essential interaction in plant cells. Although mitochondria and chloroplasts keep part of their ancestral genomes, gene transfer processes with the nuclei are continuously operating.

Mitochondria were first observed in a variety of cell types during the last decades of the nineteenth century as threads of granules previously called sarcosomes, bioblasts, or chondriocονts (Schmidt 1913). On the other hand, Nägeli (1846) discovered that chloroplasts multiplied by division in plant cells (Guilliermond and Atkinson 1941). At the beginning of the 20th century, the first reports of non-Mendelian inheritance in higher plants based on studies of variegation in higher plants were published (Correns 1908). These reports showed that few of the green-and-white variegated leaves were caused by factors inherited in a non-Mendelian manner. Further analyses of variegation in higher plants revealed that the genetic determinants for these characters were associated with chloroplasts, suggesting that these organelles may harbor genetic information. These observations led the Russian botanist Mereschkowski to first speak about the endosymbiotic theory (Mereschkowski 1905). Wallin (1923) extended this idea to the explanation about the mitochondria origin. Many textbooks describe this theory in detail, so we will not dwell on this aspect in this chapter.

Ris and Plaut (1962) demonstrated the presence of DNA in chloroplasts of the green alga *Chlamydomonas moewusii* by electron microscopy and cytochemical methods. Years later, Gibor and Granick (1964) established that chloroplasts are endowed with their own DNA complement (referred as plastome—cpDNA) and thus suggested that these organelles are semi-autonomous systems capable of self-replication and useful models for the study of differentiation. At the same time, the discovery of the 70S ribosomes within the chloroplast stroma (Stutzt and Noll 1967) set the foundations for further studies on the importance of chloroplast genomes from a functional perspective. Bedbrook and Bogorad (1976) reported the first physical map of the maize chloroplast genome, which added convincing evidence of the homogeneity and circularity of chloroplast DNA molecules. One-year later, they cloned the first chloroplast gene from this species (Bedbrook et al. 1977).

Contemporary to these discoveries were the observations reported by Nass and Nass (1963) and by Schatz et al. (1964). By using two different approaches, these authors concurrently reported for the first time that the chick embryo and the yeast mitochondria contain a significant quantity of DNA (mtDNA), respectively. Regarding higher plants, studies in the early 1960s showed that cytoplasmic male sterility (CMS) is a maternal inherited trait, bringing attention to the existence of unique DNA species within the mitochondria of plant cells in different crop species (Leaver and Gray 1982).

Regarding tomato, Palmer and Zamir (1982) reported the first studies on its chloroplast genome based on a restriction map. This map was designed through comparative restriction enzyme digestion with tobacco and Petunia cpDNA. Later on, Phillips (1985) reported a physical map generated by digestion of the cloned PstI fragments and by Southern-blot hybridization. The model consisted of a circular molecule of ~160 kb with a large inverted repeat. Simultaneously, Piechulla et al. (1985) described nine genes in the tomato chloroplast genome that are coordinately regulated during fruit ripening.

Regarding the mitochondrial DNA (mtDNA) from tomato, however, it was not until 1992 that Melcher et al. published the first physical map of the mitochondrial genome. Years later, a model of its size and organization was reported based on mtDNA digestions and hybridizations (Shikanai et al. 1998). This model proposed that the genome is structured in five subgenomic particles of different sizes with a total length of approximately 450 kb. These particles coexist in a dynamic range regulated somehow by the recombination activity of sequence hotspots.

In this chapter, we will provide an updated overview about the current knowledge of the chloroplast and mitochondrial genomes from tomato. Particularly, their structures in comparison with sequenced genomes from other Embriophytas species will be described. We will also summarize findings on the functionality of these two genomes together with their dynamic in relation to recent events of DNA exchange with the nucleus, a process which seems to remain still operative.

The Tomato Chloroplast Genome

It was not until 1986 that the first chloroplast genome from *Marchantia polymorpha* (the common liverwort) was completely sequenced providing insights into its structural organization (Ohyama et al. 1986). Since then, over hundreds of chloroplast genome sequences from different plant species have been continuously reported. After these pioneer works, in 2006, two research groups simultaneously reported the complete chloroplast genome sequence of tomato. Daniell et al. (2006) analyzed a genome sequence from a Purdue University accession (LA3023 according to the Tomato Genetic Resource Center: http://tgrc.ucdavis.edu/), while Kahlau et al. (2006) sequenced two distinct genotypes (IPA-6, a Brazilian cultivar, and Ailsa Craig [LA2838A] a European cultivar). Although both groups performed different approaches, they reported exactly the same size of 155,461 bp for all three genotypes of *Solanum lycopersicum* chloroplast genome. These results are in agreement with sizes reported for plastomes of other land plant species (Fig. 7.1b). As observed by these authors, and somehow surprisingly, the nucleotide sequences of the IPA-6 and Ailsa Craig chloroplast DNA (cpDNA) were absolutely identical. However, current information is still controversial about conservation degrees of plastome sequences between *Solanaceae* species. Whereas Clarkson et al. (2004) described very little sequence variation between *Nicotiana sylvestris* plastid genomes and its allopolyploid descendant *N. tabacum*, Daniell et al. (2006) revealed several InDels within certain coding sequences when tomato, potato, tobacco, and *Atropa* are compared.

Even though chloroplast genomes are usually represented by circular double-stranded DNA molecules, it is currently accepted that they exist as linear, concatemeric, and highly branched complex molecules (Bendich 2004). Generally, plastomes present highly conserved tetrapartite structures with two copies of large inverted repeat (IR) regions separating the large and small single copy regions (LSC and SSC). IR regions usually range from 5 to 76 kb (Palmer 1991; Sugiura 1992). In the case of the tomato plastome, two IR regions of 25 kb each separate the LSC and SSC regions of 85.6 and 18.4 kb, respectively. Compared to tobacco and potato plastomes, the tomato IR region is slightly expanded on both ends (into *rps19* and *ycf1* genes in the LSC and SSC, respectively). Besides the two large IR, tomato plastome contains also near 40 IR of 30–40 bp that are highly conserved among closer species and are located in the same genes or intergenic regions. These characteristics thus suggest a functional role. Moreover, this plastome also harbors other four IR of 57 bp, which are not found in those of potato, tobacco nor *Atropa* (Daniell et al. 2006). However, the tomato chloroplast genome is smaller than that of tobacco owing to deletions in the noncoding intergenic spacer regions (Kahlau et al. 2006; Daniell et al. 2006).

In noncoding regions, the tomato plastid contains 25 intergenic spacer regions shearing 80–100 % identity with the same regions of potato, tobacco, and *Atropa*. Only four regions are 100 % identical among species and three of them are located in IR regions. These identical variations made intergenic spacer regions useful markers for phylogenetic research studies.

Regarding gene content, the tomato chloroplast genome is more gene-dense than the

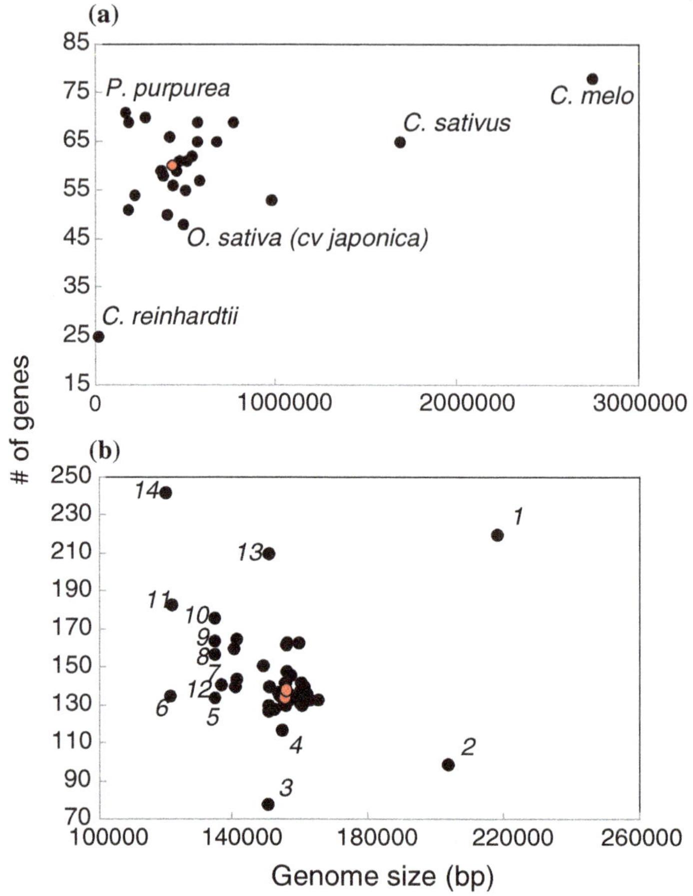

Fig. 7.1 Number of encoded genes in relation to mitochondrial (**a**) and chloroplast (**b**) genome sizes for tomato (*red circles*) and other selected taxa (*black circles*). Names of those species with a genome size and/or a gene number above or below the average ± SD are given on the graph for chloroplasts analyses (panel **a**). On panel (**b**), species are referenced as follows: *1 Pelargonium × hortorum*. *2 Chlamydomonas reinhardtii*. *3 Phaseolus vulgaris*. *4 Arabidopsis thaliana*. *5 Triticum aestivum*. *6 Marchantia polymorpha*. *7 Agrostis stolonifera*. *8 Oryza sativa*. *9 Oryza nivara*. *10 Oryza sativa* (*indica* cultivar-group). *11 Guillardia theta*. *12 Hordeum vulgare subsp. vulgare*. *13 Chlorella vulgaris*. *14 Pinus thunbergii*. Data were extracted from GenBank (www.ncbi.nlm.nih.gov/genbank/) and/or from the corresponding published paper listed in the reference section

mitochondrial (see below) and the nuclear genomes (The Tomato Genome Consortium 2012). This chloroplast genome consists of 41.7 % of noncoding regions (intergenic spacers and introns) and 58.3 % of coding regions, with 133 annotated genes. Of these 133 genes, 113 are unique and 20 were found duplicated in the IR. The same gene content and gene order is found conserved in the closest tobacco, potato, and *Atropa* species (Fig. 7.1b). Of the 113 annotated genes: 61 encode for tRNA, rRNA, ribosomal proteins, RNA polymerase, -maturase, and proteases; 47 correspond to photosynthesis-related genes; and the remaining 5 to other genes and conserved open reading frames. Table 7.2 summarizes a comparative analysis between the main features reported for all Embriophyta plastome sequences up to November, 2012.

Chloroplast Functional Genomics

Most plant plastid genomes encode proteins that function in photosynthesis. However, few of these proteins are involved in many other cellular functions: the chloroplast tRNA-Glu is required in tetrapyrrole biosynthesis (Schön et al. 1986); the plastid genome-encoded D subunit of the essential enzyme acetyl-CoA carboxylase participates in fatty acid biosynthesis (Kode et al. 2005; Kahlau and Bock 2008); and the plastid-encoded ClpP1 protease subunit is involved in plastid protein homeostasis (Shikanai et al. 2001; Kuroda and Maliga 2003). The involvement of plastid gene expression in these essential functions is probably why the loss of plastid translational activity is fatal in most plants.

Regarding the organization of the chloroplast genomes, similarly to cyanobacteria genomes, genes are clustered and arranged in operons and co-transcribed as polycistronic mRNAs and translated on 70S ribosomes. The gene processing and maturation consist of several steps such as cleavage of polycistronic mRNA, intron splicing and RNA editing by C-to-U conversions (Barkan and Goldschmidt-Clermont 2000; Bock 2000). However, higher plant plastids are far more complex than those from the prokaryotes, because the regulation of plant plastids depends on their own mechanisms and on nuclear genome "signals" influencing plastid functionality. For instance, plastid genes are transcribed specifically by plastid-encoded RNA polymerase or nuclear encoded RNA polymerase or they can even share both RNA polymerases (Allison et al. 1996; Hajdukiewicz et al. 1997; Lerbs-Mache 2000; Legen et al. 2002). More complexity is observed in the transcription factors required for promoter recognition, which are encoded by genes residing in the nuclear genome (Tanaka et al. 1996). The regulation of plastid genes is mainly at transcriptional and translational levels; however, their contributions have been scarcely discussed and remain controversial. Some studies support that transcriptional regulation is the main contribution to gene control in plastids (Pfannschmidt et al. 1999; Tullberg et al. 2000). By contrast, other studies pointed that translation constitutes the rate-limiting step in plastid gene expression (Eberhard et al. 2002).

Particularly in tomato, chloroplasts undergo peculiar drastic changes in both ultrastructure and function during fruit maturation. Among these changes, researchers have described the disappearance of the thylakoid membrane system, the degradation of chlorophyll, the appearance of plastoglobuli, and an increment in carotenoid biosynthesis that finally accumulated inside the chromoplast membranes (Rosso 1968; Harris and Spurr 1969; Egea et al. 2011). The genetic control of chloroplast during this transition has been studied for many years (Piechulla et al. 1985; Bathgate et al. 1985; Kahlau and Bock 2008). In this regard, whereas a drastic downregulation of photosynthetic genes and significant decreases in ribosomal RNAs occur, the expression of other nonphotosynthetic genes rises. Accordingly, recent studies on tomato plastid transcriptomics and proteomics have shown that photosynthetic and carbohydrate metabolism genes are strongly downregulated during fruit development (Kahlau and Bock 2008; Barsan et al. 2012). Conversely, the expression of the genetic system genes (rRNAs, tRNAs, ribosomal proteins, RNA polymerase) seems to be kept at higher levels. Interestingly, the chloroplast-to-chromoplast conversion during the ripening period is not accompanied by drastic changes in transcript abundance. Translational regulation analyses by polysome-bounded mRNA analyses showed that a strong downregulation also affects most of plastid genes in fruits in comparison with expanded leaves. During ripening, polysome association successively declines and is particularly pronounced in the photosynthesis gene group, suggesting that plastid translation is the main contribution in gene expression control during chloroplast-to-chromoplast differentiation. An exception to this was observed for the *accD* gene, which encodes an acetyl-CoA carboxylase subunit. The expression of this gene displays strong upregulation and polysome association during fruit ripening; which correlates with the high demand of lipid biosynthesis to generate a storage matrix that will accumulate carotenoids (Kahlau and Bock 2008). However, ACCD protein level decreased between mature green and ripe fruit stages, suggesting another point of regulation for this enzyme (Barsan et al. 2012). *TrnA* (encoding the tRNA-Ala) and *rpoC2* (encoding an RNA polymerase subunit) genes tended to be also upregulated during this process. In the same study, Kahlau and Bock (2008) analyzed the expression of genes predominantly transcribed by the nuclear (NEP) and plastid (PEP) encoded RNA polymerases. In their study, they found that the PEP is more intensively used in leaves, whereas transcription from the NEP promoter prevails in red fruits.

Notwithstanding the mentioned contributions to the functional role of the tomato plastid genome, knowledge about how plastid translation is regulated in fruits during the autotrophic to heterotrophic transition is scarce.

On prokaryotic-type 70S ribosomes, the plastid translation machinery consists of two subsets of RNA components. A subset comprises those components encoded by the plastid genome: the 16S rRNA of the small ribosomal subunit as well as the 23S, 5S, and 4.5S rRNAs of the large subunit. The remainder consists of the components encoded by the nuclear DNA. Although the abolishment of plastid protein biosynthesis is lethal, particular studies are focused on identifying each individual component of plastid ribosome that may not be essential (Rogalski et al. 2008). Fleischmann et al. (2011) studied candidates for non-essential plastid ribosomal proteins in tobacco. Through reverse genetic analyses, the authors revealed a previously unrecognized role of plastid translational fidelity in two developmental processes: shoot branching and leaf morphogenesis. Noteworthy in this study, the authors also suggested that the transfer of plastid ribosomal protein genes to the nucleus is greatly accelerated in non-photosynthetic lineages. Besides the common plastid ribosomal proteins, plant plastid contains plastid-specific ribosomal proteins (PSRP) not found in bacteria (Sharma et al. 2007). PSRP are encoded by the nuclear genome and the function of five of them has been recently studied (Tiller et al. 2012). In that research, the knock-down of three of these proteins decreased accumulation of the 30S or 50S subunit of the plastid ribosomes, while the others showed no change.

In general, whereas all the mentioned evidence accounts for the functional role of the tomato plastid genome, the intricate network of coregulation with the other genomes (i.e., mitochondrial and nuclear) is still obscure.

The Tomato Mitochondrial Genome

Anderson et al. (1981) reported the first complete genome sequence from a eukaryotic organelle (the human mitochondrion), and in 1997, Unseld et al. published the first complete mitochondrion genome sequence from a higher plant (*Arabidopsis thaliana*). After these groundbreaking reports, and within few decades, the advent of rapid DNA sequencing methods resulted in a profound boost over the scope and speed required for the completion of large-scale whole genome sequencing projects. As a result, in 2012, the Tomato Genome Consortium (a multinational team of scientists from 14 countries) reported a high-quality genome draft for the Heinz cultivar 1706 (LA4345 according to the Tomato Genetic Resource Center: http://tgrc.ucdavis.edu/). In this context, not only the nuclear sequence was obtained but the semi-autonomous DNA from the mitochondria (chondrome) was also sequenced, assembled and annotated.

A shotgun sequencing strategy was used to produce an assembly of the tomato mitochondrial genome. Highly purified mitochondrial DNA (mtDNA) isolated from etiolated seedlings was used as starting material to produce 4154 Sanger paired-end sequence reads with an average length of 750 nt. Shotgun clones were deposited into a dedicated database and are currently available upon request at http://www.mitochondrialgenome.org/. After trimming, clipping and filtering, high-quality ($Q_v \geq 20$) paired-reads were used as input for the assembly pipeline. In brief, an overlap-layout-consensus algorithm was chosen owing to their lengths and library features and the reads were then fed to the CAP3 Sequence Assembly Program (Huang and Madan 1999). As a result, the tomato chondrome was assembled into six scaffolds (SlmtSC_A, _V, _M, _R, _L and _B) and 164 contigs, spanning 579,717 nucleotides for the first draft of the tomato chondrome (SOLYC_MT_v1.50).

The tomato chondrome is also available for download at the Mitochondrial Genome website mentioned above. At the same time, these sequences have been deposited as a whole genome project (BioProject ID: 67471) at DDBJ/EMBL/GenBank under the accession AFYB00000000.

The version described in this chapter is the first version, AFYB01000000. Overall, the size of the final assembly is in agreement with the physical map previously reported by Shikanai et al. (1998). Furthermore, its multipartite organization (i.e., the existence of mtDNAs of varying structures) is comparable to those reported for the tobacco (Sugiyama et al. 2005) and rice (Tian et al. 2006) chondromes. In this regard, it is currently accepted that the organization of angiosperm chondromes is characterized by the presence of multipartite genome structures, which arises from high-frequency recombination via repeated sequences in the genome (Fauron and Casper 1995). A master circle (MC) model is traditionally constructed based on the restriction fragment mapping of mtDNA in higher plants, in which the total genetic information can be accommodated (Tian et al. 2006). By contrast, an extensive electron microscopy investigation has shown that the mtDNA from *Chenopodium album* cell cultures appear to consist mainly of linear molecules of various sizes, together with rosette-like and sigma-like structures, in vivo (Backert and Börner 2000). Since the relative amounts of these structures change during the course of cell growth, they may represent replication intermediates. Similar large branched molecules have also been observed in mtDNA from BY-2 tobacco cells under the light microscope (Oldenburg and Bendich 1996). Thus, there are differences between the forms of mtDNA molecules derived from genome mapping data and from microscopic observations (Sugiyama et al. 2004).

Although this discrepancy has not yet been resolved, both types of evidence indicate that the structural organization of mtDNA is highly dynamic. Furthermore, the multipartite structure can provide a redundant gene assembly and modulate the genome copy number in plant chondromes. Low-frequency ectopic recombination among multipartite structures will produce chimeras, aberrant ORFs, and novel subgenomic DNA molecules (Abdelnoor et al. 2003). Thus, multipartite structures are an important factor to consider when analyzing the scaffolds and contigs of the tomato chondrome assembly. This genomic shuffling is apparently reversible and can alter plant phenotype as suggested by two early reports of Kanazawa and Hirai (1994) and Janska et al. (1998). These authors showed that cytoplasmic male sterility (CMS) in *Nicotiana tabacum* and *Phaseolus vulgaris* species is related to the occurrence of multipartite structures, heteroplasmy (see below), and/or paternal leakage.

The origin of the tomato chondrome various scaffolds can also be related to the occurrence of heteroplasmic DNA structures. Heteroplasmy is defined as a state in which more than one mitochondrial genotype occurs in an organism. Usually, one mitotype is prevalent and the alternative one(s) are present in a very low proportion. Under such conditions, the phenotype of the organism is determined by the predominant mtDNA variant (Kmiec et al. 2006). In plants, this phenomenon has been investigated most often to clarify some mitochondrial abnormalities. For example, there are reports on CMS (Janska et al. 1998), non-chromosomal stripe mutants in maize (NCS) (Yamato and Newton 1999), the chloroplast mutator mutant in *Arabidopsis* (CHM) (Martínez-Zapater et al. 1992; Sakamoto et al. 1997) and the mitochondrial mutator system in maize (Kuzmin et al. 2005). Recent studies indicate that heteroplasmy exists also in healthy humans (Kajander et al. 2000) and wild-type plants (Arrieta-Montiel et al. 2001; Taylor et al. 2001).

Recombinations between large repeated sequences are commonly assumed to be the most important force responsible for maintaining the multipartite structure of the chondrome as a dynamic entity (Kmiec et al. 2006). These recombinations are frequent and easily reversible during plant life probably in order to fulfill their integrative role. Besides the main genome whose parts are maintained in a dynamic equilibrium by large repeated sequences, plant mitochondria contain recombinant molecules known as sublimons. These sublimons are very low in number compared to the main mitochondrial genome and are products of rare and irreversible recombinations mediated by short repeated sequences (Kmiec et al. 2006). Short repeats are common in plant mitochondrial genomes (Notsu et al. 2002; Sugiyama et al. 2005; Kubo et al. 2000; Clifton et al. 2004) and they may be originated from the insertion of reverse-transcribed copies of un-translated RNA (Gualberto et al. 1988). Another possible origin could be from the recombinational activity of oligonucleotide motifs (Woloszynska et al. 2001). As a consequence of these active recombination events mediated via both large and short repeats, two types of mtDNA of different quantitative representation coexist in one organism: the mitotype and the sublimons. The mytotype is the most predominant and creates the main genome, while the sublimons exist at a substoichiometric level. These findings suggest that chondrome heteroplasmy may also occur in the tomato cell. This is an important feature to take into account while revising the assembly results. In this vein, the tomato chondrome possesses a high number (849) of single repeats of 50 and 2200 bp. Likewise, 34 short tandem repeats (2–8) of size ranging between 15 and 100 bp were detected.

Gene Annotation

In spite of their larger size, chondromes from higher plant species do not encode many more proteins than mitochondrial genomes from other eukaryotes such as mammals. Most plant mitochondrial genomes are comprised of non-coding sequences. In *Arabidopsis*, only 20 % of the mitochondrial genome is responsible for functional genes (Unseld et al. 1997). The number of mitochondrial genes in angiosperms ranges from 25 (in the rice cultivar japonica) to 78 (in melon —*Cucumis melo*) without considering copy number (Fig. 7.1a). Most of the genes that are lost from the mitochondrion appear to have been transferred to the nuclear genome (Adams and Palmer 2003). The tomato mitochondrial genome encodes at least 36 protein-coding genes, three ribosomal RNA genes and 18 tRNA genes. These numbers are similar to those reported for other angiosperm mtDNAs, in which most of the genes encode conserved ribosomal proteins and components of the electron transport chain (complexes I–V). Furthermore, an ORF search resulted in the identification of 30 additional sequences encoding hypothetical proteins. A preliminary survey on the expression levels of these mitochondrial genes throughout tomato fruit development have indicated that many of the annotated genes are differentially expressed during this process. For instance, 23 genes belonging to the electron transport chain machinery and 11 ORFs that presented detectable levels of expression differed in their expression during fruit development (Conte et al. 2013).

Nuclear Copies of Mitochondrial DNA (NUMTs) and Nuclear Insertions of Chloroplast DNA (NUPTs)

The plastome is considered the evolutionary remnant of a cyanobacterial genome (Keeling 2010) where genetic information was transferred from the endosymbiont's genetic system to the host nuclear genome; interestingly, this transfer is still underway (reviewed in Kleine et al. 2009).

In 2012, the fully sequenced nuclear genome of tomato was published along with a comprehensive structural and comparative analysis with other Solanaceas (The Tomato Genome Consortium 2012). Similarly to other species (Timmis and Scot 1983; Stern and Palmer 1984; Blanehard and Schmidt 1995; Thorsness and Weber 1996), sequences of plastid and mitochondrial origin contribute also to the complexity of the nuclear tomato genome. These sequences have long been called "promiscuous DNA" and the idea behind this regrettable name was that they constitute a kind of mutation buffering (Conrad 1985). In mechanistic terms, the concept of plastid and mitochondrial DNA transposition to the nucleus and their subsequent integration into the nuclear genome has prevailed. In this respect, the small genomes of these organelles are also believed to be remnants after the relocation of gene function from the ancestral prokaryotes. This process has been accompanied by deletion of the endosymbiont genomes with a subsequent dependence of mitochondrial and chloroplast biogenesis on nuclear genes. Strong molecular evidence (Baldauf and Palmer 1990) suggests that such gene transfers have occurred. Furthermore, these gene transfers have also been achieved experimentally in mitochondrial (Gray et al. 1996) and chloroplast (Kanevski and Maliga 1994) systems. Both mitochondrial and chloroplast sequences homologies have been identified within the nuclear genomes of spinach (Timmis and Scot 1983; Scott and Timmis 1984; Cheung and Scott 1989), tomato (Pichersky and Tanksley 1988; Pichersky et al. 1991), tobacco (Ayliffe and Timmis 1992a, b), potato (du Jardin 1990), and members of the Chenopodiaceae family (*Beta vulgaris*, *C. album*, *Chenopodium quinoa*, *Atriplex cinerea*, and *Enchyleana tomentosa*) (Ayliffe et al. 1998).

Through different analyses, the Tomato Genome Sequencing Consortium further demonstrated the presence of DNA fragments of mitochondrial and chloroplastic origin found as insertions within the nuclear genome (NUMTs and NUPTs, respectively). In summary, 667 fragments, longer than 250 bp, were found and reported as NUPTs insertions. Furthermore, a colinearity analysis between the tomato chloroplast and the nuclear genome sequences demonstrated that 492 fragments could be true insertions with a plastome origin. In addition, two noteworthy long colinear insertions were found inserted in chromosomes 2 and 11. Conversely, the tobacco nuclear genome contains multiple chloroplast DNA integrants (i.e., >100 copies of a single plastid sequence), which can be in excess of 18 kb (Ayliffe and Timmis 1992a, b).

Following the endosymbiont theory (Margulis and Bermudes 1985), the mitochondrion and its genome are the remnants of a free-living eubacteria ancestor (probably an extant α-proteobacterium). Therefore, this ancestor was engulfed by a eukaryotic host cell and, as a result, established a symbiotic relationship with it (Gray 1999). The host provided the nuclear genome and most of the endosymbiont genes were either lost or transferred to the nuclear genome at an early stage in evolution. Thus, very little of the original gene pool is found in modern mtDNA. In this regard, many features distinguish the mtDNAs of higher plants from those of animals and other organisms (Sugiyama et al. 2004). Although the transfer of mitochondrial genes to the nucleus and their functional activation ceased in the common ancestor of animals, mitochondrial gene loss, and gene transfer have been an ongoing and frequent process in flowering plants (Palmer et al. 2000). Extensive Southern-blot analyses of 280 genera of flowering plants have provided a global view of gene loss in plant mtDNA (Adams et al. 2000). In addition, the possible mechanisms of DNA transfer between organelles with closed membrane systems and the integration of the DNA into the host genome have been reviewed by Kurland and Andersson (2000). The different chondromes in land plants have significantly expanded in size compared with those of green algae. Land plants evolved from green algae belonging to the *Charophyceae* (Graham et al. 2000). By comparisons of completely sequenced mtDNAs, *Chara vulgaris* was recently inferred to be the last common ancestor of green algae and land plants (Turmel et al. 2003). *Chara* possesses a densely packed mitochondrial genome with a gene content similar to that of its *Marchantia* counterpart (Oda et al. 1992). This led Turmel et al. (2002a, b) to infer that the growth in mtDNA size in *Marchantia* occurred by the enlargement of intergenic spacers because of frequent duplications and substitutions during evolution from *Charophytes* to *Bryophytes*. The subsequent size increase of angiosperm chondromes during evolution from bryophytes occurred both by further enlargement of spacer regions owing to frequent duplications and by the frequent capture of sequences from the chloroplast and nuclear genomes (Marienfeld et al. 1999). Of these incoming DNAs, only plastome-tRNA genes have gained functions in angiosperm chondrome-DNA (Joyce and Gray 1988). Furthermore, the contribution of frequent recombination and transposition of many different classes of retrotransposons to the mitochondrial genome expansion of land plants is at most 15 %.

Thus, the origin of most unique sequences (~50 %) in plant mtDNA is not known (Sugiyama et al. 2004). The chondrome size variation is exceptionally wide among higher plants, ranging from the smallest 208 kb estimated for white mustard (*Brassica hirta*; Palmer and Herbon 1987) to the largest that are believed to be over 2400 kb in muskmelon (*C. melo*; Ward et al. 1981) (Fig. 7.1a). Such an extensive expansion is attributable to two major factors: protein-coding redundancy and a high level of mitochondrial DNA recombination that results in extraneous DNA integration (Mackenzie and McIntosh 1999). Altogether, these findings have allowed researchers to establish that fragments of mitochondrial DNA are integrated into the nuclear genomes of many organisms including numerous animal and plant species (Bensasson et al. 2001; Timmis et al. 2004). These sequences are named NUMTs (pronounced "new mights"), an abbreviated term for "nuclear mitochondrial DNA," and describe any transfer or "transposition" of cytoplasmic mtDNA sequences into the separate nuclear genome of a eukaryotic organism (Lopez et al. 1994). As whole genome sequencing projects accumulate, more and more NUMTs have been detected in many diverse eukaryotic organisms (see http://www.pseudogene.net for a list of examples). Although no evidence of recent mtDNA transfer into metazoan nuclei has been reported, this process is still ongoing in plants. Current studies indicate that escape of the genetic material from organelles to the nucleus occurs much more frequently than generally believed (Timmis et al. 2004). Computational analyses comparing the tomato mitochondrial and nuclear assemblies revealed 111 locally collinear blocks (LCB) on the chondrome, which are collinear with the nuclear sequence. Of these LCB, 72 (~197 kb) were inferred to be NUMTs. The analysis showed NUMTs of varied number, size, and position, ranging between zero and seven on chromosomes 2 and 5, respectively, and with the highest number (21) detected over chromosome 11. Fluorescence in situ hybridization (FISH) of mtDNA generally supported this in silico analysis. Whether this kind of instability of the chondrome (called "molecular poltergeists" by Hazkani-Covo et al. 2010) has direct consequences over the tomato plant fitness is still an open question.

Chloroplast and Mitochondrial Genomes Comparisons Across Green Species

As an additional resource of the tomato genome project, a mitochondrial database (www.mitochondrialgenome.org) was built and made available to facilitate exchanging information about chondrome genomes. This tool allows flexible BLAST searches and comparisons of more than 47 mitochondrial genomes from *Viridiplantae* species that are currently available, including the different versions of the tomato chondrome assembly. Nucleotide sequences of all clones included in the tomato chondrome assembly are available to be downloaded from the same website and, if necessary, these clones can also be requested for research purposes.

Similarly, the Chloroplast Genome Database (http://chloroplast.cbio.psu.edu/, Cui et al. 2006) offers data from more than 100 plastomes of land plants; which allows the search of genes, by using their annotated names, as well as flexible BLAST searches. This database also allows researchers to download protein and nucleotide sequences extracted from a selected chloroplast genome and to browse the putative protein families (tribes).

Among many different applications, these resources allow very general descriptions of the main features founds in the up to date known plastomes and chondromes. Tables 7.1 and 7.2 summarize the main features of these mitochondria and chloroplast genomes, respectively.

Comparatively, the size disparity between the *Viridiplantae* species chondromes appears to reflect a dynamic history of expansion and possibly contractions of several regions, such as intergenic and/or repetitive regions. Indeed, these disparities could be explained by the loss or acquisition of nuclear and chloroplastic sequences. However, gene content analyses of all Embryophyta chondromes showed that these genomes share the complete core gene set of the

Table 7.1 Main features of mitochondrial genomes from *Viridiplantae* species

Species (common name)	Genome size(a)/structure	Main features	References
Chaetosphaeridium globosum	56,574 bp 1 master circle	48.3 kb of coding sequence: 85 % of total size. 8.3 kb intergenic sequences acquired by horizontal transfer from phage or bacterial DNA	Turmel et al. (2002a, b)
Chara vulgaris (stonewort)	67,737 bp 1 master circle	High density of coding sequences (90.7 %), 14 group-I introns and 13 group-II introns account for 38.5 % of total size. Poor in repeated sequence elements. No evidence for editing sites	Turmel et al. (2003)
Chlamydomonas eugametos	22,897 bp single circular molecule	Densely packed coding sequences. 9 group-I introns, two large direct repeats and short repetitive sequences. G C-rich repetitive elements with putative post-transcriptional regulation functions	Denovan-Wright et al. (1998)
Chlorokybus atmophyticus	201,763 bp 1 master circle	41.4 % of conserved gene sequences. 6 group-I introns and 14 group-II introns. Repeats represent 7.5 %	Lemieux et al. (2007)
Mesostigma viride	42,424 bp 1 master circle	86.6 % of conserved gene sequences. 4 group-I introns and 3 group-II introns (evidence of acquired by lateral transfer). Two regions of overlapping genes	Turmel et al. (2002a, b)
Nephroselmis olivacea	45,223 bp 1 master circle	78 % of conserved gene sequences. 4 group-I introns vertically inherited from a green algal ancestor. 4 potential transcriptional units	Turmel et al. (1999a, b)
Oltmannsiellopsis viridis	56,761 bp 1 master circle	68.7 % of coding sequences, intergenic regions with a large number of repeated elements. 3 introns (2 of group-I and 1 of group II) acquired by horizontal transfer	Pombert et al. (2006)
Ostreococcus tauri	44,237 bp 1 master circle	Genes encompass 93 % of the genome. A unique duplicated region encodes genes	Robbens et al. (2007)
Pedinomonas minor	25,137 bp 1 master circle	Reduced set of genes, packed in 60 % of the genome. A single intron of group II. 9 kb of repeated sequences	Turmel et al. (1999a, b)
Physcomitrella patens	105,340 bp 1 master circle	Many genes encoded by the clockwise strand. 2 group-I introns and 25 group-II introns. Putative RNA editing sites	Terasawa et al. (2007)
Scenedesmus obliquus	42,919 bp 1 master circle	60.6 % of identified gene sequences. Deviant genetic code (standard sense is used as stop). Repetitive sequences in intergenic regions	Nedelcu et al. (2000)
Pseudendoclonium akinetum	95,880 bp 1 master circle	Coding genes in 47.4 % of the genome sequences. 7 group-I type introns. Repeated elements with recombinant activity: evolutive role	Pombert et al. (2004)
Chlamydomonas reinhardtii	15,758 bp 1 linear molecule	Densely packed coding regions. Universal genetic code. No introns. Low fraction of intergenic DNA. Presence of inverted terminal repeats	Popescu and Lee (2007)

(continued)

Table 7.1 (continued)

Species (common name)	Genome size(a)/structure	Main features	References
Cycas taitungensis	414,903 bp 1 master circle and alternative circular molecules	10.1 % of coding sequences. 20–25 group-II introns, no group-I introns. Presence of Bpu elements (transposable-like elements) in non-coding regions. Abundant RNA editing sites	Chaw et al. (2008)
Oriza rufipogon (rice)	559,045 bp, circular molecule	6 copies of repeat regions with active recombination activity. Presence of RNA editing sites. Presence of sequences of chloroplast origin	Sun et al. (2002)
Oryza sativa (rice)	490,520 bp, circular map	Presence of 6.3 and 13.4 % of plastid and nuclear sequences	Notsu et al. (2002)
Tripsacum dactyloides	704,100 bp circular molecules	Large amount of plastidic DNA sequences (transfer of exogenous DNA?)	Wang et al. (2012)
Sorghum bicolour (sorghum)	468,628 bp circular molecules	Partial genome assemblies of three *Sorghum bicolor* genotypes	Zheng et al. (2011)
Triticum aestivum (wheat) cv. Chinese spring	452,528 bp 1 master circle and subgenomic molecules	16.7 % of coding sequences. Many genes present at multiple-copy. Direct and inverted repeats with intramolecular recombination function. Chloroplast-derived sequences. Presence of retroelements	Ogihara et al. (2005)
Solanum lycopersicum (tomato)	581,837 bp 6 scaffolds and subgenomic molecules	Presence of single repeats and short tandem repeats. 22 introns and 92 nuclear sequences of mitochondrial origin (NUMTs)	TGSC (2012)
Nicotiana tabacum (tobacco)	430,597 bp master circle with a multipartite organization	9.9 % of coding sequences. Homologous recombination via short direct repeats. 17 and 6 *cis*- and *trans*-splicing group-II introns, respectively. Presence of retrotransposon of nuclear origin	Sugiyama et al. (2005)
Vitis vinífera (grape)	773,279 bp Master circle and subgenomic particles	4.98 % of gene encoding sequences. Few genes present partial pseudo copies. Large genome size by expansion of spacer sequences. HTG with chloroplast and nuclear DNA	Goremykin et al. (2009)
Carica papaya (papaya)	476,890 bp circular DNA	Partial non-annotated genome	Yu et al. (2009)
Ferrocalamus rimosivaginus (bamboo)	432,839 bp, 1 circular molecule	8.9 % of gene encoding sequences. 22 group-II introns, 6 trans-spliced. Few large repeats	Ma et al. (2012)
Bamboosa oldhamii (Giant timber bamboo)	509,941 bp circular molecule	No information available	GenBank Acc EU365401
Zea luxurians	539,368 bp 1 circular molecule	8.6 % of gene encoding sequences. Homologous recombination between direct repeats. Presence of NUMTs	Darracq et al. (2010)
Zea perennis	570,354 bp 1 circular molecule	8.5 % of gene encoding sequences. Homologous recombination between direct repeats. Presence of NUMTs	

(continued)

Table 7.1 (continued)

Species (common name)	Genome size(a)/structure	Main features	References
Zea mays subsp. *mays* (maize)	569,630 bp 1 circular molecule, alternative physical structures	8.4 % of gene encoding sequences. 22 group-II introns, 7 are trans-spliced. Homologous recombination between direct repeats. 2 large insertion of chloroplast DNA. Chondrome sequences in the nuclear genome. No evidences of NUMTs	Clifton et al. (2004)
Citrullus lanatus	379,236 single circular genome	45.9 % of gene encoding sequences.19 *cis*- and 5 *trans*-spliced group-II introns and 1 group-II intron (derived from horizontal transfer). 14 synthenic gene clusters. Chloroplast origin sequences and nuclear-derived retroelements	Alverson et al. (2010)
Cucurbita pepo (pumpkin)	982,833 bp single circular genome	16.6 % of gene encoding sequences.19 *cis*- and 5 *trans*-spliced group-II introns. Nuclear-derived retroelements. Proliferation of small repeats	
Cucumis sativus (cucumber)	1,555,935, 83,817; and 44,840 bp 3 circular autonomous particles and a large pool of sub-stoichiometric forms	18 *cis*- and 5 *trans*-spliced group-II introns. One single group-I intron. NUMTs represent 1/3 of the chondrome genome. 36 % of the chondrome corresponds to repetitive sequences with recombination activity	Alverson et al. (2011a, b)
Cucumis melo (melon)	2,738,402 bp 6 scaffolds (multipartite?)	Only 1.7 % of gene encoding sequences. 17 duplicated genes. 20 *cis*- and 1 *trans*-spliced introns. High proportion of repetitive sequences and 47 % of the chondrome corresponds to NUMTs. HTG explains it large size	Rodríguez-Moreno et al. (2011)
Pleurozia purpurea (Purple Spoonwort)	168,526 bp 1 master circle	52 % of gene encoding sequences. 7 group-I and 24 group-IIintrons. Small number of RNA editing events. 4 repeat sequences	Wang et al. (2009); Li et al. (2009)
Megaceros aenigmaticus	184,908 bp 1 master circle	16, 34 and 50 % of introns, exons and intergenic spacer sequences. 30 group-I introns. Few RNA editing events. It genome is a remnant of transition stage between Charopythes and land plants	
Phaeoceros laevis	209,482 bp 1 linear molecule	36.5, 10.9 and 52.6 % of introns, exons intergenic spacer sequences. RNA-editing detected in 54 genes. 64 cis-spliced group-II introns	Xue et al. (2010)
Arabidopsis thaliana	366.924 bp Variable number of different molecules	10 % of gene encoding sequences. 62 % of the chondrome has no clear origin and function. Some are highly similar to chloroplast and nuclear sequences. 2 large repeats active in recombination events	Unseld et al. (1997)

(continued)

Table 7.1 (continued)

Species (common name)	Genome size(a)/structure	Main features	References
Brassica napus (rape)	223,412 bp 1 master circle and 2 subgenomic particles	17.4 % of gene encoding sequence. Presence of direct repeats, active in intramolecular recombination. 19 *cis-* and 5 *trans*-splices group-II introns. RNA editing sites. Presence of plastid and nuclear-derived sequences	Chen et al. (2011)
Beta vulgaris (sugar beet)	368,799 bp 1 single circular molecule	11.3 % of gene encoding sequence. 14 *cis-* and 6 *trans*-spliced introns of group II. Presence of plastid and nuclear sequences, product of DNA transfer events. RNA editing sites. Three-copy of recombining-repeats and short repeats	Kubo et al. (2000)
Marchantia polymorpha (liverwort)	121,025 bp 1 single circular molecule	25 group-II and 7 group-I introns. No foreign DNA fragments, no recombination to generate subgenomes, no RNA editing system	Ohyama (1996)
Vigna radiata (bean)	401,262 bp single circular molecule	16.4 % of gene encoding sequence. 17 *cis-* and 5 *trans*-spliced group-II introns. Few and small repeats: 1 with recombining activity. Chloroplast and nuclear-derived-DNA in intergenic regions	Alverson et al. (2011a, b)
Daucus carota (carrot)	281,132 bp 2 putative master circles	20 % of gene encoding sequence. 19 group-II introns, 7 of which are *trans*-spliced. Large inverted and direct repeats. Gene loss by transfer to the nuclear genome	Iorizzo et al. (2012)
Ricinus communiis (castor bean)	502,773 bp Circular map	Horizontal gene transfer to the nuclear genome	Rivarola et al. (2011)
Polytomella capuana	12,998 bp 1 linear molecule with inverted repeats	2 transcriptional clusters represent 82 % of total size. Two conformations for telomeric repeats: open and closed. Presence of short inverted repeat elements	Smith and Lee (2008)
Polytomella parva	13,500 and 3500 bp 2 linear molecules and small subgenomic circular particles	Coding regions compactly organized. Intron free and arranged into two size clusters. Inverted repeats sequences are involved in the multipartite organization of the genome	Fan and Lee (2002)

electron transport chain complexes I, III and IV. Exceptions are the chondromes of *Pleurozia purpurea*, *Phaeoceros laevis*, *Megaceros aenigmaticus*, *Mesostigma viride,* and *M. polymorpha* which lack the *nad7* gene. Besides, the chondromes of *Pseudendoclonium akinetum* lacks the *nad9* gene and that from *Oryza rufipogon* lacks 4 genes of complex I (*nad1*, *nad2*, *nad4* and *nad5*) and the *cox3* gene (complex IV). Although an incomplete annotation cannot be ruled out, this might reflect an important gene loss in the chondromes of these species.

Regarding genes of the other complexes (II, V, cytochrome *C* biogenesis and rRNAs—rps and rpl), a wide range of situations can be found. Whereas in some species they are all encoded by the chondrome, for others these complexes are completely absent. A conspicuous example is the

Table 7.2 Main features of plastid genomes from *Viridiplantae* species

Species (common name)	Genome size(a)/structure	Main features	References
Nuphar advena (spatterdock)	160,866 bp. Two IR: 25,835 bp. LSC: 90,379 bp, SSC: 18,817 bp	60.2 % of gene encoding sequences. 113 annotated genes. Eighteen of these genes contain introns including two genes, clpP and ycf3, each with two introns, and one gene, rps12, also composed of three exons, but with the 5′ exon separated from the two 3′ exons	Raubeson et al. (2007)
Ranunculus macranthus	155,129 bp. Two IR: 25,791 bp. LSC: 84,638 bp, SSC: 18,909 bp	62 % of gene encoding sequences. 113 annotated genes. Eighteen of these genes contain introns including two genes, clpP and ycf3, each with two introns, and one gene, rps12, also composed of three exons, but with the 5’ exon separated from the two 3’ exons	
Vitis vinifera (grape)	160,928 bp. Two IR: 26,358 bp. LSC: 89,147 bp SSC:19,065 bp	57.5 % of gene encoding sequences. 113 annotated genes. Seventeen 17 intron-containing genes, 15 and 2 contain one and two introns, respectively. Phylogenies support Vitaceae as the earliest-diverging lineage of rosids	Jansen et al. (2006)
Triticum aestivum L. cv. Chinese Spring (wheat)	134,545 bp. Two IR: 20,703 bp. LSC: 80,349 bp. SSC: 12,790 bp	The same gene content of rice and maize. Structural divergence indicates that wheat and rice are related more closely to each other than to maize	Ogihara et al. (2002)
Solanum tuberosum (potato)	155,312 bp. Two IR: 25,595 bp. LSC: 85,749. SSC: 18,373 bp	130 annotated genes. Eighteen genes contain one or two introns, and few tRNA are encoded within these introns. Four introns are located in IR and one intron in SSC	Chung et al. (2006)
Acorus calamus (calamus)	153,821 bp. Two IR	112 annotated genes. accD and ycf15 genes are missed	Goremykin et al. (2005)
Adiantum capillus-veneris	150,568 bp. Two IR: 23,447 bp. LSC: 82,282 bp. SSC: 21,392 bp	118 annotated genes: 85 protein-encoding, 29 tRNAs and 4 rRNAs	Wolf et al. (2003)
Amborella trichopoda	162,686 bp. Two IR	132 annotated genes: 114 individual gene species and 18 genes duplicated in the inverted repeats	Goremykin et al. (2003)
Agrostis stolonifera (Common Bent)	136,584 bp. Two IR: 21,649 bp. LSC: 80,546 bp. SSC: 12,740 bp	53.6 % of gene encoding sequences. 131 annotated genes (113 different and 18 duplicated in the IR). 30 distinct tRNAs encoded	Saski et al. (2007)
Anthoceros formosae (Hornwort)	161,162 bp. Two IR: 15,744 bp. LSC: 107,503 bp. SSC: 22,171 bp	112 annotated genes: 76 protein, 32 tRNA and 4 rRNA genes	Kugita (2003)
Arabidopsis thaliana	154,478 bp. Two IR: 26,264 bp. LSC: 84,170 bp. SSC: 17,780 bp	128 annotated genes: a total of 87 potential protein-coding genes including 8 genes duplicated in the inverted repeat regions, 4 rRNA and 37 tRNA genes	Sato et al. (1999)
Atropa belladonna	156,688 bp. Two IR: 25,906 bp. LSC: 86,868 bp. SSC: 18,008 bp	113 annotated genes and arranged in an identical order as tobacco. Intron numbers and positions are highly conserved	Schmitz–Linneweber et al. (2002)

(continued)

Table 7.2 (continued)

Species (common name)	Genome size(a)/structure	Main features	References
Chlamydomonas reinhardtii	203,395 bp. Two IR: 22,211 bp	112 annotated genes: 72 *bona fide* protein-coding genes, 30 tRNA genes, 10 rRNA and 30 tRNA genes	Maul et al. (2002)
Chlorella vulgaris	150,613 bp. Two IR. LSC: 80,873 bp SSC: 78,100 bp	62 % of gene encoding sequences. 124 annotated genes: 71 protein genes, 33 tRNA genes and 10 putative ORFs	Wakasugi et al. (1997)
Citrus sinensis (Sweet Orange)	160,129 bp. Two IR: 26,996 bp. LSC: 87,744 bp. SSC: 18,393 bp	57.3 % of gene encoding sequences. 89 protein-coding genes, 4 rRNAs and 30 distinct tRNAs	Bausher et al. (2006)
Cucumis sativus cultivar Baek (cucumber)	155,527 bp. Two IR: 25,187 bp. LSC: 86,879 bp. SSC: 18,274 bp	55.8 % of gene encoding sequences. 76 protein-coding genes, 30 tRNA genes, 4 rRNA genes, and 3 conserved ORFs	Kim et al. (2006)
Daucus carota (wild carrot)	155,911 bp. Two IR: 27,051 bp. LSC: 84,242 bp. SSC: 17,567 bp	56.4 % of gene encoding sequences. 115 unique genes and 21 duplicated ones within the IR. 4 rRNAs, 30 distinct tRNA genes and 18 intron-containing genes	Ruhlman et al. (2006)
Eucalyptus globules (blue gum)	160,286 bp. Two IR: 26 393 bp. LSC: 89,012 bp. SSC: 18,488 bp	128 annotated genes: 112 individual gene species and 16 duplicated ones within the IR. 78 protein-coding genes, 30 tRNAs, 4 rRNAs	Steane (2005)
Glycine max (soybean)	152,218 bp. Two IR: 25,574 bp. LSC: 83,175 bp. SSC: 17,895 bp	60 % of gene encoding sequences. 130 annotated genes: 111 unique genes and 19 are duplicated within the IR. 30 distinct tRNAs	Saski et al. (2005)
Gossypium barbadense (cotton)	160,317 bp. Two IR: 25,591 bp. LSC: 88,841 bp. SSC: 20,294 bp	131 annotated genes: 116 unique genes and 15 are duplicated within the IR. 37 distinct tRNAs	Ibrahim et al. (2006)
Gossypium hirsutum (cotton)	160,301 bp. Two IR: 25,608 bp. LSC: 88,816 bp. SSC: 20,269 bp	56.5 % of gene encoding sequences. 131 annotated genes: 112 are unique and 19 are duplicated within the IR. 4 rRNAs and 30 distinct tRNA genes	Lee et al. (2006)
Guillardia theta	121,524 bp. Two IR: 4,900 bp. LSC: 96,300 bp. SSC: 15,400 bp	90 % of gene encoding sequences. 183 annotated genes: 66 are protein-encoding, 30 tRNA, 44 rRNAs, 3 translation factors, 8 genes encoding components of the transcriptional machinery and 26 additional *ycfs*	Douglas and Penny (1999)
Hordeum vulgare subsp. Vulgare (barley)	136,462 bp. Two IR: 21,579 bp. LSC: 80,600 bp. SSC: 12,704 bp	56.7 % of gene encoding sequences. 131 annotated genes. 113 are protein-encoding and 18 of these are duplicated within the IR. 30 distinct tRNAs	Saski et al. (2007)
Jasminum nudiflorum (jasmine)	165,121 bp. Two IR: 29,486 bp. LSC: 92,877 bp. SSC: 13,272 bp	57 % of gene encoding sequences. 113 unique genes: 80 protein-coding genes, 30 tRNAs and 4 rRNAs	Lee et al. (2007)
Lotus japonicus	150,519 bp. Two IR: 25,156 bp. LSC: 81,936 bp. SSC: 18,27 bp	84 annotated genes: 77 are unique species and 7 are duplicated within the IR. 37 tRNA genes. Two copies of rRNA gene clusters (16S-23S-4.5S-5S)	Kato et al. (2000)
Solanum lycopersicum cultivar IPA-6	155,461 bp. Two IR: 25,608 bp. LSC: 85,882 bp. SSC: 18,363 bp	58.8 % of gene encoding sequences. 114 unique genes, 30 tRNA, 4 rRNA genes	Kahlau et al. (2006)

(continued)

Table 7.2 (continued)

Species (common name)	Genome size(a)/structure	Main features	References
Cultivar LA3023 (tomato)	155,461 bp. Two IR: 25,611 bp. LSC: 85,876 bp. SSC: 18,363 bp	58.3 % of gene encoding sequences. 133 annotated genes: 113 are unique and 20 are duplicated within the IR. 30 distinct tRNAs	Daniell et al. (2006)
Marchantia polymorpha (liverwort)	121,024 bp. Two IR: 10,058 bp. LSC: 81,095 bp. SSC: 19,813 bp	136 annotated genes: 103 encode stable RNA or proteins. 32 species of tRNA	Ohyama et al. (1986)
Nandina domestica	156,599 bp. Two IR: 26,062 bp. LSC: 85473 bp. SSC: 19002 bp	128 annotated genes: 70 protein-encoding, 30 tRNAs and 4 rRNA genes	Moore et al. (2006)
Nicotiana sylvestris (woodland tobacco)	155,941 bp. Two IR: 25,342 bp. LSC: 86,684 bp. SSC: 18,573 bp	146 annotated genes. Identical gene organization of that of *N. tabacum*, except for one ORF	Yukawa et al. (2006)
Nicotiana tabacum (cultivated tobacco)	155,844 bp. Two IR: 25,339 bp. LSC: 86,684 bp. SSC: 18482 bp	146 annotated genes. 39 different proteins, 4 rRNAs, 30 tRNAs and 11 putative ORFs	Shinozaki et al. (1986)
Nicotiana tomentosiformis	155,745 bp. Two IR: 25,429 bp. LSC: 86,392 bp. SSC: 18,495 bp	146 annotated genes. Identical gene organization of that of *N. tabacum,* except for 4 ORFs and 1 pseudogene	Yukawa et al. (2006)
Oryza sativa (rice)	134,525 bp. Two IR: 20,799 bp	121 annotated genes: 30 tRNAs and 4 rRNA genes	Hiratsuka et al. (1989)
Panax ginseng (Chinese ginseng)	156,318 bp. Two IR: 26,071 bp. LSC: 86,106 bp. SSC: 18,070 bp	58 % of gene encoding sequences. 131 annotated genes: 75 peptide-encoding genes, 30 tRNA genes, 4 rRNA genes and 5 putative ORFs	Kim and Lee (2004)
Pelargonium x hortorum (geranium)	217,942 bp. Two IR: 75,741 bp. LSC: 59,710 bp. SSC: 6,750 bp	51.5 % of gene encoding sequences. 160 annotated genes: 76 unique protein genes (39 of which are duplicated within the IR) 4 rRNA genes (all of which are duplicated within the IR), and 29 tRNA genes (8 are duplicated within the IR)	Chumley et al. (2006)
Phalaenopsis Aphrodite	148,964 bp. Two IR: 25,732 bp. LSC: 85,957 bp. SSC: 11,543 bp	110 annotated genes: 76 protein-encoding genes, 4 rRNA genes and 30 tRNA genes	Chang et al. (2006)
Phaseolus vulgaris (bean)	150,285 bp. Two IR: 26,426 bp. LSC: 79,824 bp. SSC: 17,610 bp	59.6 % of gene encoding sequences. 127 annotated genes: 75 unique protein genes, 30 tRNA genes and 4 rRNA genes	Guo et al. (2007)
Pinus thunbergii (Japanese black pine)	119,707 bp. LSC: 65,696 bp SSC: 53,021 bp	127 annotated genes: 70 coding-protein genes, 41 tRNA genes and 4 rRNA genes	Wakasugi et al. (1994)
Saccharum officinarum (sugar cane)	141,182 bp. Two IR: 22,795 bp. LSC: 83,048 bp. SSC: 12,544 bp	The number, gene content and order of the functional chloroplast genes are identical to those of rice, maize and wheat	Asano et al. (2004)
Solanum bulbocastanum (wild potato)	155,371 bp. Two IR: 25,588 bp. LSC: 85,814 bp. SSC: 18,381 bp	59.6 % of gene encoding sequences. 133 annotated genes: 113 unique genes, 30 distinct tRNAs and 4 rRNA genes	Daniell et al. (2006)
Sorghum bicolor (sorghum)	140,754 bp. Two IR: 22,782 bp. LSC: 82,688 bp. SSC: 12,502 bp	52.1 % of gene encoding sequences. 131 annotated genes: 113 unique genes, 30 distinct tRNAs and 4 rRNA genes	Saski et al. (2007)

(continued)

Table 7.2 (continued)

Species (common name)	Genome size(a)/structure	Main features	References
Piper cenocladum (pepper)	160,624 bp. Two IR: 27,039 bp. LSC: 87,668 bp. SSC: 18,878 bp	130 annotated genes. 113 unique genes, 30 distinct tRNAs and 4 rRNA genes	Cai et al. (2006)
Platanus occidentalis (sycamore)	161,791 bp. Two IR: 25,066 bp. LSC: 92150 bp. SSC: 19509 bp	129 annotated genes.79 protein-coding genes, 30 tRNA genes and 4 rRNA genes	Moore et al. (2006)
Drimys granadensis	160,604 bp. Two IR: 26,649 bp. LSC: 88,685 bp. SSC: 18,621 bp	50.1 % of gene encoding sequences. 131 annotated genes: 79 protein-coding genes, 30 tRNA genes and 4 rRNA genes	Cai et al. (2006)

case of the green alga *Ostreococcus tauri,* which harbors two copies of the *nad4*L, *cob*, *cox1* and *atp8* genes in its mitochondrial genome. Furthermore, many species (*A. thaliana, B. vulgaris* subsp. *vulgaris, Oriza sativa* subsp. *Indica, O. sativa* subsp. *japonica, Sorghum bicolor, Tripsacum dactiloides, Zea luxurians, Zea mays mays, Z. mays parviglumis, Zea perennis, Ferrocalamus rimosivaginus, Bamboosa oldhamii, Silene latifolia* and *Vigna radiate*) harbour the complete set of cytochrome *C* biogenesis genes (*ccm*C, *ccm*FC, *ccm*FN, *ccm*B) but, by contrast, they lack the *sdh3* and *sdh4* genes of the protein complex II. On the other hand, other species (*Chaetosporidium globosum, C. vulgaris, Chlorokybus atmophyticus, M. polymorpha and M. viride*) contain all of the complex II genes but they lack the cytochrome c biogenesis genes. Only *N. tabacum, P. purpurea, Vitis vinífera, Physcomitrella patens, Carica papaya, Ricinus communiis,* and *S. lycopersicum* harbor the complete set of genes for these two complexes encoded in their mitochondrial genomes. The rest of the analyzed chondromes showed disparity regarding the complex II and cytochrome *C* biogenesis encoding genes.

As for the different encoded ATP synthase subunits (complex V), it is also very variable among Embryophyta species. Similarly, ribosomal coding genes are all well conserved in some species (i.e. *M. polymorpha*-16 in total, *P. purpurea*-16 in total and *Cycas taitungensis*-18 in total), whereas in others, most of them are absent (as for example for *S. latifolia* and *B. vulgaris*). In this regard, it should be noted that *V. vinífera* chondrome encodes for the highest number of rRNA genes (29) among all analyzed species, being 17 of them of chloroplastic origin.

Phylogenetic Analyses

Conservation of gene content and a relatively slow rate of nucleotide substitution in protein-coding genes have made the chloroplast genome an ideal focus for studies of plant evolutionary history (Martin et al. 1998; Adachi et al. 2000; De Las Rivas et al. 2002). However, several criteria should be taken into account for these kind of analyses such as exclusion of: (i) species with non-annotated sequences, (ii) missing genes in their annotated genomes, and (iii) protein-encoding sequences that are not present across the chosen species.

Figure 7.2a shows a phylogenetic tree performed by comparing the sequences of 50 orthologous proteins from 50 species of the *Viridiplantae* clade. The clusters of different species match with the current accepted plant classification, thus, confirming the strong association between chloroplast protein modification and the plant speciation. Noteworthy in this respect, *S. lycopersicum* clustered closer to *Solanum bulbunocastum* than to *Solanum tuberosum* and, altogether, *S. lycopersicum* clustered with *Atropa* and *Nicotiana* species (Fig. 7.2a). Clarkson et al. (2004) described a very low degree of sequence variation between

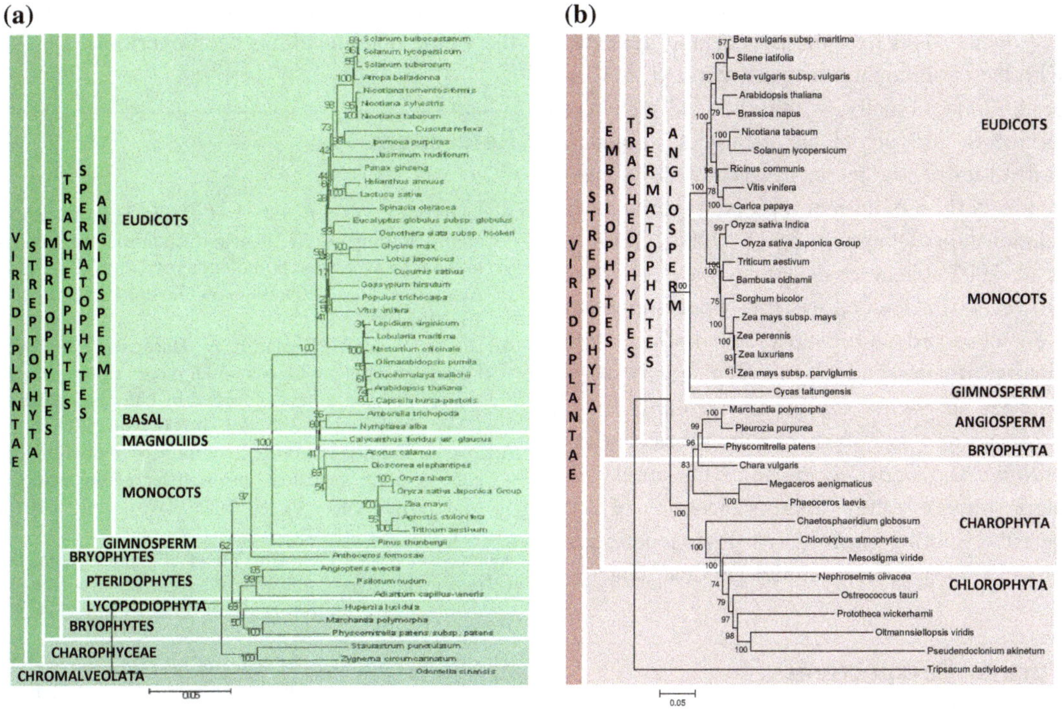

Fig. 7.2 Evolutionary relationships of taxa assessed with chloroplast (**a**) and mitochondrial (**b**) protein sequences. The evolutionary history was inferred using the neighbor-joining method (Saitou and Nei 1987). The optimal tree with the sum of branch lengths is shown (1.51474643 in **a** and 2.79018422 in **b**). Percentages of replicate trees in which the associated taxa clustered together in the bootstrap test (500 replicates) are shown next to the branches (Felsenstein 1985). Trees are drawn to scale, with branch lengths in the same units as those of the evolutionary distances used to infer the phylogenetic tree. The evolutionary distances were computed using the *p*-distance method (Nei and Kumar 2000) and are in the units of the number of amino acid differences per site. Analyses involved 50 and 35 amino acid sequences for **a** and **b**, respectively. All ambiguous positions were removed for each sequence pair. There were a total of 3275 and 5856 positions in the final dataset for **a** and **b**, respectively. Evolutionary analyses were conducted by using the MEGA5 software package (Tamura et al. 2011)

the plastid genomes of *N. sylvestris* and its allopolyploid descendant *N. tabacum*. By contrast, Daniell et al. (2006) revealed a significant number of InDels within certain coding sequences between tomato, potato, tobacco and *Atropa*.

The closest phylogenetic position to tomato within the *Viridiplantae*, as inferred from mitochondrial genomic data, appears to be *N. tabacum* (Fig. 7.2b). In this sense, the nearest species to these last two are *Vitis vinifera* and *C. papaya* (from the order Vitales and Brassicales), which are connected by *Ricinus communis* (Malpighiales order).

As expected, an analysis based on the neighbor-joining method strongly supports the placement of most of the included taxa with Chlorophycean green algae separated from Streptophyta taxa. Within Streptophyta, the only Gymnosperm included in the analysis (*C. taitungensis*) appeared as the ancestor of all Angiosperm species, showing that Gymnosperms are the earliest-diverging lineage among the Streptophyta. *M. polymorpha* and *P. purpurea,* which are placed as the early diverging lineages of land plants, are the exceptions regarding Angiosperms. Thus, they represent the ancestral type of mtDNA. This hypothesis is in line with the finding that the mitochondrial genome of these species closely related to protists, in both gene content and order (Wang et al. 2009). This analysis also shows that gene loss, especially those encoding ribosomal proteins, seems to have occurred after the Angiosperms lineage divergence. This hypothesis is also in agreement with the evolutionary analysis reported by Chaw et al. (2008). Finally, within the land plant taxa, monocots and dicots are clearly separated. In general terms, the reconstructed tree is in accordance with the current accepted phylogenetic relationships (Pombert et al. 2004; Terasawa et al. 2007; Chaw et al. 2008; Ma et al. 2012). However, in few cases, low bootstrap values were observed within taxa with known phylogenetic relations, such as *Zea* genus (61 % between *Z. mays* subsp. *marviglumis* and *Z. luxurians* and 79 % between the Brassicaceae *A. thaliana* and *Brassica napus*). This observation alerts about the appropriateness of the neighbor-joining method for phylogenetic relations based on mitochondrial genome data.

Further Perspectives and Applications

Outcomes from whole genome sequencing projects of crop plant species exponentially increase the available information needed to understand the incidence of plastid genome modification in plant evolution and plant speciation. Particularly in tomato, post-genomic, and functional genomics tools can help elucidating how the transition of chloroplasts to chromoplasts occurs during the ripening of fruits. However, little is still known about the regulation of gene transcription and protein translation as well as of the flow of information between the nucleus and the chloroplast. Knowing the intricate connections between the nucleus and chloroplast is the challenge for the future and will probably introduce an improvement in crops. These organelles are fundamental for the production of a wide variety of metabolites for the food industry as well as for the adaptation of plants to stressful conditions.

Even less understood is the function and regulation of the evolutionary mosaics that represent plant mitochondrial genomes. Solid evidence supports the acquisition (and loss) of genetic information (and possible even active genes) from several distinct sources in the course of evolution. However, the impact of these events at the whole plant level has been overlooked.

References

Abdelnoor RV, Yule R, Elo A et al (2003) Substoichiometric shifting in the plant mitochondrial genome is influenced by a gene homologous to MutS. Proc Natl Acad Sci USA 100:5968–5973. doi:10.1073/pnas.1037651100

Adachi J, Waddell PJ, Martin W, Hasegawa M (2000) Plastid genome phylogeny and a model of amino acid substitution for proteins encoded by chloroplast DNA. J Mol Evol 50:348–358. doi:10.1007/s002399910038

Adams K, Palmer J (2003) Evolution of mitochondrial gene content: gene loss and transfer to the nucleus. Mol Phylogenet Evol 29:380–395. doi:10.1016/S1055-7903(03)00194-5

Adams KL, Daley DO, Qiu YL et al (2000) Repeated, recent and diverse transfers of a mitochondrial gene to the nucleus in flowering plants. Nature 408:354–357. doi:10.1038/35042567

Allison LA, Simon LD, Maliga P (1996) Deletion of rpoB reveals a second distinct transcription system in plastids of higher plants. EMBO J 15:2802–2809

Alverson AJ, Wei X, Rice DW et al (2010) Insights into the evolution of mitochondrial genome size from complete sequences of *Citrullus lanatus* and *Cucurbita pepo* (Cucurbitaceae). Mol Biol Evol 27:1436–1448. doi:10.1093/molbev/msq029

Alverson AJ, Rice DW, Dickinson S et al (2011a) Origins and recombination of the bacterial-sized multichromosomal mitochondrial genome of cucumber. Plant Cell. doi:10.1105/tpc.111.087189

Alverson AJ, Zhuo S, Rice DW et al (2011b) The mitochondrial genome of the legume *Vigna radiata* and the analysis of recombination across short mitochondrial repeats. PLoS One 6:e16404

Anderson S, Bankier A, Barrell B et al (1981) Sequence and organization of the human mitochondrial genome. Nature 290:457–465

Arrieta-Montiel M, Lyznik A, Woloszynska M et al (2001) Tracing evolutionary and developmental implications of mitochondrial stoichiometric shifting in the common bean. Genetics 158:851–864

Asano T, Tsudzuki T, Takahashi S et al (2004) Complete nucleotide sequence of the sugarcane (*Saccharum officinarum*) chloroplast genome: a comparative analysis of four monocot chloroplast genomes. DNA Res 11:93–99

Ayliffe MA, Timmis JN (1992a) Tobacco nuclear DNA contains long tracts of homology to chloroplast DNA. Theor Appl Genet 85–85:229–238. doi:10.1007/BF00222864

Ayliffe MA, Timmis JN (1992b) Plastid DNA sequence homologies in the tobacco nuclear genome. Mol Gen Genet 236:105–112

Ayliffe MA, Scott NS, Timmis JN (1998) Analysis of plastid DNA-like sequences within the nuclear genomes of higher plants. Mol Biol Evol 15:738–745

Backert S, Börner T (2000) Phage T4-like intermediates of DNA replication and recombination in the mitochondria of the higher plant *Chenopodium album* (L.). Curr Genet 37:304–314

Baldauf SL, Palmer JD (1990) Evolutionary transfer of the chloroplast tufA gene to the nucleus. Nature 344:262–265

Barkan A, Goldschmidt-Clermont M (2000) Participation of nuclear genes in chloroplast gene expression. Biochimie 82:559–572

Barsan C, Zouine M, Maza E et al (2012) Proteomic analysis of chloroplast-to-chromoplast transition in tomato reveals metabolic shifts coupled with disrupted thylakoid biogenesis machinery and elevated energy-production components. Plant Physiol 160:708–725. doi:10.1104/pp.112.203679

Bathgate B, Purton ME, Grierson D, Goodenough PW (1985) Plastid changes during the conversion of chloroplasts to chromoplasts in ripening tomatoes. Planta 165:197–204. doi:10.1007/BF00395042

Bausher MG, Singh ND, Lee S-B et al (2006) The complete chloroplast genome sequence of *Citrus sinensis* (L.) Osbeck var "Ridge Pineapple": organization and phylogenetic relationships to other angiosperms. BMC Plant Biol 6:21. doi:10.1186/1471-2229-6-21

Bedbrook JR, Bogorad L (1976) Endonuclease recognition sites mapped on *Zea mays* chloroplast DNA. Proc Natl Acad Sci USA 73:4309–4313

Bedbrook JR, Kolodner R, Bogorad L (1977) *Zea mays* chloroplast ribosomal RNA genes are part of a 22,000 base pair inverted repeat. Cell 11:739–749

Bendich AJ (2004) Circular chloroplast chromosomes: the grand illusion. Plant Cell 16:1661–1666. doi:10.1105/tpc.160771

Bensasson D, Zhang D-X, Hartl DL, Hewitt GM (2001) Mitochondrial pseudogenes: evolution's misplaced witnesses. Trends Ecol Evol 16:314–321. doi:10.1016/S0169-5347(01)02151-6

Blanehard JL, Schmidt GW (1995) Pervasive migration of organellar DNA to the nucleus in plants. J Mol Evol 41:397–406

Bock R (2000) Sense from nonsense: how the genetic information of chloroplasts is altered by RNA editing. Biochimie 82:549–557

Cai Z, Penaflor C, Kuehl JV et al (2006) Complete plastid genome sequences of Drimys, Liriodendron, and Piper: implications for the phylogenetic relationships of magnoliids. BMC Evol Biol 6:77. doi:10.1186/1471-2148-6-77

Chang C-C, Lin H-C, Lin I-P et al (2006) The chloroplast genome of Phalaenopsis aphrodite (Orchidaceae): comparative analysis of evolutionary rate with that of grasses and its phylogenetic implications. Mol Biol Evol 23:279–291. doi:10.1093/molbev/msj029

Chaw S-M, Shih AC-C, Wang D et al (2008) The mitochondrial genome of the gymnosperm *Cycas taitungensis* contains a novel family of short interspersed elements, Bpu sequences, and abundant RNA editing sites. Mol Biol Evol 25:603–615. doi:10.1093/molbev/msn009

Chen J, Guan R, Chang S et al (2011) Substoichiometrically different mitotypes coexist in mitochondrial genomes of *Brassica napus* L. PLoS One 6:e17662

Cheung WY, Scott NS (1989) A contiguous sequence in spinach nuclear DNA is homologous to three separated sequences in chloroplast DNA. Theor Appl Genet 77:625–633

Chumley TW, Palmer JD, Mower JP et al (2006) The complete chloroplast genome sequence of *Pelargonium × hortorum*: organization and evolution of the largest and most highly rearranged chloroplast genome of land plants. Mol Biol Evol 23:2175–2190. doi:10.1093/molbev/msl089

Chung H-J, Jung JD, Park H-W et al (2006) The complete chloroplast genome sequences of *Solanum tuberosum* and comparative analysis with Solanaceae species identified the presence of a 241-bp deletion in cultivated potato chloroplast DNA sequence. Plant Cell Rep 25:1369–1379. doi:10.1007/s00299-006-0196-4

Clarkson JJ, Knapp S, Garcia VF et al (2004) Phylogenetic relationships in *Nicotiana* (Solanaceae) inferred from multiple plastid DNA regions. Mol Phylogenet Evol 33:75–90. doi:10.1016/j.ympev.2004.05.002

Clifton SW, Minx P, Fauron CM et al (2004) Sequence and comparative analysis of the maize NB mitochondrial genome. Plant Physiol 136:3486–3503. doi:10.1104/pp.104.044602.3486

Conrad M (1985) The mutation buffering concept of biomolecular structure. J Biosci 8:669–679

Conte M, López M, Lichtenstein G, Carrari F (2013) Mitochondrial and ripening transcriptome analyses during tomato fruit development and ripening. In: 8th International Conference for Plant Mitochondrial Biology ICPMB 2013. Rosario, Argentina

Correns VCL (1908) Vererbungsversuche mit blass (gelb)grtinen und buntblittrigen Sippen bei Mirabilis-jalapa, *Urtica pilulifera* und. *Lunaria annua*. Zeitschrift für Induktive Abstammungs und Vererbungslehre 1:291–329

Cui L, Veeraraghavan N, Richter A et al (2006) Chloroplast DB: the chloroplast genome database. Nucleic Acids Res 34:D692–D696. doi:10.1093/nar/gkj055

Daniell H, Lee S-B, Grevich J et al (2006) Complete chloroplast genome sequences of *Solanum bulbocastanum*, *Solanum lycopersicum* and comparative analyses with other Solanaceae genomes. TAG Theor Appl Genet 112:1503–1518. doi:10.1007/s00122-006-0254-x

Darracq A, Varré J-S, Touzet P (2010) A scenario of mitochondrial genome evolution in maize based on rearrangement events. BMC Genom 11:233. doi:10.1186/1471-2164-11-233

De Las Rivas J, Lozano JJ, Ortiz AR (2002) Comparative analysis of chloroplast genomes: functional annotation, genome-based phylogeny, and deduced evolutionary patterns. Genome Res 12:567–583

Denovan-Wright EM, Nedelcu AM, Lee RW (1998) Complete sequence of the mitochondrial DNA of *Chlamydomonas eugametos*. Plant Mol Biol 36:285–295

Douglas SE, Penny SL (1999) The plastid genome of the cryptophyte alga, Guillardia theta: complete sequence and conserved synteny groups confirm its common ancestry with red algae. J Mol Evol 48:236–244

Du Jardin P (1990) Homologies to plastid DNA in the nuclear and mitochondrial genomes of potato. Theor Appl Genet 79:807–812. doi:10.1007/BF00224249

Eberhard S, Drapier D, Wollman F-A (2002) Searching limiting steps in the expression of chloroplast-encoded proteins: relations between gene copy number, transcription, transcript abundance and translation rate in the chloroplast of *Chlamydomonas reinhardtii*. Plant J Cell Mol Biol 31:149–160

Egea I, Bian W, Barsan C et al (2011) Chloroplast to chromoplast transition in tomato fruit: spectral confocal microscopy analyses of carotenoids and chlorophylls in isolated plastids and time-lapse recording on intact live tissue. Ann Bot 108:291–297. doi:10.1093/aob/mcr140

Fan J, Lee RW (2002) Mitochondrial genome of the colorless green alga *Polytomella parva*: two linear DNA molecules with homologous inverted repeat Termini. Mol Biol Evol 19:999–1007

Fauron C, Casper M (1995) The maize mitochondrial genome: dynamic, yet functional. Trends Genet TIG 11:228–235

Felsenstein J (1985) Confidence limits on phylogenies: an approach using the bootstrap. Evolution 39:783–791

Fleischmann TT, Scharff LB, Alkatib S et al (2011) Nonessential plastid-encoded ribosomal proteins in tobacco: a developmental role for plastid translation and implications for reductive genome evolution. Plant Cell 23:3137–3155. doi:10.1105/tpc.111.088906

Gibor A, Granick S (1964) Plastids and mitochondria: inheritable systems: do plastids and mitochondria contain a chromosome which controls their multiplication and development? Science 145:890–897. doi:10.1126/science.145.3635.890

Goremykin VV, Hirsch-Ernst KI, Wolfl S, Hellwig FH (2003) Analysis of the *Amborella trichopoda* chloroplast genome sequence suggests that amborella is not a basal angiosperm. Mol Biol Evol 20:1499–1505. doi:10.1093/molbev/msg159

Goremykin VV, Holland B, Hirsch-Ernst KI, Hellwig FH (2005) Analysis of *Acorus calamus* chloroplast genome and its phylogenetic implications. Mol Biol Evol 22:1813–1822. doi:10.1093/molbev/msi173

Goremykin VV, Salamini F, Velasco R, Viola R (2009) Mitochondrial DNA of *Vitis vinifera* and the issue of rampant horizontal gene transfer. Mol Biol Evol 26:99–110. doi:10.1093/molbev/msn226

Graham LE, Cook ME, Busse JS (2000) The origin of plants: body plan changes contributing to a major evolutionary radiation. Proc Natl Acad Sci USA 97:4535–4540

Gray MW (1999) Evolution of organellar genomes. Curr Opin Genet Dev 9:678–687

Gray RE, Law RHP, Devenish RJ, Nagley P (1996) Allotopic expression of mitochondrial ATP synthase genes in nucleus of *Saccharomyces cerevisiae*. In: Attardi GM, Chomyn A (eds) Mitochondrial biogenesis and genetics, part B. Academic Press, London, pp 369–389

Gualberto JM, Wintz H, Weil JH, Grienenberger JM (1988) The genes coding for subunit 3 of NADH dehydrogenase and for ribosomal protein S12 are present in the wheat and maize mitochondrial genomes and are co-transcribed. Mol Gen Genet MGG 215:118–127

Guilliermond A, Atkinson LMR (1941) The cytoplasm of the plant cell. In: Frans Verdoorn (ed) A new series of plant science books

Guo X, Castillo-Ramírez S, González V et al (2007) Rapid evolutionary change of common bean (*Phaseolus vulgaris* L.) plastome, and the genomic diversification of legume chloroplasts. BMC Genom 8:228. doi:10.1186/1471-2164-8-228

Hajdukiewicz PT, Allison LA, Maliga P (1997) The two RNA polymerases encoded by the nuclear and the plastid compartments transcribe distinct groups of genes in tobacco plastids. EMBO J 16:4041–4048. doi:10.1093/emboj/16.13.4041

Harris WM, Spurr AR (1969) Chromoplasts of tomato fruits. II. The red tomato. Am J Bot 56:380–389

Hazkani-Covo E, Zeller RM, Martin W (2010) Molecular poltergeists: mitochondrial DNA copies (numts) in sequenced nuclear genomes. PLoS Genet 6:e1000834. doi:10.1371/journal.pgen.1000834

Hiratsuka J, Shimada H, Whittier R, et al. (1989) The complete sequence of the rice (*Oryza sativa*) chloroplast genome: intermolecular recombination between distinct tRNA genes accounts for a major plastid DNA inversion during the evolution of the cereals. Mol Gen Genet 217(2–3):185–194

Huang X, Madan A (1999) CAP3: A DNA sequence assembly program. Genome Res 9:868–877

Ibrahim RIH, Azuma J-I, Sakamoto M (2006) Complete nucleotide sequence of the cotton (*Gossypium barbadense* L.) chloroplast genome with a comparative analysis of sequences among 9 dicot plants. Genes Genet Syst 81:311–321

Iorizzo M, Senalik D, Szklarczyk M et al (2012) De novo assembly of the carrot mitochondrial genome using next generation sequencing of whole genomic DNA provides first evidence of DNA transfer into an angiosperm plastid genome. BMC Plant Biol 12:61. doi:10.1186/1471-2229-12-61

Jansen RK, Kaittanis C, Saski C et al (2006) Phylogenetic analyses of *Vitis* (Vitaceae) based on complete chloroplast genome sequences: effects of taxon sampling and phylogenetic methods on resolving relationships among rosids. BMC Evol Biol 6:32. doi:10.1186/1471-2148-6-32

Janska H, Sarria R, Woloszynska M et al (1998) Stoichiometric shifts in the common bean mitochondrial genome leading to male sterility and spontaneous reversion to fertility. Plant Cell 10:1163–1180

Joyce PBM, Gray MW (1988) Nucleotide sequence of a wheat mitochondrial glutamine tRNA gene. Nucleic Acids Res 16:1210

Kahlau S, Bock R (2008) Plastid transcriptomics and translatomics of tomato fruit development and chloroplast-to-chromoplast differentiation: chromoplast gene expression largely serves the production of a single protein. Plant Cell 20:856–874. doi:10.1105/tpc.107.055202

Kahlau S, Aspinall S, Gray JC, Bock R (2006) Sequence of the tomato chloroplast DNA and evolutionary comparison of solanaceous plastid genomes. J Mol Evol 63:194–207. doi:10.1007/s00239-005-0254-5

Kajander OA, Rovio AT, Majamaa K et al (2000) Human mtDNA sublimons resemble rearranged mitochondrial genomes found in pathological states. Hum Mol Genet 9:2821–2835. doi:10.1093/hmg/9.19.2821

Kanazawa A, Hirai A (1994) Reversible changes in the composition of the population of mtdnas during dedifferentiation and regeneration in tobacco. Genetics 138:865–870

Kanevski I, Maliga P (1994) Relocation of the plastid rbcL gene to the nucleus yields functional ribulose-1,5-bisphosphate carboxylase in tobacco chloroplasts. Proc Natl Acad Sci USA 91:1969–1973

Kato T, Kaneko T, Sato S et al (2000) Complete structure of the chloroplast genome of a legume, Lotus japonicus. DNA Res 7:323–330

Keeling PJ (2010) The endosymbiotic origin, diversification and fate of plastids. Philos Trans R Soc Lond B Biol Sci 365:729–748. doi:10.1098/rstb.2009.0103

Kim K-J, Lee H-L (2004) Complete chloroplast genome sequences from Korean ginseng (*Panax schinseng* Nees) and comparative analysis of sequence evolution among 17 vascular plants. DNA Res 11:247–261

Kim J-S, Jung JD, Lee J-A et al (2006) Complete sequence and organization of the cucumber (*Cucumis sativus* L. cv. Baekmibaekdadagi) chloroplast genome. Plant Cell Rep 25:334–340. doi:10.1007/s00299-005-0097-y

Kleine T, Maier UG, Leister D (2009) DNA transfer from organelles to the nucleus: the idiosyncratic genetics of endosymbiosis. Annu Rev Plant Biol 60:115–138. doi:10.1146/annurev.arplant.043008.092119

Kmiec B, Woloszynska M, Janska H (2006) Heteroplasmy as a common state of mitochondrial genetic information in plants and animals. Curr Genet 50:149–159. doi:10.1007/s00294-006-0082-1

Kode V, Mudd EA, Iamtham S, Day A (2005) The tobacco plastid accD gene is essential and is required for leaf development. Plant J Cell Mol Biol 44:237–244. doi:10.1111/j.1365-313X.2005.02533.x

Kubo T, Nishizawa S, Sugawara A et al (2000) The complete nucleotide sequence of the mitochondrial genome of sugar beet (*Beta vulgaris* L.) reveals a novel gene for tRNA(Cys)(GCA). Nucleic acids research 28:2571–2576

Kugita M (2003) The complete nucleotide sequence of the hornwort (*Anthoceros formosae*) chloroplast genome: insight into the earliest land plants. Nucleic Acids Res 31:716–721. doi:10.1093/nar/gkg155

Kurland CG, Andersson SG (2000) Origin and evolution of the mitochondrial proteome. Microbiol Mol Biol Rev MMBR 64:786–820

Kuroda H, Maliga P (2003) The plastid clpP1 protease gene is essential for plant development. Nature 425:86–89. doi:10.1038/nature01909

Kuzmin EV, Duvick DN, Newton KJ (2005) A mitochondrial mutator system in maize. Plant Physiol 137:779–789. doi:10.1104/pp.104.053611.1

Leaver CJ, Gray MW (1982) Mitochondrial genome organization and expression in higher plants. Annu Rev Plant Physiol 33:373–402. doi:10.1146/annurev.pp.33.060182.002105

Lee S-B, Kaittanis C, Jansen RK et al (2006) The complete chloroplast genome sequence of *Gossypium hirsutum*: organization and phylogenetic relationships to other angiosperms. BMC Genom 7:61. doi:10.1186/1471-2164-7-61

Lee H-L, Jansen RK, Chumley TW, Kim K-J (2007) Gene relocations within chloroplast genomes of Jasminum and Menodora (Oleaceae) are due to multiple, overlapping inversions. Mol Biol Evol 24:1161–1180. doi:10.1093/molbev/msm036

Legen J, Kemp S, Krause K et al (2002) Comparative analysis of plastid transcription profiles of entire plastid chromosomes from tobacco attributed to wild-type and PEP-deficient transcription machineries. Plant J Cell Mol Biol 31:171–188

Lemieux C, Otis C, Turmel M (2007) A clade uniting the green algae *Mesostigma viride* and *Chlorokybus atmophyticus* represents the deepest branch of the Streptophyta in chloroplast genome-based phylogenies. BMC Biol 5:2. doi:10.1186/1741-7007-5-2

Lerbs-Mache S (2000) Regulation of rDNA transcription in plastids of higher plants. Biochimie 82:525–535

Li L, Wang B, Liu Y, Qiu Y-L (2009) The complete mitochondrial genome sequence of the hornwort *Megaceros aenigmaticus* shows a mixed mode of conservative yet dynamic evolution in early land plant mitochondrial genomes. J Mol Evol 68:665–678. doi:10.1007/s00239-009-9240-7

Lopez JV, Yuhki N, Masuda R et al (1994) Numt, a recent transfer and tandem amplification of mitochondrial DNA to the nuclear genome of the domestic cat. J Mol Evol 39:174–190

Ma P-F, Guo Z-H, Li D-Z (2012) Rapid sequencing of the bamboo mitochondrial genome using Illumina technology and parallel episodic evolution of organelle genomes in grasses. PLoS One 7:e30297

Mackenzie S, McIntosh L (1999) Higher plant mitochondria. Plant Cell 11:571–586

Margulis L, Bermudes D (1985) Symbiosis as a mechanism of evolution: status of cell symbiosis theory. Symbiosis (Philadelphia, PA) 1:101–124

Marienfeld J, Unseld M, Brennicke A (1999) The mitochondrial genome of *Arabidopsis* is composed of both native and immigrant information. Trends Plant Sci 4:495–502

Martin W, Stoebe B, Goremykin V et al (1998) Gene transfer to the nucleus and the evolution of chloroplasts. Nature 393:162–165. doi:10.1038/30234

Martínez-Zapater JM, Gil P, Capel J, Somerville CR (1992) Mutations at the *Arabidopsis* CHM locus promote rearrangements of the mitochondrial genome. Plant Cell 4:889–899. doi:10.1105/tpc.4.8.889

Maul JE, Lilly JW, Cui L et al (2002) The *Chlamydomonas reinhardtii* plastid chromosome: islands of genes in a sea of repeats. Plant Cell 14:2659–2679. doi:10.1105/tpc.006155.present

Mereschkowski C (1905) Über Natur und Ursprung der Chromatophoren im Pflanzenreiche. Biol Centralbl 25:593–604

Moore MJ, Dhingra A, Soltis PS et al (2006) Rapid and accurate pyrosequencing of angiosperm plastid genomes. BMC Plant Biol 6:17. doi:10.1186/1471-2229-6-17

Nägeli C (1846) Über Polysiphonia und Herposiphonia. Zeitschrift für wissenschaftliche Botanik 4:207–256

Nass S, Nass MM (1963) Intramitochondrial fibers with DNA characteristics. II. Enzymatic and other hydrolytic treatments. J Cell Biol 19:613–629

Nedelcu AM, Lee RW, Lemieux C et al (2000) The complete mitochondrial DNA sequence of *Scenedesmus obliquus* reflects an intermediate stage in the evolution of the green algal mitochondrial genome. Genome Res 10:819–831

Nei M, Kumar S (2000) Molecular evolution and phylogenetics. Oxford University Press, Oxford

Notsu Y, Masood S, Nishikawa T et al (2002) The complete sequence of the rice (*Oryza sativa* L.) mitochondrial genome: frequent DNA sequence acquisition and loss during the evolution of flowering plants. Mol Genet Genomics MGG 268:434–445. doi:10.1007/s00438-002-0767-1

Oda K, Yamato K, Ohta E et al (1992) Gene organization deduced from the complete sequence of liverwort *Marchantia polymorpha* mitochondrial DNA. A primitive form of plant mitochondrial genome. J Mol Biol 223:1–7

Ogihara Y, Isono K, Kojima T et al (2002) Structural features of a wheat plastome as revealed by complete sequencing of chloroplast DNA. Mol Genet Genomics MGG 266:740–746. doi:10.1007/s00438-001-0606-9

Ogihara Y, Yamazaki Y, Murai K et al (2005) Structural dynamics of cereal mitochondrial genomes as revealed by complete nucleotide sequencing of the wheat mitochondrial genome. Nucleic Acids Res 33:6235–6250. doi:10.1093/nar/gki925

Ohyama K (1996) Chloroplast and mitochondrial genomes from a liverwort, *Marchantia polymorpha*: gene organization and molecular evolution. Biosci Biotechnol Biochem 60:16–24

Ohyama K, Fukuzawa H, Kohchi T et al (1986) Chloroplast gene organization deduced from complete sequence of liverwort *Marchantia polymorpha* chloroplast DNA. Nature 322:572–574. doi:10.1038/322572a0

Oldenburg DJ, Bendich AJ (1996) Size and structure of replicating mitochondrial DNA in cultured tobacco cells. Plant Cell 8:447–461. doi:10.1105/tpc.8.3.447

Palmer JD (1991) CHAPTER 2—plastid chromosomes: structure and evolution. In: Molecular The (ed) Plastids IVBT-TMB of biology of plastids. Academic Press, London, pp 5–53

Palmer JD, Herbon LA (1987) Unicircular structure of the *Brassica hirta* mitochondrial genome. Curr Genet 11:565–570

Palmer JD, Zamir D (1982) Chloroplast DNA evolution and phylogenetic relationships in Lycopersicon. Proc Natl Acad Sci USA 79:5006–5010

Palmer JD, Adams KL, Cho Y et al (2000) Dynamic evolution of plant mitochondrial genomes: mobile genes and introns and highly variable mutation rates. Proc Natl Acad Sci USA 97:6960–6966

Pfannschmidt T, Nilsson A, Tullberg A et al (1999) Direct transcriptional control of the chloroplast genes psbA and psaAB adjusts photosynthesis to light energy distribution in plants. IUBMB Life 48:271–276. doi:10.1080/713803507

Phillips AL (1985) Restriction map and clone bank of tomato plastid DNA. Curr Genet 10:147–152

Pichersky E, Tanksley SD (1988) Chloroplast DNA sequences integrated into an intron of a tomato nuclear gene. Mol Gen Genet 215:65–68

Pichersky E, Logsdon JM, McGrath JM, Stasys RA (1991) Fragments of plastid DNA in the nuclear genome of tomato: prevalence, chromosomal location, and possible mechanism of integration. Mol Gen Genet 225:453–458

Piechulla B, Imlay KRC, Gruissem W (1985) Plastid gene expression during fruit ripening in tomato. Plant Mol Biol 5:373–384

Pombert J-F, Otis C, Lemieux C, Turmel M (2004) The complete mitochondrial DNA sequence of the green alga *Pseudendoclonium akinetum* (Ulvophyceae) highlights distinctive evolutionary trends in the chlorophyta and suggests a sister-group relationship between the Ulvophyceae and Chlorophyceae. Mol Biol Evol 21:922–935. doi:10.1093/molbev/msh099

Pombert J-F, Lemieux C, Turmel M (2006) The complete chloroplast DNA sequence of the green alga *Oltmannsiellopsis viridis* reveals a distinctive quadripartite architecture in the chloroplast genome of early diverging ulvophytes. BMC Biol 4:3. doi:10.1186/1741-7007-4-3

Popescu CE, Lee RW (2007) Mitochondrial genome sequence evolution in *Chlamydomonas*. Genetics 175:819–826. doi:10.1534/genetics.106.063156

Raubeson LA, Peery R, Chumley TW et al (2007) Comparative chloroplast genomics: analyses including new sequences from the angiosperms *Nuphar advena* and *Ranunculus macranthus*. BMC Genomics 8:174. doi:10.1186/1471-2164-8-174

Ris H, Plaut W (1962) Ultrastructure of DNA-containing areas in the chloroplast of *Chlamydomonas*. J Cell Biol 13:383–391

Rivarola M, Foster JT, Chan AP et al (2011) Castor bean organelle genome sequencing and worldwide genetic diversity analysis. PLoS One 6:e21743

Robbens S, Derelle E, Ferraz C et al (2007) The complete chloroplast and mitochondrial DNA sequence of *Ostreococcus tauri*: organelle genomes of the smallest eukaryote are examples of compaction. Mol Biol Evol 24:956–968. doi:10.1093/molbev/msm012

Rodríguez-Moreno L, González VM, Benjak A et al (2011) Determination of the melon chloroplast and mitochondrial genome sequences reveals that the largest reported mitochondrial genome in plants contains a significant amount of DNA having a nuclear origin. BMC Genom 12:424. doi:10.1186/1471-2164-12-424

Rogalski M, Schöttler MA, Thiele W et al (2008) Rpl33, a nonessential plastid-encoded ribosomal protein in tobacco, is required under cold stress conditions. Plant Cell 20:2221–2237. doi:10.1105/tpc.108.060392

Rosso SW (1968) The ultrastructure of chromoplast development in red tomatoes. J Ultrastruct Res 25:307–322

Ruhlman T, Lee S-B, Jansen RK et al (2006) Complete plastid genome sequence of *Daucus carota*: implications for biotechnology and phylogeny of angiosperms. BMC Genom 7:222. doi:10.1186/1471-2164-7-222

Saitou N, Nei M (1987) The neighbor-joining method: a new method for reconstructing phylogenetic trees. Mol Biol Evol 4:406–425

Sakamoto W, Tan S-H, Murata M, Motoyoshi F (1997) An unusual mitochondrial atp9-rpl16 cotranscript found in the maternal distorted leaf mutant of *Arabidopsis thaliana*: implication of GUG as an initiation codon in plant mitochondria. Plant Cell Physiol 38:975–979

Saski C, Lee S-B, Daniell H et al (2005) Complete chloroplast genome sequence of *Gycine max* and comparative analyses with other legume genomes. Plant Mol Biol 59:309–322. doi:10.1007/s11103-005-8882-0

Saski C, Lee S-B, Fjellheim S et al (2007) Complete chloroplast genome sequences of *Hordeum vulgare*, *Sorghum bicolor* and *Agrostis stolonifera*, and comparative analyses with other grass genomes. TAG Theor Appl Genet 115:571–590. doi:10.1007/s00122-007-0567-4

Sato S, Nakamura Y, Kaneko T et al (1999) Complete structure of the chloroplast genome of *Arabidopsis thaliana*. DNA Res 6:283–290

Schatz G, Haslbrunner E, Tuppy H (1964) Deoxyribonucleic acid associated with yeast mitochondria. Biochem Biophys Res Commun 15:127–132

Schmidt EW (1913) Pflanzliche Mitochondrien. Progressus rei botanicae 4:164–183

Schmitz-Linneweber C, Regel R, Du TG et al (2002) The plastid chromosome of *Atropa belladonna* and its comparison with that of *Nicotiana tabacum*: the role of RNA editing in generating divergence in the process of plant speciation. Mol Biol Evol 19:1602–1612

Schön A, Krupp G, Gough S et al (1986) The RNA required in the first step of chlorophyll biosynthesis is a chloroplast glutamate tRNA. Nature 322:281–284

Scott NS, Timmis JN (1984) Homologies between nuclear and plastid DNA in spinach. Theor Appl Genet 67:279–288

Sharma MR, Wilson DN, Datta PP et al (2007) Cryo-EM study of the spinach chloroplast ribosome reveals the structural and functional roles of plastid-specific ribosomal proteins. Proc Natl Acad Sci USA 104:19315–19320. doi:10.1073/pnas.0709856104

Shikanai T, Kaneko H, Nakata S et al (1998) Mitochondrial genome structure of a cytoplasmic hybrid between tomato and wild potato. Plant Cell Rep 17:832–836. doi:10.1007/s002990050493

Shikanai T, Shimizu K, Ueda K et al (2001) The chloroplast clpP gene, encoding a proteolytic subunit of ATP-dependent protease, is indispensable for chloroplast development in tobacco. Plant Cell Physiol 42:264–273

Shinozaki K, Ohme M, Tanaka M et al (1986) The complete nucleotide sequence of the tobacco chloroplast genome: its gene organization and expression. EMBO J 5:2043–2049

Smith DR, Lee RW (2008) Mitochondrial genome of the colorless green alga *Polytomella capuana*: a linear molecule with an unprecedented GC content. Mol Biol Evol 25:487–496. doi:10.1093/molbev/msm245

Steane DA (2005) Complete nucleotide sequence of the chloroplast genome from the Tasmanian bluegum, *Eucalyptus globulus* (Myrtaceae). DNA Res 12:215–220. doi:10.1093/dnares/dsi006

Stern DB, Palmer JD (1984) Extensive and widespread homologies between mitochondrial DNA and chloroplast DNA in plants. Proc Natl Acad Sci USA 81:1946–1950

Stutzt B, Noll H (1967) Polysomies in plants: evidence for three classes of ribosomal RNA in nature. Proc Natl Acad Sci USA 57:774–781

Sugiura M (1992) The chloroplast genome. In: Schilperoort R, Dure L (eds) 10 Years plant molecular biology. Springer, Netherlands, pp 149–168

Sugiyama Y, Watase Y, Nagase M et al (2004) Timing of tRNA gene transfer from chloroplast to mitochondrion revealed by genomic analysis of dicotyledonous plant mitochondria. Endocytobiosis Cell Res 15:77–86

Sugiyama Y, Watase Y, Nagase M et al (2005) The complete nucleotide sequence and multipartite organization of the tobacco mitochondrial genome: comparative analysis of mitochondrial genomes in higher plants—supp info. Mol Genet Genomics 272:303–315

Sun Q, Wang K, Yoshimura A, Doi K (2002) Genetic differentiation for nuclear, mitochondrial and chloroplast genomes in common wild rice (*Oryza rufipogon* Griff.) and cultivated rice (*Oryza sativa* L.). TAG Theor Appl Genet 104:1335–1345. doi:10.1007/s00122-002-0878-4

Tamura K, Peterson D, Peterson N et al (2011) MEGA5: molecular evolutionary genetics analysis using maximum likelihood, evolutionary distance, and maximum parsimony methods. Mol Biol Evol 28:2731–2739. doi:10.1093/molbev/msr121

Tanaka K, Oikawa K, Ohta N et al (1996) Nuclear encoding of a chloroplast RNA polymerase sigma subunit in a red alga. Science (New York, NY) 272:1932–1935

Taylor DR, Olson MS, McCauley DE (2001) A quantitative genetic analysis of nuclear-cytoplasmic male sterility in structured populations of *Silene vulgaris*. Genetics 158:833–841

Terasawa K, Odahara M, Kabeya Y et al (2007) The mitochondrial genome of the moss *Physcomitrella patens* sheds new light on mitochondrial evolution in land plants. Mol Biol Evol 24:699–709. doi:10.1093/molbev/msl198

The Tomato Genome Consortium (2012) The tomato genome sequence provides insights into fleshy fruit evolution. Nature 485:635–641. doi:10.1038/nature11119

Thorsness PE, Weber ER (1996) Escape and migration of nucleic acids between chloroplasts, mitochondria, and the nucleus. Int Rev Cytol 165:207–234

Tian X, Zheng J, Hu S (2006) The rice mitochondrial genomes and their variations. Plant Physiol 140:401–410. doi:10.1104/pp.105.070060.Palmer

Tiller N, Weingartner M, Thiele W et al (2012) The plastid-specific ribosomal proteins of *Arabidopsis thaliana* can be divided into non-essential proteins and genuine ribosomal proteins. Plant J Cell Mol Biol 69:302–316. doi:10.1111/j.1365-313X.2011.04791.x

Timmis JN, Scot SN (1983) Sequence homology between spinach nuclear and chloroplast genomes. Nature 305:65–67

Timmis JN, Ayliffe MA, Huang CY, Martin W (2004) Endosymbiotic gene transfer: organelle genomes forge eukaryotic chromosomes. Nat Rev Genet 5:123–135. doi:10.1038/nrg1271

Tullberg A, Alexciev K, Pfannschmidt T, Allen JF (2000) Photosynthetic electron flow regulates transcription of the psaB gene in pea (*Pisum sativum* L.) chloroplasts through the redox state of the plastoquinone pool. Plant Cell Physiol 41:1045–1054. doi:10.1093/pcp/pcd031

Turmel M, Lemieux C, Burger G et al (1999a) The complete mitochondrial DNA sequences of *Nephroselmis olivacea* and *Pedinomonas minor*. Two radically different evolutionary patterns within green algae. Plant Cell 11:1717–1730

Turmel M, Otis C, Lemieux C (1999b) The complete chloroplast DNA sequence of the green alga *Nephroselmis olivacea*: insights into the architecture of ancestral chloroplast genomes. Proc Natl Acad Sci USA 96:10248–10253

Turmel M, Otis C, Lemieux C (2002a) The complete mitochondrial DNA sequence of *Mesostigma viride* identifies this green alga as the earliest green plant divergence and predicts a highly compact mitochondrial genome in the ancestor of all green plants. Mol Biol Evol 19:24–38

Turmel M, Otis C, Lemieux C (2002b) The chloroplast and mitochondrial genome sequences of the charophyte *Chaetosphaeridium globosum*: insights into the timing of the events that restructured organelle DNAs within the green algal lineage that led to land plants. Proc Natl Acad Sci USA 99:11275–11280. doi:10.1073/pnas.162203299

Turmel M, Otis C, Lemieux C (2003) The mitochondrial genome of *Chara vulgaris*: insights into the mitochondrial DNA architecture of the last common ancestor of green algae and land plants. Plant Cell 15:1888–1903. doi:10.1105/tpc.013169.these

Unseld M, Marienfeld JR, Brandt P, Brennicke A (1997) The mitochondrial genome of *Arabidopsis thaliana* contains 57 genes in 366,924 nucleotides. Nat Genet 15:57–61

Wakasugi T, Tsudzuki J, Ito S et al (1994) Loss of all ndh genes as determined by sequencing the entire chloroplast genome of the black pine *Pinus thunbergii*. Proc Natl Acad Sci USA 91:9794–9798

Wakasugi T, Nagai T, Kapoor M et al (1997) Complete nucleotide sequence of the chloroplast genome from the green alga *Chlorella vulgaris*: the existence of genes possibly involved in chloroplast division. Proc Natl Acad Sci USA 94:5967–5972

Wallin IE (1923) The University of Chicago. Am Nat 57:255–261

Wang B, Xue J, Li L et al (2009) The complete mitochondrial genome sequence of the liverwort *Pleurozia purpurea* reveals extremely conservative mitochondrial genome evolution in liverworts. Curr Genet 55:601–609. doi:10.1007/s00294-009-0273-7

Wang D, Rousseau-Gueutin M, Timmis JN (2012) Plastid sequences contribute to some plant mitochondrial genes. Mol Biol Evol 29:1707–1711. doi:10.1093/molbev/mss016

Ward BL, Anderson RS, Bendich AJ (1981) The mitochondrial genome is large and variable in a family of plants (Cucurbitaceae). Cell 25:793–803

Wolf PG, Rowe CA, Sinclair RB, Hasebe M (2003) Complete nucleotide sequence of the chloroplast genome from a leptosporangiate fern, *Adiantum capillus-veneris* L. DNA Res 10:59–65

Woloszynska M, Kieleczawa J, Ornatowska M et al (2001) The origin and maintenance of the small repeat in the bean mitochondrial genome. Mol Genet Genomics 265:865–872. doi:10.1007/s004380100481

Xue J-Y, Liu Y, Li L et al (2010) The complete mitochondrial genome sequence of the hornwort *Phaeoceros laevis*: retention of many ancient pseudogenes and conservative evolution of mitochondrial genomes in hornworts. Curr Genet 56:53–61. doi:10.1007/s00294-009-0279-1

Yamato KT, Newton KJ (1999) Heteroplasmy and homoplasmy for maize mitochondrial mutants: a rare

homoplasmic nad4 deletion mutant plant. J Hered 90:369–373. doi:10.1093/jhered/90.3.369
Yu Q, Tong E, Skelton RL et al (2009) A physical map of the papaya genome with integrated genetic map and genome sequence. BMC Genom 10:371. doi:10.1186/1471-2164-10-371
Yukawa M, Tsudzuki T, Sugiura M (2006) The chloroplast genome of *Nicotiana sylvestris* and *Nicotiana tomentosiformis*: complete sequencing confirms that the *Nicotiana sylvestris* progenitor is the maternal genome donor of *Nicotiana tabacum*. Mol Genet Genomics MGG 275:367–373. doi:10.1007/s00438-005-0092-6
Zheng L-Y, Guo X-S, He B et al (2011) Genome-wide patterns of genetic variation in sweet and grain sorghum (*Sorghum bicolor*). Genome Biol 12:R114. doi:10.1186/gb-2011-12-11-r114

Assembly and Application to the Tomato Genome

8

Jifeng Tang, Erwin Datema, Antoine Janssen and Roeland C.H.J. van Ham

Abstract

The computational process of reconstructing a genome by assembling large amounts of raw sequencing data into long DNA fragments poses great challenges. This chapter illustrates current genome sequencing technologies and assembly algorithms by example of the tomato genome sequencing project. Over the last decade, "Next Generation Sequencing" technologies have placed great emphasis on efficient library preparation, high throughput and long read length. These developments have pushed the evolution of genome assembly approaches from greedy overlap-layout-consensus approaches that were used to assemble Sanger sequences, to de Bruijn graph and string graph approaches that are currently in use to assemble these new types of sequencing data produced in large volume. Nonetheless, many species still lack a high-quality, gold-standard genome sequence as genome assembly is still far from a solved problem. Several approaches have been developed to estimate the quality of assembled genome sequences and to perform so-called genome finishing, a complicated and costly procedure to complete the unresolved regions of the genome. We expect that within this decade sequencing technologies will undergo another dramatic improvement, resulting in "Third Generation Sequencing" technologies with which chromosomes and genomes can be sequenced in their entirety with high accuracy. Plant breeding will benefit enormously from this development, providing breeders with the tools, data and understanding to design new traits and varieties from natural and induced genetic variation in an entirely rationalized and economical manner, and much beyond our current capabilities. The tomato genome described here was sequenced within an international collaboration and its completion spanned almost a decade.

J. Tang · E. Datema · A. Janssen · R.C.H.J. van Ham (✉)
Keygene N.V., Agro Business Park 90, 6708 PW Wageningen, The Netherlands
e-mail: roeland.van-ham@keygene.com

M. Causse et al. (eds.), *The Tomato Genome*, Compendium of Plant Genomes,
DOI 10.1007/978-3-662-53389-5_8

The novel sequencing technologies that were invented and commercialized during the course of this effort resulted in the generation of multiple types of sequence datasets. This in turn required development and application of state-of-the-art bioinformatics approaches to process the vast and varied datasets in order to produce a near-complete and high quality genome assembly.

Keywords
Tomato · Genome sequence · Genome assembly · Genome finishing

Genome Sequencing and Assembly

Introduction

Since the elucidation of the structure of DNA in 1953 by Watson and Crick, scientists have put great efforts in unravelling the structure and composition of genomes. In the 1970s, the first DNA sequencing technologies were developed that allowed reconstruction of the precise order of nucleotides within a DNA molecule. Among these, the Sanger sequencing method (Sanger and Nicklen 1977) became the most successful technology and which ultimately enabled the sequencing of the entire genome of a species. In 2001, the first human genome sequence was published which required three billion US dollars and 10 years of work by a large international consortium.

In recent years, various novel sequencing technologies have been developed and successfully applied in whole genome sequencing, notably including the 454 pyrosequencing and the Illumina technologies. Collectively these are called "Next Generation Sequencing (NGS) technologies". Although NGS technologies have made whole genome sequencing less laborious and several orders of magnitude faster and cheaper, the computational process of reconstructing a genome by assembling very large amounts of raw sequencing data into long DNA fragments, such as chromosomes, still poses great challenges. The root of this problem lies, on the one hand, in the complexity of the genome sequence itself, which is often highly repetitive over short and long distances (polyploidy) and heterozygous to varying extents. On the other hand, the sequence reads from which a genome needs to be reconstructed are extremely short compared to the size of a genome and they typically contain experimental errors, which hampers the process of identifying unambiguous overlaps between short fragments.

Plant species in particular have highly complex genomes comprising, many and often large repeats and high rates of heterozygosity. The largest eukaryotic genome known to date was identified in a plant (*Paris japonica*; PELLICER et al. 2010): it's ~150 Gb haploid genome is almost 50 times larger than the human genome. Many plant species are polyploid (Meyers and Levin 2006) and carry large gene families and abundant pseudogenes in their genome, resulting from genome duplication events and proliferation of transponsons. For example, the maize genome consists of at least 75 % repetitive sequences, most of which are mobile DNA elements (Meyers et al. 2001; Schnable et al. 2009). Besides these, some genomes or genomic regions have high GC-biases. All of these factors confound the process of genome sequencing and assembly.

All current sequencing technologies have their limitations either in read length, base accuracy or throughput. The Illumina sequencing technology can produce extremely large numbers of sequence reads per run (2 billion single or paired-end reads) with a relatively low error rate (below ~0.4 %) (Quail et al. 2012), while the read length with a maximum of 250 nucleotides is relatively short. The 454 sequencing technology (Margulies et al. 2005) has a much lower throughput and generates reads with up to 1000 nucleotides but homopolymer sequencing errors pose a problem. In contrast to these 2 s generation sequencing technologies, the recently developed single molecule PacBio sequencing technology can generate relatively long reads (~8.5 kb on average), but the throughout is relatively low and the sequencing error rate is very high (~13 %) (Quail et al. 2012).

The limitations of the sequencing technologies primarily impact the power to assemble repetitive regions of a genome. To help overcome these problems, paired-end and mate-pair sequencing methods, with which reads from both ends of a DNA fragment can be generated, have been developed by Illumina and 454. The two reads of a pair with a known approximate distance can bridge repetitive regions. Despite this improvement, the huge amounts of reads generated by the NGS platforms pose computational challenges in their own right. They require extensive IT resources in terms of storage capacity and compute power (memory and processors). In recent years, a number of software programs have been developed to analyze genomes from very large data volumes and different genome sequencing and assembly strategies have been devised that specifically address biological, experimental and computational challenges.

In this chapter we describe current genome sequencing and assembly strategies, and genome quality evaluation and finishing methods. Furthermore we illustrate some of these in the context of the international tomato reference genome assembly project. In addition, we also provide an outlook on the future developments in genome sequencing technologies.

Genome Sequencing Strategies

Genome sequencing is a technology to reveal the order of nucleotides of each chromosome in the genome of a species. The two major genome sequencing strategies are whole genome shotgun (WGS) sequencing and clone-based sequencing. Besides these two strategies, recently a few local sequencing and assembly strategies have shown advantages of reducing the complexity of a genome in whole genome sequencing approaches.

In the WGS approach, a large number of copies of genomic DNA is randomly sheared into smaller and partially overlapping fragments. Typically, the sheared fragments are separated by size by running them in a gel after which fragments with a required size are extracted from the gel and purified. These fragments are then sequenced using any of the current sequencing technologies. Based on the overlaps between the sequences generated from the fragments, the short sequences can be assembled into longer contiguous sequences, so-called "contigs".

Clone-based sequencing strategies divide the whole genome into a number of large overlapping fragments that can be inserted into an appropriated cloning vector, such as Bacterial Artificial Chromosomes (BACs), cosmids, fosmids or Yeast Artificial Chromosomes (YACs). The overlapping fragments are generated by random shearing or by partially digesting the

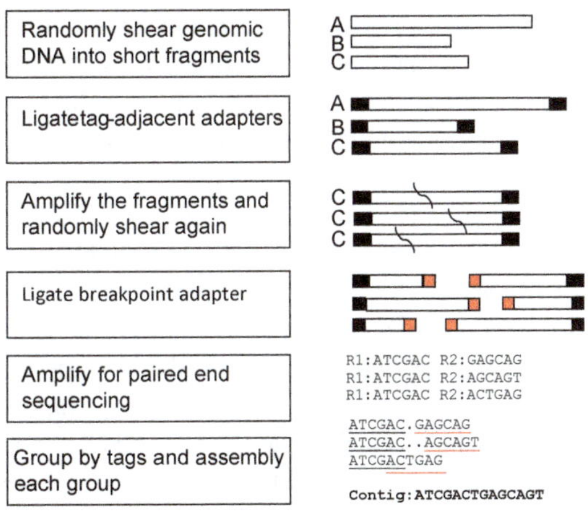

Fig. 8.1 A local sequencing and assembly strategy according to Hiatt et al. (2010). Genomic DNA is randomly sheared into short fragments, indicated "*A*", "*B*" and "*C*"; tag-adjacent adapters are ligated to "*A*", "*B*" and "*C*"; the fragments are amplified and randomly sheared again (examplified for fragment "*C*"); breakpoint adapters are ligated to the breakpoints and the fragments are amplified from the tag-adjacent and breakpoint adapters for Illumina sequencing to generate paired-end reads (*R*1 and *R*2); paired reads are grouped by *R*1 (*black underlines*) and assembled

DNA with a restriction enzyme, in both approaches followed by selection of fragments with a particular size range compatible with the cloning vector. After insertion into the cloning vector, the constructs are transformed into a host organism, such as *E. coli*, in which they are replicated and stored. Each host cell contains one unique cloned fragment of the original genome and together they represent a genomic library. All clones in the library can be characterized using a fingerprinting technology, e.g., the Whole Genome Profiling method (WGP^{TM}) (van Oeveren et al. 2011). Based on the fingerprint information, the clones can be assembled into contigs using software, such as FPC (Soderlund et al. 1997), to produce a physical map of the underlying genome. From the physical map, a minimal tiling path can be designed that will comprise the minimum number of clones covering the maximum part of the genome. Clones from the minimal tiling path are then selected for sequencing, which for each clone individually can be done using the WGS approach.

Local sequencing and assembly strategies partition whole genome sequencing and assembly into small regions, which reduces the overall assembly problem into many small subproblems. One of these approaches is the tag-directed sequencing method developed by Hiatt et al. (2010), in which genomic DNA is randomly partitioned into small fragments ($\sim$500 bp), followed by ligation of tag-adjacent adapters on both sides of the fragments. The approach is illustrated in Fig. 8.1. In more detail, the fragments are amplified and sheared, the breakpoints are ligated to a breakpoint-adjacent adapter. Segments between the tag-adjacent adapter and breakpoint-adjacent adapter are amplified for sequencing using Illumina paired-end sequencing. One read of a pair corresponds to the tag and the other read of the pair to the breakpoint read from the random shearing. The breakpoint reads are grouped by their tag reads and each group can be assembled into either end of the fragments. As an alternative for random shearing, another approach was developed in which genomic DNA is partitioned using restriction enzymes to create a series of reduced representation libraries from different fragment sizes (Young et al. 2010). Besides these two strategies, Keygene developed a paired-end WGP approach, an integrated strategy of local sequencing and assembly based on a BAC library and a physical mapping approach using WGP (van Oeveren et al. 2011). The approach differs from the previously described approaches in that the BAC libraries are digested by one or a few restriction enzyme(s) independently and the local assemblies can be linked to the WGP map directly.

In general, the WGS approach entails massively parallel sequencing of overlapping DNA fragments and assembly of these into longer contigs. In comparison with the clone-based sequencing approach and the local sequencing and assembly approach, WGS is relatively straightforward and usually more cost-effective. However, the assembly step of WGS is often greatly hampered by repetitive and complex regions in a genome. While generating the recombinant clones is relatively slow, labour intensive and expensive in the clone-based method, sequence assembly per se is relatively straightforward, because it is much less affected by the repetitiveness and complexity of the genome. Like the clone-based approach, the local sequencing and assembly approach requires relatively labour intensive sample preparation steps in comparison with the WGS approach. Local assembly, however, requires little computational resources in terms CPU, memory and I/O (read and write throughput) per *individual* assembly, but it needs to be massively parallelized in order to handle the total set of *individual* assemblies.

Library Preparation Protocols

The NGS era brought several new or improved library preparation protocols. Illumina has further developed the paired-end sequencing protocol, which was originally invented by Sanger. The Sanger paired-end sequencing method is a clone-based approach, which requires a cloning step with ligating adaptor sequences containing restriction sites for endonucleases. The Illumina paired-end sequencing protocol is clone-free and generates read pairs from fragments with approximately fixed distances, typically shorter than 1 kb. A special application of paired-end reads is to use them to generate so-called "pseudoreads". For example, paired-end reads of 2 × 100 bp produced from fragments with a size of approximately 180 bp can be used to generate pseudoreads of 180 bp, based on an overlap of approximately 20 bp. A program called "FLASH" (Magoč and Salzberg 2011) was developed to construct such pseudoreads. In contrast, 454 paired-end sequencing can generate read pairs with an approximate distance of up to 20 kb. In the protocol, the two ends of the fragments are connected with a bioitin labelled linker. The circularized DNA fragments are randomly sheared and biotin-labelled fragments are selected. The selected fragments are sequenced using the Roche 454 sequencer, which produces single reads covering the linker sequences. The linker sequences can be identified and the single reads then can be split into two reads to represent a pair corresponding to the ends of the original fragment.

To generate read pairs from fragments lager than 1 kb, Illumina developed a so-called "mate pair" protocol. The original Illumina mate-pair protocol used biotin to label both sides of the fragments. Like in the 454 protocol, the fragments need to be circularized and randomly sheared and the biotin-labelled fragments are selected for paired-end sequencing. This approach turned out to cause problems, because without linkers, the exact sequences corresponding to fragment ends could not be identified. To solve the issue, an identifiable junction sequence called "cre-lox" was introduced to link the two ends of fragments and a bioinformatics tool called DeLoxer can be used to identify and remove the cre-lox adapters (Van Nieuwerburgh et al. 2012). Similar to the cre-lox method, a recently released Illumina-Nextera mate-pair protocol uses identifiable junction sequences, which can be identified and processed using bioinformatics tools from the Biopieces pipeline (www.biopieces.org) or AdapterRemoval (Lindgreen 2012).

Compared to single shotgun reads, paired-end reads and mate-pair reads not only have double read length, more importantly the approximate distances of read pairs from a library can be used to constrain the assembly. Read pairs can be forced to be assembled only within expected distances, which can greatly improve the assembly of repetitive regions (Wetzel et al. 2011). Almost all recently developed assemblers make use of paired-end and mate-pair reads to constrain the assembly process, including the widely used prgrams SOAPdenovo (Luo et al. 2012) and ALLPATHS-LG (Gnerre et al. 2011).

Sequence Pre-Processing

Sequences generated with different library preparation protocols and different sequencing technologies all contain sequencing errors. It is well known that 454 reads contain homopolymer errors, while Illumina reads have a low quality at the end and both ends of Sanger reads are normally of low quality. Reptile (Yang et al. 2010) is one of various programs that can correct sequencing errors. It builds an index of so-called k-mers, short overlapping substrings occurring in the original sequence reads. It uses these to identify k-mers containing erroneous bases and constructs multiple sequence alignments based on the k-mers and their neighbouring k-mers for error corrections. The tool Quake (Kelley et al. 2010) also enables read error correction based on k-mers and does so by making use of base quality information in the process. Instead of using k-mers, the tool HiTEC (Ilie et al. 2011) builds a suffix array for all reads and uses statistical analysis to find and correct errors.

Besides sequencing errors, both 454 reads and Illumina reads contain clone duplications. These pertain to exact copies of the same fragment, which are amplified and sequenced hundreds to millions times. These reads originate from only one initial DNA fragment, but take up more space than one read and accordingly more computational resources during downstream bioinformatics analysis.

Sequences generated from (BAC) clones can be contaminated with sequences derived from the cloning vector and from the bacterial host *E. coli*. Such reads need to be trimmed before further analysis. Besides this, in plant genome sequencing, the extracted genomic DNA may contain chloroplast and mitochondrial DNA, even after purification of nuclear DNA. Because chloroplast and mitochondrial genomes have sequences homologous to nuclear DNA, they may result in misassembles of the nuclear genome, and it is therefore common practice to remove these sequences prior to assembly.

Genome Assembly Methods

Sequence assembly is the approach to reconstruct genome sequences using a set of short sequences that can be generated by shotgun sequencing technologies. Advances in these sequencing technologies have driven the rapid development of new sequence assembly tools. All tools rely on the assumption that sequence reads derived from the same locus on the chromosome should be identical. The identical and overlapping parts between the reads are used to stitch them together. However, in reality, this assumption does not always hold. Sequence reads derived from the same locus can contain different bases due to the occurrence of sequencing errors, and sequences from different loci can be (nearly) identical because they stem from duplicated copies of a locus or from highly identical repetitive sequences. In order to deal with sequencing errors, repetitive sequences and at the same time taking into account sequencing characteristics from different technology platforms, various assembly algorithms have been developed.

The earliest algorithm used in sequence assembly employed a simple 'greedy' approach (Bonfield et al. 1995; Sutton et al. 1995; Ewing and Green 1998) in which shortest common super-sequences (Timkovsky 1993) in a set of random, short overlapping sequence reads are reconstructed by "greedily" joining those reads that are most similar to each other. This approach was soon followed by the overlap-layout-consensus (OLC) method (Myers et al. 2000) in which similarity relations between reads are represented by a graph. Sequence reads are represented by nodes and their overlaps with other reads are edges. The assembly problem can thus be transformed to the problem of finding a path through the graph that contains all the nodes, which can be solved by the application of more common graph theory techniques. Both greedy and OLC approaches were designed for assembly of the relatively long sequence reads ($\sim$1000 bp) that are produced by the classical Sanger sequencing method (Bonfield et al. 1995; Sutton et al. 1995; Ewing and Green 1998; Myers et al. 2000). Both algorithms require an all-versus-all pair-wise comparison between sequence reads and

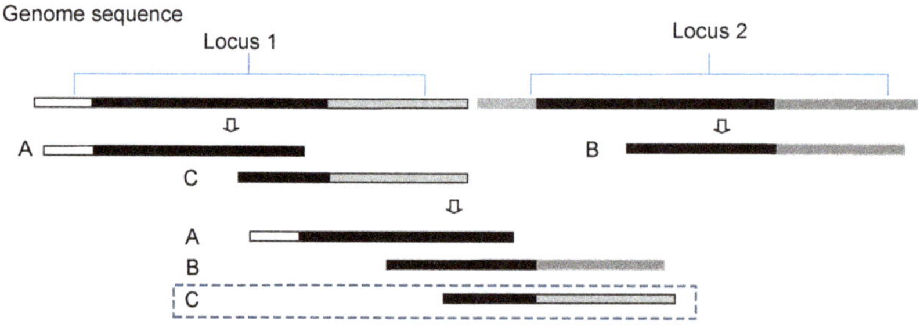

Fig. 8.2 An example of misassembly in the greedy assembly approach. From a duplicated region in the genome (locus 1 and 2, indicated by *solid black blocks*) three reads were produced through shotgun sequencing: reads *A* and *C* derive from locus 1, and read *B* from locus 2. The greedy algorithm first merges sequence *A* and *B* together because they share the longest overlap, to the exclusion of sequence *C*, which has a shorter overlap with sequence *A*. The contig generated from sequence *A* and *B* represents a misassembly of the two loci

use the overlap information between sequences for constructing the assembly. However, compared to classical Sanger sequencing, the datasets produced by next generation sequencing technologies consist of much shorter reads that come in volumes that prohibit computation of all-versus-all similarity and overlap. This computational limitation has driven the development of entirely new assembly algorithms, among which the "de Bruijn graph" method is the most used and powerful one (Pevzner et al. 2001). The de Bruijn graph approach is similar to the OLC approach in that it relies on the construction of a graph, starting from a population of sequence reads. Unlike OLC, however, in which the graph is based on sequence overlap information of the entire read, the de Bruijn graph uses *k*-mers to construct a path through a graph. *k*-mers are substrings with a fixed length of *k* nucleotides. A de Bruijn graph is therefore also called "*k*-mer" graph. For each distinct k-mer the frequency in the total population of reads is counted. Redundancy in the read population is compressed which enables computation on the much larger datasets produced by NGS technologies. The recently developed "string graph" method takes advantage of both the de Bruijn graph and the OLC approach (Myers 2005; Simpson and Durbin 2010; Gonnella and Kurtz 2012; Simpson and Durbin 2012). The initial string graph uses intervals between two sequences as nodes and boundaries between two sequences as edges.

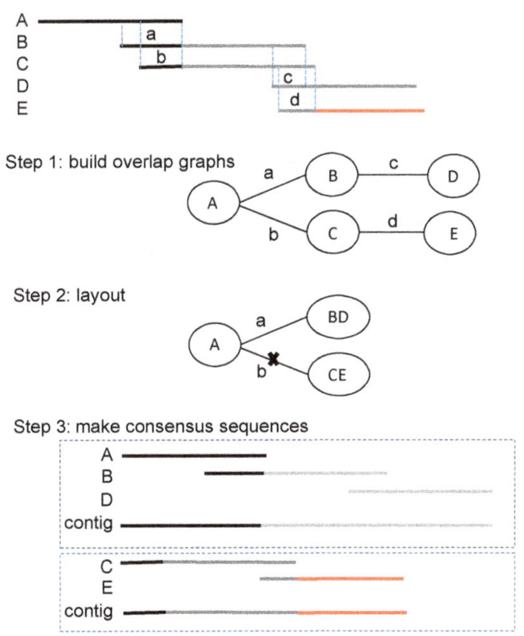

Fig. 8.3 Overlap-layout-consensus (OLC) approach. The overlap between sequence *A* and *B* is indicated by "*a*", the overlap between *A* and *C* by "*b*", the overlap between sequence *B* and *D* by "*c*" and the overlap between sequence *C* and *E* by "*d*". On basis of overlaps, a graph is built with nodes representing sequences and edges representing overlaps (step 1). In step 2, nodes *B* and *D* connected by a unique edge are compressed into the single node *BD*, and similarly node *C* and *E* are compressed in the node CE. Next, the optimal paths are identified: path *ABD* and path *CE*. In the final step, the consensus sequence is called based on multiple sequence alignment of sequences in each path

Greedy algorithms provide the simplest method for genome assembly. They use the overlap information between sequences which are calculated from pair-wise comparisons. Sequences with the best overlaps are joined first, and iteratively merge with other sequences as long as the sequences do not conflict with the already constructed assembly group (see Fig. 8.2). The consensus sequences, also called contigs (contiguous sequences), are computed based on the sequence depth from the merged reads in each assembly group. If per-base sequence-quality information is available, the quality scores are used also for the computation of the consensus sequences. A number of the earliest available assembly software tools were developed based on the greedy algorithm, including widely used tools such as Phrap, Cap3 and TIGR assembler (Bonfield et al. 1995; Ewing and Green 1998; Huang and Madan 1999; Luo et al. 2012). More recently developed genome assemblers based on a greedy algorithm include SSAKE and VCAKE (Warren et al. 2007; Jeck et al. 2007). Although each of these tools implemented the algorithm differently, they all employ two important parameters in the assembly process: the minimum overlap length and the minimum similarity between the overlapping regions of reads. Since the greedy algorithm merges the sequences with the best overlaps, it may not lead to a global optimal solution. Especially sequences coming from repetitive regions may be misassembled together. Figure 8.2

Table 8.1 Assembly programs

Program name	Approach used	Input data	References
Phrap	Greedy algorithm	FASTA reads with or without quality information, Sanger ACE file	Ewing and Green (1998)
Cap3	Greedy algorithm	FASTA reads with or without quality information	Huang and Madan (1999)
Newbler	Overlap-layout-consensus algorithm	FASTA reads with the limitation of max read length of 1999nt, 454 sff files	Margulies et al. (2005)
Celera assembler	Overlap-layout-consensus algorithm	Fastq files generated from Illumina and PacBio, 454 sff files	Myers et al. (2000)
ALLPATHS-LG	De Bruijn graph algorithm	Illumina Fastq files, PacBio reads in FASTA	Gnerre et al. (2011)
SOAPdenovo	De Bruijn graph algorithm	Illumina Fastq files, FASTA reads	Luo et al. (2012)
Velvet	De Bruijn graph algorithm	Illumina Fastq files	Zerbino and Birney (2008)
Abyss	De Bruijn graph algorithm	Illumina Fastq files	Simpson et al. (2009)
CLC bio	De Bruijn graph algorithm	Illumina Fastq files	A commercial assembler released by CLC

illustrates this problem. To avoid the influence of repetitive sequences, many tools allow masking of repetitive sequences before they take part in the assembly. The tool "RepeatMasker" (Smit and Green 1996) is frequently used for this task.

Unlike the greedy algorithm, the overlap-layout-consensus (OLC) approach uses the overlap information between all sequences, and initially builds a global overlap graph. OLC assembly normally includes three steps after pair-wise comparison between all sequences. The first step is to build the overlap graph based on all overlap information. In the graph, nodes are sequences and edges represent the overlapping part between sequences (see Fig. 8.3, step 1). As long as the overlaps between sequences comply with the customizable constraints of "minimum overlap length" and "minimum similarity of the overlap", all edges between the sequences will be present. The genome assembly problem then boils down to finding the minimum number of paths through the graph that visit all nodes only once. The solution can be easy but can also be "NP-hard", and thus difficult to solve. If the genome is completely devoid of repetitive sequences, then only the start and end nodes have a single edge while all other nodes have two edges. The graph can be traversed by visiting each node only once. However, in the real world, genomes are usually complex and contain many repetitive sequences of varying lengths and similarities. Finding the optimal paths through a graph built from such a genome is often highly problematic. In fact, the goal of the second step in the OLC approach is to find the optimal paths that cover and traverse all nodes in the graph. It uses a hierarchical approach to compress the overlap graph first. The nodes with unique edges are compressed into one node (see Fig. 8.3, step 2) and the optimal paths can be identified from the compressed graph. However, due to the influence of repetitive sequences, the optimal paths may not be easily discovered and it is often necessary to mask repetitive sequences before constructing an overlap graph. In the final step the consensus sequences are calculated based on all sequences present in the same path. A multiple sequence alignment is usually constructed based on these sequences, and the most reliable bases are discovered from the multiple sequence alignment. If the quality information is available, the base quality will be taken into account. The Celera assembler (Myers et al. 2000) represents one of the most powerful implementations of the OLC method (Table 8.1).

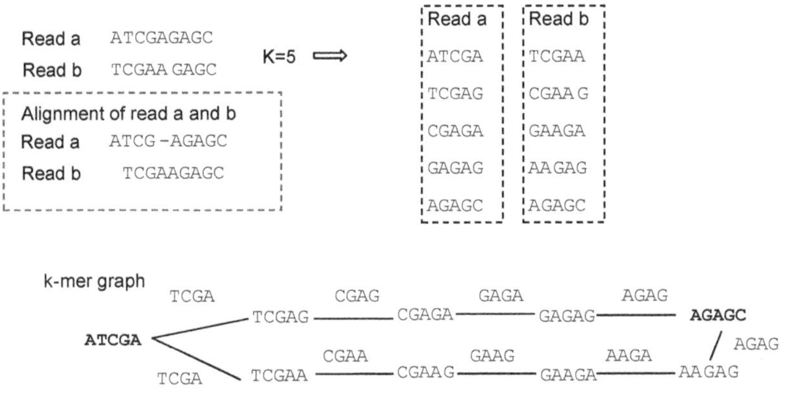

Fig. 8.4 K-mer graph approach. Read *a* and *b* represent two sequences from the same locus, in which read *b* appears to have an insertion (base "*A*" at position 4) based on alignment of the reads. Using a k-mer size of 5, both reads *a* and *b* can be split into five sub-reads, from which a k-mer graph can be constructed, with k-mers as nodes and overlaps as edges. In the graph shown, two paths can be defined that share the same start-node and end-node, and that create a "bubble" caused by the insertion

The Newbler assembler (Margulies et al. 2005) is the most recent tool that implemented this approach and was developed primarily for assembly of 454 sequence datasets.

The *k*-mer graph (de Bruijn graph) method was developed recently to deal with short reads and the extremely large numbers of sequence reads generated by the next generation sequencing technologies. The method partitions all reads into even shorter k-mers (see Fig. 8.4). A k-mer graph is then constructed, considering the k-mers as nodes and overlaps of k-1 nucleotides as edges. Based on the k-mer graph, paths can then be easily read out from a node to the next node following the edges. However, k-mers containing sequencing errors, polymorphisms between alleles, repetitive sequences and sequences derived from homeologous chromosomes in polyploid species, result in "bubbles" or "branches" in the graph that break the linear paths. The frequency distribution of k-mers is often used to identify and remove or "pinch" the bubbles in the graph. This works reasonably well for bubbles caused by k-mers containing sequencing errors because these have significantly lower frequencies than k-mers with correct bases. However, bubbles caused by polymorphisms between alleles, repetitive sequences and homeologous chromosomes cannot be solved in a similarly straightforward way by k-mer frequencies. Most programs therefore break-up paths as soon as these bubbles are encountered. The consensus sequences resulting from the linear paths are then filled up by ambiguous bases (N's) in gaps. The gap size is estimated from the approximate distances of paired-end and mate-pair reads. Some programs use an internal local assembly at the two sides of a gap to close or reduce the gap. The local assemblies use paired-end and mate-pair reads, which have one read of read pairs mapping on the flanking regions of the gap, and the other reads of the pairs are subsequently assembled. Since the k-mer approach partitions a sequence read into tens of k-mers, the full sequence information of the read is lost, although this approach reduces the graph size and memory usage. Some k-mer graph-based assemblies keep the full sequence information and trace them back in the k-mer graph a posteriori and remove some branches caused by small repetitive sequences, relying on the fact that branches are often supported by the original sequence reads.

K-mer size is the most important parameter for k-mer graph-based assemblers. In practise, the k-mer only takes an odd value. Smaller k-mer sizes will compress the data more and generate a smaller k-mer graph, while more bubbles or branches may occur in the graph. In contrast,

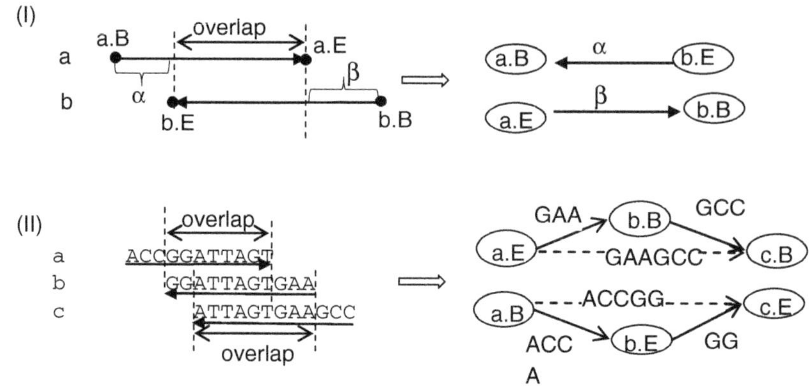

Fig. 8.5 String graph approach. **I** In *left panel*, the overlap between reads *a* and *b* is indicated with the orientation of the overlap (*arrow* header); the overhangs of read *a* and *b* are called "α" and "β". Each read has two ends (read *a* has ends of *a.B* and *a.E*, and read *b* has ends of *b.B* and *b.E*), which are nodes in the string graph shown in the *right panel*. Overhangs of "α" and "β" are two bidirected edges, which are defined according to the rules in the paper of Myers (2005). **II** In *left panel*, read *a* and read *b* overlap each other, read *b* and *c* overlap each other and the overlap between read *a* and *c* exists, but is not necessary to take as an edge, because the relationship between read *a* and *c* can be presented by read $a \rightarrow b \rightarrow c$. The *right figure* shows the bidirectional string graph. The edges between *a.E* and *c.B* and between *a.B* and *c.E* are transitive edges (*dashed lines*)

larger k-mer sizes generate more unique k-mers and larger k-mer graphs with less branches, but they require more memory to store the graph. Whatever k-mer size is used, a substring derived from a sequence read that contains sequencing errors will result in a unique or rare k-mer. Therefore, error correction in the k-mer graph-based approach is an important step.

String graph-based assembly represents a new and recently developed approach (Myers 2005). It uses a novel way to compress the sequence reads and their overlap information. Unlike overlap graphs, nodes in the string graph are ends of reads, instead of whole sequences; edges are overhangs between two overlapping sequences, instead of overlaps (Fig. 8.5a). Because the input data for de novo assembly normally has a very high sequence coverage, storing overhangs instead of overlaps can significantly reduce memory usage. Transitive edges stored in overlap graphs are reduced in string graph (Myers 2005), which also demonstrates in Fig. 8.5b. In this way, the string graph uses less memory than the overlap graph and also contains full sequence information which is otherwise lost in the k-mer graph. The recently developed string graph assembly method SGA uses Ferragina–Manzini index (FM-index) derived from the Burrows–Wheeler transform to efficiently construct the string graph (Simpson and Durbin 2010). This was shown to greatly reduce compute time and memory usage. In addition, the FM-index based compressed data structure was optimized and the assembler employs an error correction method prior to assembly. SGA is to date the only assembler that successfully introduced a string graph-based approach. Simpson and Durbin (2012) showed that SGA produced a comparable result to several widely used assemblers, including SOAPdenovo (Luo et al. 2012), Abyss (Simpson et al. 2009) and Velvet (Zerbino and Birney 2008), using less memory, but relatively long processing time.

Genome assembly has been shown to represent an NP-hard problem. It is difficult to develop algorithms that can solve such problems and all available solutions have their disadvantages and limitations. A greedy algorithm always joins the sequences with the best overlaps, and does not take into account global sequence information; artefacts can therefore arise through local assembly solutions that disregard the more global optimal assembly. Besides, a greedy algorithm requires pair-wise comparison between all sequences, which requires extensive compute time. To address this, some implementations have been parallelized, for example, PCAP as the parallelized modification of cap3 (Huang and Madan 1999). The down side of this solution is that the risk of building assemblies that are optimal only locally is aggravated. Another way to handle this is to use a greedy algorithm based assembler only in conjunction with a suitable experimental sequencing design. For example, the Phrap assembler was used to assemble Illumina reads that were generated in a local sequencing strategy design (Hiatt et al. 2010), also see section on genome sequencing strategies. In the study, Phrap successfully assembled tag-directed Illumina sequences and generated multiple local assemblies of on average 500 nt. Similar to the greedy assembly approach, the OLC approach requires all-versus-all pair-wise comparisons to be performed and which uses all overlap information between sequences and builds a global graph. In addition, it requires large amounts of memory to handle the overlap graphs. In the implementation of OLC in the Celera assembler, the pair-wise comparison is parallelized and the overlap graph is compressed. The k-mer graph algorithm does not need all-vs-all pair-wise comparisons, but it is very sensitive to sequencing errors. Most assemblers have implemented an error correction function to correct sequencing errors prior to assembly. The error correction methods usually use k-mer frequencies. K-mers with low frequencies are nearly identical in sequence to k-mers with high frequencies, indicating that the low frequency k-mers contain sequencing errors, which can be corrected using the high frequency k-mers. Some k-mer based assemblers have been implemented with a sequencing error correction function, for example, ALLPATHS-LG (Gnerre et al. 2011) and SOAPdenovo (Luo et al. 2012). In case assemblers are used that do not comprise an error correction module, it is often recommended to perform error correction directly on the input sequences, prior to feeding them to the assembler. The string graph algorithm is very promising as it can handle both short and long reads efficiently in terms of memory usage. However the implementation needs to be optimized for practical cases to become more efficient in terms of compute time.

Many assemblers were originally designed to handle a single type of data, but recently some have been further developed to handle mixed types of data. These are so-called "hybrid" assemblers with a notable example provided by the Celera assembler. It was designed originally to assemble Sanger reads using the overlap-layout-consensus approach. It was later upgraded to also handle 454, Illumina and most recently, PacBio sequences. Each of the types of data needs to be transformed into a special format which can be recognized by the assembler. Another example is the Newbler assembler, which was first developed to assemble 454 reads, alone or together with classical Sanger reads provided in FASTA format. Newbler was extended to also use Illumina reads. The "hybrid" assembly strategy is to take advantages of the different types of data into one assembly. Illumina reads have short length and modest sequencing error rate ($\sim$0.4 %) and can be generated easily and cheap to extremely high sequence depth. PacBio sequences on the other hand are relatively long, but have a relatively high sequencing error rate ($\sim$13 %). Combining Illumina and PacBio sequences may result in longer and more accurate consensus sequences than can be achieved with each of these types alone, although technical challenges to combine the two types of data remain (Au et al. 2012). A possible solution is implemented in AHA, which is a hybrid assembler developed by PacBio to assemble both PacBio reads and Illumina reads.

Scaffolding

Scaffolding is the process of joining a set of disconnected contigs into continuous long sequences using the distance information contained in paired-end (Roach et al. 1995) and mate-pair reads to link, order and orient the contigs. Almost all recently developed assemblers have built-in functions to use paired-end and mate-pair reads generated from 454 or Illumina for assembly and scaffolding. In addition, a few software tools have been developed that use paired-end and mate-pair reads only for the purpose of scaffolding, for example, Bambus (Pop et al. 2004), SSPACE (Boetzer and Pirovano 2012), MIP (Salmela et al. 2011) and GRASS (Gritsenko et al. 2012). The tools ERANGE (Mortazavi et al. 2008) and L_RNA_scaffolder (Xue et al. 2013) enable scaffolding of contigs by integrating RNA-seq reads into the assembly.

Beyond read-distance based scaffolding (using the distance information of paired-end and mate-pair reads), long range information derived from a physical map, such as sequence-based whole genome physical maps (van Oeveren et al. 2011), optical map and/or genetic maps can be used for higher-level scaffolding and the construction of pseudomolecules that aim to represent entire chromosomes.

Evaluation of Assembly Quality

Different assemblers can be validated and compared with the use of high-quality or gold standard, real or simulated input data and references. The Assemblathon (Earl et al. 2011; Vezzi et al. 2012; Bradnam et al. 2013) and GAGE (Salzberg et al. 2012; Vezzi et al. 2012) projects provide such datasets. The comparisons have demonstrated differences between assemblers and their results and they provide guidance for selecting assemblers that can be expected to be most suitable for specific classes of genome complexity, experimental designs, and types of input data. This highlights the need for evaluation of the quality of an assembly, despite factors such as lack of a good reference genome or absence of a physical or genetic map hampering such evaluations.

Commonly used measures for the description and quality of a genome assembly include the total number of bases assembled (genome coverage), the absolute number of contigs and scaffolds, their average size, and measures such as the N50 size and N50 index of contigs and scaffolds. In essence, these are measures for the continuity and coverage of an assembly, but not necessarily for assembly quality as a measure for the accuracy of the genome reconstruction. In particular the N50 size and index, as statistical assessment for the continuity of a genome assembly, are used in many genome publications and are often compared among genome assemblies. To calculate them, the contigs or scaffolds are sorted by decreasing size. The cumulative size of the contigs or scaffolds that account for more than 50 % of the total assembled size is then calculated and the smallest contig or scaffold in that set is called the N50 contig or scaffold size. The number of contigs or scaffolds accounting for more than 50 % of the total assembly size is called the N50 index. Although powerful as a means to compare assemblies, the N50 size and index should always be considered together with other measures such as the total number of assembled bases and the average contig size (Nagarajan and Pop 2013).

Although assembly quality cannot be accurately evaluated in the absence of a good reference, a few approaches can be used to find reflections of assembly quality. One of these is to map the input sequences back onto the assembled genome. Based on the assumption that the genome was sampled perfectly randomly and provided that sufficient depth of sequencing was performed, the distribution of sequence depth of the mapped reads then provides an indication of assembly quality. Misassembles due to repetitive sequences, haplotype diversity or sequencing errors will lead to aberrant patterns of mapped read depth and representations of the underlying genome. Regions covered evenly and close to the average coverage are likely to have been assembled correctly. On the other hand, regions covered by significantly higher numbers of mapped reads than average indicate a collapse of repeats and regions covered shallowly indicate possible misassembles. If the input data is paired-end or mate-pair, the orientations and distances of the pairs can be used as good indications for assembly quality. Wrong orientations indicate misassembled rearrangements and distances between read pairs that deviate too much from the expected size indicate mis-scaffolding. A further approach to assess assembly quality is to evaluate the integrity with which independently acquired long sequences are mapped to the assembly. EST/transcript sequences, assembled BAC sequences and PacBio sequences can be used to this end if these sequences are accurate. Partial mapping or unmapped sequences indicate incomplete or misassembled regions. Genomic sequences or transcripts from closely related species may also be used to evaluate assembly quality. However, results from such comparisons should be dealt with utmost caution as conflicts need not always represent misassembles but may reflect true differences between two species. If orthologous regions between two closely related species (co-linear regions) differ much more than expected, this may indicate that the assembly quality should be improved.

In the AllPathsLG assembler, a basic assembly validation module and reference-based validation module have been implemented. The modules use the mate-pair reads that are mapped back to the assembly and if available, a reference sequence to detect misassembled contigs and scaffolds and corrects these. Several independent assembly quality assessment tools have been developed. One example is AMOSvalidate (Phillippy et al. 2008), a tool that checks the consistency of an assembly based on the mapped sequence depth and orientations and distances of mapped pairs. In the Assemblathon project (Earl et al. 2011), a probabilistic method for evaluating the assembly (GAV), was used. Recently a tool called "CGAL" (Rahman and Pachter 2013) was developed that uses a likelihood based approach to assess the assembly quality. The likelihood calculation evaluates the uniformity of sequence coverage of the assembly and takes into account errors in sequences, the insert size distribution of the paired-end reads and the unassembled sequences.

Genome Finishing

Almost all currently available de novo assembled crop genomes contain tens of thousands of contigs and scaffolds (Bevan and Uauy 2013). The scaffolds in turn often contain large numbers of gaps. As a result, a typical de novo assembled genome is in fact a rather fragmented and incomplete representation of the underlying genome. This incompleteness greatly impacts downstream genome analyses, such as gene prediction, annotation, variation extraction, whole genome comparison, etc. To contain this impact, not only the upfront generation of a good input dataset and the selection of a suitable assembler are required: the quality of a genome assembly can also be improved after assembly through dedicated "genome finishing" operations.

To reconstruct a complete genome, sequence gaps that remain after assembly should be closed and unlinked contigs should be connected. The traditional approach for gap closure is the chromosome walking approach. This normally included designing primers surrounding gaps and selecting (BAC) clones spanning gaps using primer hybridizations. The selected clones and BACs are then sequenced and merged with the draft sequences (Frohme et al. 2001). Garber et al. (2009) developed a protocol to design primers surrounding gaps, PCR amplicons based on these primers, and 454 sequencing to sequence the products. However, the PCR approach did not work for long repetitive sequences. Recently an approach for gap filling was proposed that uses paired-end and mate-pair sequences (Boetzer and Pirovano 2012; Luo et al. 2012). One read of a pair is mapped to the vicinity of a gap and a local assembly is then executed using unmapped reads in an iterative manner. A drawback of this approach is that local assembly can easily introduce novel misassembles. Because gaps are often surrounded by repetitive regions or regions that are difficult to sequence (high GC content) or assemble, the repetitive sequences from different loci can be misassembled together. The distances of paired-end and mate-pair sequence reads should be used in these cases to constrain the local assemblies. Nevertheless, the approach has been shown to be effective for filling, especially, of smaller sized gaps and it has been accommodated in several assemblers, for instance SOAPdenovo (Luo et al. 2012).

Gaps with sizes larger than the paired-end and mate-pair library insert size are difficult to close because no physical fragments in the library span the gaps. Long sequences produced by a platform such as PacBio have demonstrated their great advantages for gap filling in the study of English et al. (2012). The authors developed the software tool PBjelly, which finds PacBio long sequences that map to either or both ends of a gap and then assembles these sequences to obtain high-quality consensus sequences. PBjelly allows iterative assembly of local regions that surround gaps, until the gaps are closed or no sequences can be mapped for extending the local assemblies. After the process, some gaps can be completely or partly closed; other gaps may not be addressed at all, depending on the sequencing depth, read quality and reference quality. Whole genome shotgun sequencing using the PacBio platform is still relatively expensive, which hampers its application in gap filling. If there is a BAC library available, selecting BACs from a minimum-tiling path may be a good approach for gap filling.

Besides gapfilling, base accuracy in the assembled sequences is also very relevant to consider, although it is often hard to assess because it requires availability of either additional, high-quality sequence data or very high read depths. SEQual (Ronen et al. 2012) is a tool for correcting errors (indels and substitutions) in assembled contigs based on deep read coverage.

In general, developing an effective genome finishing approach remains a considerable challenge and the costs attached to it may significantly exceed the cost of generating the initial whole genome draft sequence.

The Tomato Genome Assembly

Introduction

In this section we illustrate how several of the genome sequencing and assembly technologies described in the previous section have successfully been applied in the sequencing of the tomato reference genome (The Tomato Genome Consortium 2012).

The tomato genome sequencing project was started in 2004 by the Tomato Genome Sequencing Consortium, a multinational team of scientists from 14 countries. The project was launched just after the human and rice genome sequencing projects were completed. At that time, Sanger sequencing using a BAC based physical map strategy was the dominant approach for genome sequencing and assembly of large and complex eukaryotic genomes. The initial plan was to follow that approach and to sequence only the 220 Mb gene-rich euchromatic regions of the genome (estimated size 900 Mb). Over the course of the project, next generation sequencing technologies such as 454, Illumina and SOLiD demonstrated their power and advantages for genome sequencing and assembly over classical approaches, including greatly enhanced throughput and low cost. Soon several plant genomes were sequenced and assembled successfully entirely based on next generation sequencing technologies (Margulies et al. 2005). Therefore, the tomato genome consortium changed the original plan and turned from BAC by BAC Sanger sequencing to a complementary whole genome shotgun sequencing approach using a comprehensive next generation sequencing strategy: 454, Illumina and Solid sequencing technologies were used for whole genome shotgun sequencing and as well some BAC sequencing. This posed a new challenge with respect to genome assembly. At that time, an assembly tool that could take the advantages from all datasets, did not exist and a custom approach was developed in the project. A backbone assembly was produced from 454 and Sanger sequences. The assembly was further scaffolded using BAC and fosmid end sequences. Illumina and Solid shotgun sequences, assembled BAC contigs and a second de novo assembly from the 454 and Sanger reads were used for gap filling and base correction. After this the assembled genome was integrated with the genetic map, two physical maps and BAC fluorescence in situ hybridization (FISH) information. Based on the integrations, structural inconsistencies were discovered and resolved. Furthermore, alien sequences (bacterial contamination and organeller) were identified and removed. The last step was an evaluation of the quality of the genome sequence.

Table 8.2 Tomato sequencing data generated and raw read coverage, using an estimated genome size of a 900 Mbp

Sequencing technologies	Data type	Raw read coverage	Total raw read coverage
Sanger	3 kb paired-end	3.3×	3.6×
	40 kb fosmid ends	0.1×	
	120 kb BAC ends	0.2×	
454	Shotgun	15×	31×
	3 kb mate-pair	8×	
	8 kb mate-pair	4×	
	20 kb mate-pair	3×	
Illumina	300 bp paired-end	70×	82×
	2 kb mate-pair	3×	
	3 kb mate-pair	3×	
	4 kb mate-pair	3×	
	5 kb mate-pair	3×	
SOLiD	Shotgun* (from 7 kb mate-pair)	22×	140×
	1 kb mate-pair	21×	
	4 kb mate-pair	31×	
	8 kb mate-pair	66×	

Shotgun*: was from 7 kb mate-pair, one reads in pairs were generated and the other reads were not

Data Pre-Processing

In the project, a comprehensive whole genome shotgun dataset was generated using Sanger, 454, Illumina and Solid sequencing technologies. Table 8.2 lists the data generated using the various technologies and library types, and provides the read coverage attained using an estimated genome size of 900 Mbp. Sanger sequences generated from BACs and BAC/fosmid end clones containing vector and *E. coli* contaminations were identified and removed by the tool Lucy2 (Li and Chou 2004) and phred/cross_match (Ewing and Green 1998). Furthermore, chloroplast and mitochondrial sequences were identified and removed based on NCBI blast searching. Besides, duplicate reads and sequencing errors and poor quality regions in all types of data were identified and removed.

Read duplicates and homopolymer errors were encountered in the 454 sequencing reads, and caused problems in the assembly. Because tools to pre-process and remove such data were not available at that time, custom built scripts were used for this processing. Homopolymers were compressed and identical reads were identified and the longest reads were retained. After this, homopolymers in the remaining reads were uncompressed. In total, about 26 % of 454 reads were discarded. The majority of discarded reads were clonal duplicates, which were most likely caused by PCR amplification in the library preparation and the emulsion PCR during sequencing. More than twice as many reads were discarded from the 20 kb insert mate-pair libraries than from other insert size libraries. Moreover, for libraries with the same insert size fragments, most reads were discarded from libraries that were sequenced to the highest depth.

The SOLiD reads were trimmed from low quality regions and the remaining reads were then improved using the SOLiD Accuracy Enhance Tool which is based on the error correction algorithm in Euler assembler (Pevzner et al. 2001). Similar to Sanger and 454 reads, read duplicates were identified and removed. Because of the large data volume, instead of read comparison, SOLiD reads were first aligned to the first draft genome, which was generated based on 454 and Sanger reads. Reads that mapped on the same genomic position were identified and only one representative read per position was kept.

De Novo Genome Assembly

The filtered genomic 454 reads (total 23.2× coverage) and Sanger sequences (total 3.5× coverage) from BAC/fosmids ends and 3 k paired-ends, together with previously sequenced BACs were assembled using Newbler v2.3. In the assembly, Sanger sequences were treated as shotgun reads. This assembly resulted in a total of 782 Mb assembled sequence, partitioned in 3,761 scaffolds with an average length of 208 kb. According to the estimated genome size of 900 Mb, the scaffolds covered 87 % of the estimated genome. More than 90 % of the 454 and Sanger sequences were used in the assembly, and the majority of the unassembled reads were identified as repetitive reads by Newbler, whereas a small fraction was treated as singletons and outliers. This indicates that the assembled sequences nearly cover the complete genome.

The filtered 50 bp SOLiD (paired) reads were used for base error correction. These reads were aligned using PASS (Campagna et al. 2009), and putative indels and substitution errors were identified based on the alignment. The substitution errors were corrected if at least three reads confirmed the substitution and if 90 % of all aligned reads supported the substitution. Since most of the indels were expected to occur in homopolymer regions, the filtered SOLiD reads that mapped to homopolymer regions were remapped without allowing mismatches. The correct lengths of homopolymers were thus defined if at least 80 % of the aligned reads were identical on length, which resulted in 42,481 putative errors were corrected. The Illumina reads were used in a second round of base error correction which resulted in a total of 84,344 corrected errors.

Furthermore, the assembled sequences were checked for contamination with *E. coli*, cloning vectors, chloroplast or mitochondrial sequences, which were missed in the pre-processing steps. In total, 17 scaffolds spanning 87 kb were identified by this analysis and removed from the assembly.

After base correction and removing contaminating sequences, structural inconsistencies in the assembly were assessed through comparisons with the genetic map and the WGP physical map. Scaffolds matching to marker sequences located on multiple chromosomes were marked as chimeric scaffolds. This was examined further based on the matched WGP tags on the scaffolds. The breakpoints in the scaffolds were discovered by manual inspection, which resulted in 22 breakpoints in 20 scaffolds.

As the Celera assembler at that moment in time was upgraded to enable handling of 454 data, a second de novo assembly was produced on the same dataset using that assembler. This assembly resulted in a lower number of contigs with larger sizes compared with the first assembly, while the number of scaffolds in the assembly was twice as high as that obtained in the Newbler assembly. Highly contiguous contigs from the second assembly were used to fill gaps in the first assembly. The sequences surrounding gaps were used to blast against the Celera contigs. Using this approach, 3095 gaps in the Newbler assembly were filled with sequences from the Celera assembly.

The improved scaffolds were further scaffolded using 135,271 BAC and 64,722 fosmid end sequences using Bambus (Pop et al. 2004). This resulted in a remarkably small N95 index of the assembly of 73 scaffolds.

Previously assembled BAC sequences were integrated into the assembly based on a Megablast analysis. The high-quality BAC contigs were used to replace the matched regions of the assembled sequences if the BAC contigs were uniquely matched to the assembled sequences. As a result, 2597 gaps within scaffolds were closed. During the process, some small scaffolds were replaced with their linked assembled BAC sequences. After the process, the final assembly contained 91 scaffolds spanning 760 Mb.

Map Integration

In the project, two physical maps, a SNaPshot fingerprinting and Keygene's WGP map, were generated for scaffolding the de novo assembly. Subsequently, the scaffolds were assigned to their corresponding positions on the chromosomes through integration with a high-density genetic map and a genome-wide cytogenetic map.

Through the matched sequence-based WGP tags on the scaffolds, the scaffolds were linked to WGP contigs. Compared to the sequence assembly of 95 % of all bases into only 73 WGS scaffolds, 95 % of the BACs assembled into 1674 WGP contigs and 1217 SNaPshot contigs. The sequence assembly was thus more contiguous than the physical maps, and the contribution of the physical maps to the final assembly was modest. In total, six pairs of scaffolds were linked through the physical maps: two pairs through the WGP map, and four pairs through the SNaPshot map.

Subsequently, the scaffolds were ordered and oriented on the chromosomes based on the genetic and cytogenetic maps. The maker sequences from the genetic map and available BAC sequences of BACs used in FISH were aligned to the assembled scaffolds using BLASTN. Through the matched marker sequences and BAC sequences, the scaffolds were assigned to their corresponding chromosomes. The scaffolds were oriented based on the order of the matched markers along the genetic map and BAC-FISH. If conflicts on the orientations between genetic map and BAC-FISH data were found, the information deriving from the genetic map was used as leading. Following this approach, 53 scaffolds covering 594 Mb were assigned to chromosomes with orientation, while 38 scaffolds were assigned to chromosomes without orientation. As a result, the final integrated assembly consisted of 12 chromosomal pseudomolecules spanning 760 Mb deriving from 91 scaffolds and the remaining 22 Mb assembled bases could not be anchored to any chromosomes.

Evaluation of the Assembly

The structural correctness of the final assembly was examined using the alignment of SOLiD mate-pair sequences and Sanger BAC and fosmid end sequences. The mapped Sanger read pairs at the expected distance and the distribution of the mapped SOLiD mate-pairs provided insight into the structural correctness of the assembly. In total, less than 0.1 % of the Sanger BAC and fosmid end sequences showed inconsistencies with the assembly. Only 34 putative misassembled regions were identified using SOLiD 1 kb and 8 kb mate-pairs. Moreover, WGP BAC contigs were aligned to the 12 assembled pseudomolecules through WGP tags. Approximately 97 % of the BAC contigs were collinear with the pseudomolecules.

The per-base accuracy was evaluated through the alignment of the assembled Sanger BAC contigs against the final assembly. 117 Mb of non-redundant BAC contigs were mapped to the final assembly. The matched BAC contigs were realigned accurately to the corresponding regions of the final assembly using BLASTZ (Schwartz et al. 2003). The examination revealed one substitution error per 29.4 kb and one indel error per 6.4 kb. Furthermore, the WGP tags and 265,234 tomato ESTs were aligned to the pseudomolecules. Approximately 98 % of both WGP tags and ESTs were aligned to the genome sequences with 100 % identity and at least 97 % identity, respectively. This illustrates the very high-quality, correctness, consistency and near-completeness of the tomato genome assembly.

Summary

The tomato genome sequencing project exploited datasets generated by various sequencing methods and technology platforms. The very large sizes of BAC and whole genome shotgun datasets and the availability of long-jump reads (454 and Sanger) provided great sequencing depths, base continuity and their extraordinarily high quality. Besides these datasets, two physical maps, a high-density genetic map and a cytogenetic map contributed to the quality of the final assembly. On one hand, the comprehensive datasets covered almost all information that was necessary to reconstruct the genome to a very high coverage and contiguity; on the other hand, it caused a complex bioinformatics problem of finding the optimal route and order for exploiting the data. The resulting quality of the tomato reference genome generated by the tomato genome consortium is currently among the highest on the list of sequenced crop genomes (Bevan and Uauy 2013).

Outlook

DNA sequencing technology has come to play a most essential role in nearly all aspects of research related to human health and food (improving animal/plant breeding). From the 1970s, the technology has witnessed an extraordinary pace of development, in particular in the last decade with the appearance of next generation techniques. All of these developments have made sequencing several orders of magnitude cheaper, easier and faster. We are now able to produce draft sequences of relatively complex genomes to useful accuracy, contiguity and coverage. However, many challenges remain in order to achieve completeness of genome sequences, in particular for highly complex, repetitive, polyploid and heterozygous plant genomes. In the near future, one crucially important factor in tackling the complexity of the genome sequencing and assembly problem is likely to come from "sequence read length".

None of the currently available technologies is able to produce high accurate reads of such a length that they can encompass most naturally occurring repeat stretches in genomes. The parameter of read length is currently championed by the single molecule sequencing technology developed by Pacific Biosciences. However, although read lengths of >25 kb can be achieved, this technology still suffers from a high sequencing error rate and is relatively costly. We expect that in the next 5–10 years, sequencing technolgies will dramatically improve and novel technologies will appear with which chromosomes and genomes can be sequenced in their entirity, with high accuracy and nearly complete contiguity and coverage. Long read lengths will enable us to reduce the problem of genome assembly from jigsaw puzzles with more than a million pieces to puzzles with less than a thousand pieces, and high read accuracy will reduce the complexity of the puzzle. Push-button sequencing and assembly of complex plant genomes will usher in a new era in plant genomics. High-quality reference genome sequences will improve all downstream analyses, such as genome annotation, the inference of gene regulatory networks, and the effects of sequence variation and haplotypes on the expression of relevant phenotypes. Plant breeding will benefit enormously from this development, providing breeders with the tools, data and understanding to design new traits and varieties from natural and induced genetic variation in an entirely rationalized and economical manner, and much beyond our current capabilities.

Acknowledgments The WGP™ technology is protected by patents and patent applications owned by Keygene N.V. WGP is a trademark of Keygene N.V.

References

Au KF, Underwood JG, Lee L, Wong WH (2012) Improving PacBio long read accuracy by short read alignment. PLoS One 7:e46679. doi:10.1371/journal.pone.0046679

Bevan MW, Uauy C (2013) Genomics reveals new landscapes for crop improvement. Genome Biol 14:206. doi:10.1186/gb-2013-14-6-206

Boetzer M, Pirovano W (2012) Toward almost closed genomes with GapFiller. Genome Biol 13:R56. doi:10.1186/gb-2012-13-6-r56

Bonfield JK, Smith KF, Staden R (1995) A new DNA sequence assembly program. Nucleic Acids Res 23:4992–4999

Bradnam KR, Fass JN, Alexandrov A et al (2013) Assemblathon 2: evaluating de novo methods of genome assembly in three vertebrate species. Gigascience 2:10. doi:10.1186/2047-217X-2-10

Campagna D, Albiero A, Bilardi A et al (2009) PASS: a program to align short sequences. Bioinformatics 25:967–968. doi:10.1093/bioinformatics/btp087

Earl D, Bradnam K, St. John J et al (2011) Assemblathon 1: a competitive assessment of de novo short read assembly methods. Genome Res 21:2224–2241. doi:10.1101/gr.126599.111

English AC, Richards S, Han Y et al (2012) Mind the gap: upgrading genomes with Pacific Biosciences RS long-read sequencing technology. PLoS One 7: e47768. doi:10.1371/journal.pone.0047768

Ewing B, Green P (1998) Base-calling of automated sequencer traces using phred. II. Error probabilities. Genome Res 8:186–194. doi:10.1101/gr.8.3.175

Frohme M, Camargo AA, Czink C et al (2001) Directed gap closure in large-scale sequencing projects. Genome Res 11:901–903. doi:10.1101/gr.179401

Garber M, Zody MC, Arachchi HM et al (2009) Closing gaps in the human genome using sequencing by synthesis. Genome Biol 10:R60. doi:10.1186/gb-2009-10-6-r60

Gnerre S, Maccallum I, Przybylski D et al (2011) High-quality draft assemblies of mammalian genomes from massively parallel sequence data. Proc Natl Acad Sci USA 108:1513–1518. doi:10.1073/pnas.1017351108

Gonnella G, Kurtz S (2012) Readjoiner: a fast and memory efficient string graph-based sequence assembler. BMC Bioinform 13:82. doi:10.1186/1471-2105-13-82

Gritsenko AA, Nijkamp JF, Reinders MJT, de Ridder D (2012) GRASS: a generic algorithm for scaffolding next-generation sequencing assemblies. Bioinformatics 28:1429–1437. doi:10.1093/bioinformatics/bts175

Hiatt JB, Patwardhan RP, Turner EH et al (2010) Parallel, tag-directed assembly of locally derived short sequence reads. Nat Methods 7:119–122. doi:10.1038/nmeth.1416

Huang X, Madan A (1999) CAP3: a DNA sequence assembly program. Genome Res 9:868–877. doi:10.1101/gr.9.9.868

Ilie L, Fazayeli F, Ilie S (2011) HiTEC: accurate error correction in high-throughput sequencing data. Bioinformatics 27:295–302. doi:10.1093/bioinformatics/btq653

Jeck WR, Reinhardt JA, Baltrus DA et al (2007) Extending assembly of short DNA sequences to handle error. Bioinformatics 23:2942–2944. doi:10.1093/bioinformatics/btm451

Kelley DR, Schatz MC, Salzberg SL (2010) Quake: quality-aware detection and correction of sequencing errors. Genome Biol 11:R116. doi:10.1186/gb-2010-11-11-r116

Li S, Chou H-H (2004) LUCY2: an interactive DNA sequence quality trimming and vector removal tool. Bioinformatics 20:2865–2866. doi:10.1093/bioinformatics/bth302

Lindgreen S (2012) AdapterRemoval: easy cleaning of next generation sequencing reads. BMC Res Notes 5:337. doi:10.1186/1756-0500-5-337

Luo R, Liu B, Xie Y et al (2012) SOAPdenovo2: an empirically improved memory-efficient short-read de novo assembler. Gigascience 1:18. doi:10.1186/2047-217X-1-18

Magoč T, Salzberg SL (2011) FLASH: fast length adjustment of short reads to improve genome assemblies. Bioinformatics 27:2957–2963. doi:10.1093/bioinformatics/btr507

Margulies M, Egholm M, Altman WE et al (2005) Genome sequencing in microfabricated high-density picolitre reactors. Nature 437:376–380. doi:10.1038/nature04726

Meyers BC, Tingey SV, Morgante M (2001) Abundance, distribution, and transcriptional activity of repetitive elements in the maize genome. Genome Res 11:1660–1676. doi:10.1101/gr.188201

Meyers LA, Levin DA (2006) On the abundance of polyploids in flowering plants. Evolution 60:1198–1206. doi:10.1111/j.0014-3820.2006.tb01198.x

Mortazavi A, Williams BA, McCue K et al (2008) Mapping and quantifying mammalian transcriptomes by RNA-Seq. Nat Methods 5:621–628. doi:10.1038/nmeth.1226

Myers EW (2005) The fragment assembly string graph. Bioinformatics 21(Suppl 2):ii79–ii85. doi:10.1093/bioinformatics/bti1114

Myers EW, Sutton GG, Delcher AL et al (2000) A whole-genome assembly of Drosophila. Science 287:2196–2204. doi:10.1126/science.287.5461.2196

Nagarajan N, Pop M (2013) Sequence assembly demystified. Nat Rev Genet 14:157–167. doi:10.1038/nrg3367

Pellicer J, Fay MF, Leitch IJ (2010) The largest eukaryotic genome of them all? Bot J Linn Soc 164:10–15. doi:10.1111/j.1095-8339.2010.01072.x

Pevzner PA, Tang H, Waterman MS (2001) An Eulerian path approach to DNA fragment assembly. Proc Natl Acad Sci USA 98:9748–9753. doi:10.1073/pnas.171285098

Phillippy AM, Schatz MC, Pop M (2008) Genome assembly forensics: finding the elusive mis-assembly. Genome Biol 9:R55. doi:10.1186/gb-2008-9-3-r55

Pop M, Kosack DS, Salzberg SL (2004) Hierarchical scaffolding with Bambus. Genome Res 14:149–159. doi:10.1101/gr.1536204

Quail MA, Smith M, Coupland P et al (2012) A tale of three next generation sequencing platforms: comparison of Ion Torrent, Pacific Biosciences and Illumina MiSeq sequencers. BMC Genom 13:341. doi:10.1186/1471-2164-13-341

Rahman A, Pachter L (2013) CGAL: computing genome assembly likelihoods. Genome Biol 14:R8. doi:10.1186/gb-2013-14-1-r8

Roach JC, Boysen C, Wang K, Hood L (1995) Pairwise end sequencing: a unified approach to genomic mapping and sequencing. Genomics 26:345–353. doi:10.1016/0888-7543(95)80219-C

Ronen R, Boucher C, Chitsaz H, Pevzner P (2012) SEQuel: improving the accuracy of genome assemblies. Bioinformatics 28:i188–i196. doi:10.1093/bioinformatics/bts219

Salmela L, Mäkinen V, Välimäki N et al (2011) Fast scaffolding with small independent mixed integer programs. Bioinformatics 27:3259–3265. doi:10.1093/bioinformatics/btr562

Salzberg SL, Phillippy AM, Zimin A et al (2012) GAGE: A critical evaluation of genome assemblies and assembly algorithms. Genome Res 22:557–567. doi:10.1101/gr.131383.111

Sanger F, Nicklen S (1977) DNA sequencing with chain-terminating. Biochemistry 74:5463–5467

Schnable PS, Ware D, Fulton RS et al (2009) The B73 maize genome: complexity, diversity, and dynamics. Science 80(326):1112–1114. doi:10.1126/science.1178534

Schwartz S, Kent WJ, Smit A et al (2003) Human-mouse alignments with BLASTZ. Genome Res 13:103–107. doi:10.1101/gr.809403

Simpson JT, Durbin R (2012) Efficient de novo assembly of large genomes using compressed data structures. Genome Res 22:549–556. doi:10.1101/gr.126953.111

Simpson JT, Durbin R (2010) Efficient construction of an assembly string graph using the FM-index. Bioinformatics 26:i367–i373. doi:10.1093/bioinformatics/btq217

Simpson JT, Wong K, Jackman SD et al (2009) ABySS: a parallel assembler for short read sequence data. Genome Res 19:1117–1123. doi:10.1101/gr.089532.108

Smit A, Green P (1996) RepeatMasker. http://ftp.genome.washington.edu/RM/RepeatMasker.html

Soderlund C, Longden I, Mott R (1997) FPC: a system for building contigs from restriction fingerprinted clones. Comput Appl Biosci 13:523–535

Sutton GG, White O, Adams MD, Kerlavage AR (1995) TIGR assembler: a new tool for assembling large shotgun sequencing projects. Genome Sci Technol 1:9–19. doi:10.1089/gst.1995.1.9

Timkovsky V (1993) On the approximation of shortest common non-subsequences and supersequences. Technical Report

The Tomato Genome Consortium (2012) The tomato genome sequence provides insights into fleshy fruit evolution. Nature 485:635–641. doi:10.1038/nature11119

Van Nieuwerburgh F, Thompson RC, Ledesma J et al (2012) Illumina mate-paired DNA sequencing-library preparation using Cre-Lox recombination. Nucleic Acids Res 40:e24. doi:10.1093/nar/gkr1000

Van Oeveren J, de Ruiter M, Jesse T et al (2011) Sequence-based physical mapping of complex genomes by whole genome profiling. Genome Res 21:618–625. doi:10.1101/gr.112094.110

Vezzi F, Narzisi G, Mishra B (2012) Reevaluating assembly evaluations with feature response curves: GAGE and assemblathons. PLoS ONE 7:e52210. doi:10.1371/journal.pone.0052210

Warren RL, Sutton GG, Jones SJM, Holt RA (2007) Assembling millions of short DNA sequences using SSAKE. Bioinformatics 23:500–501. doi:10.1093/bioinformatics/btl629

Wetzel J, Kingsford C, Pop M (2011) Assessing the benefits of using mate-pairs to resolve repeats in de novo short-read prokaryotic assemblies. BMC Bioinform 12:95. doi:10.1186/1471-2105-12-95

Xue W, Li J-T, Zhu Y-P et al (2013) L_RNA_scaffolder: scaffolding genomes with transcripts. BMC Genom 14:604. doi:10.1186/1471-2164-14-604

Yang X, Dorman KS, Aluru S (2010) Reptile: representative tiling for short read error correction. Bioinformatics 26:2526–2533. doi:10.1093/bioinformatics/btq468

Young AL, Abaan HO, Zerbino D et al (2010) A new strategy for genome assembly using short sequence reads and reduced representation libraries. Genome Res 20:249–256. doi:10.1101/gr.097956.109

Zerbino DR, Birney E (2008) Velvet: algorithms for de novo short read assembly using de Bruijn graphs. Genome Res 18:821–829. doi:10.1101/gr.074492.107

Annotation of the Tomato Genome

9

Stephane Rombauts

Abstract

The annotation of the tomato genome performed by the iTAG consortium (international Tomato Annotation Group) relied on a pipeline operating as a distributed, worldwide network of resources and experts. It used SGN (http://solgenomics.net/) as a central data repository and exchange node. For the iTAG pipeline, used for tomato and potato, we relied on software, as it has besides its own ab initio prediction capabilities, also an extended flexibility to integrate and combine a high diversity of extrinsic data, and other prediction results from other software. Transcript data of numerous origins were mapped on the genome sequence using several software. The detailed procedure is described.

Keywords

Tomato · Genome annotation · EUGENE · EST · Rnaseq

Introduction

The tomato (*Solanum lycopersicum*) genome sequencing was initiated at the very beginning of the technological revolution that is now referred to as the 'Next Generation Sequencing' (NGS), with 454 piro-sequencing on the forefront. The original project (SOL) for sequencing the tomato genome, described still a sequencing strategy using BAC sequences (bacterial artificial chromosomes), sanger-technology and minimal tiling paths to cover the whole genome. But the efforts, to put together the whole genome of the Heinz 1708 variety of tomato would be of very little use if no annotation would be provided.

The annotation of a genome sequence is the step that brings raw sequence data to a level of biological knowledge (Yandell and Ence 2012) that researchers need for designing experiments and breeders to shape new and better tomato varieties. The annotation of a genome can be very broad, aiming at anchoring all possible available information onto a genome. Here, with the tomato genome, we aimed primarily at providing the best possible gene structures, with a high quality, human readable, functional description for mostly protein coding genes. Both topics are distinct, with the

S. Rombauts (✉)
Univ Ghent VIB, 9052 Ghent, Belgium
e-mail: stephane.rombauts@psb.vib-ugent.be

M. Causse et al. (eds.), *The Tomato Genome*, Compendium of Plant Genomes,
DOI 10.1007/978-3-662-53389-5_9

functional description relying on the quality of the gene exon–intron structure, and the structural modeling of genes relying on the quality of the assembly.

Programs for structural gene prediction are available since almost two decades now, coinciding with the first whole genomes sequenced. Many programs exist and a number of them were developed in the frame of a particular genome project and therefore sometimes dedicated to a (limited) number of organisms. From these, only a handful survived the one project for which they were developed and remain being used, with varying popularity. Among those are FgenesH (Salamov and Solovyev 2000) at Softberry, still commonly used by institutions like the Joint Genome Institute (JGI), GeneMark (different types, Lukashin and Borodovsky 1998), Besemer et al. (2001) maintained at Georgia Tech USA, GeneID by CRG (Barcelona, Parra et al. (2000), AUGUSTUS (Stanke et al. 2006), EuGene (INRA, Toulouse and PSB, VIB-UGent (Schiex et al. 2001; Foissac et al. 2008) and the ENSEMBL pipeline run at the EBI (to name a few, see Table 9.1). Software like CONRAD (DeCaprio et al. 2007), and others, though proposing interesting innovations, found too few followers or were forgotten. With this aspect in mind, one could agree that it is not necessarily the best program that gathers all, but that it is the people that know how to use a particular program at best that sustain the software, and the interest of those same people to work on new genomes. A software is being used by a community is mostly linked to its ease of use, flexibility and mostly ability to train for new organisms. As the nGASP competition showed (Coghlan et al. 2008), no software is outperforming all other software for all genomes, while the number of genomes annotated by one or the other software, indicates that the ease of use is clearly the dominating factor leading to adopting or not a program. The major element, influencing popularity of software, is the training, as some software can be trained more or less easily. It is even so that training software remains the main bottleneck.

Whatever software gathers the highest popularity; they still cannot be used as such. Even if all living creatures (as far as we know) use the same DNA to describe their gene repertoire, each uses its own dialect. Therefore, software needs to be trained, and features specific for an organism need to be captured adequately in the models necessary for ab initio gene prediction. In some cases, closely related organisms can take advantage of prior work by recycling previously built models and parameter settings, but it is likely that specifically trained

Table 9. 1 Underlying architecture for ab initio gene-modeling

Positional weight matrices (PWM)	The simplest MMs are homogeneous zero order MMs which assume that each base occurs independently with a given frequency. Such simple models are often used for non-coding regions
Weight array model (WAM)	An inhomogeneous higher order MM capable of capturing potential dependencies between adjacent positions of a signal
Three-periodic Markov model	Characterize coding sequence. Coding regions are defined by three MMs, one for each position inside a codon
Interpolated Markov model (IMM)	IMMs combine statistics from several MMs, from order zero to a given order k (typically k = 8), according to the information available
Hidden Markov model (HMM)	HMMs allow for insertions and deletions and so variation in signal length
Generalized Hidden Markov model (GHMM)	GHMMs allow a string, rather than a single symbol, as the output of a state
Semi-Markov conditional random field (SMCRF)	A more flexible variation of GHMM which allows a wider range of biological features to be incorporated with fewer technical concerns
Evolutionary Hidden Markov model (EHMM)	EHMMs model molecular evolution as a Markov process in two dimensions: a substitution process over time at each site in the aligned genomes, which is guided by a phylogenetic tree; and a process by which the rate of evolution changes from one site to the next

From Picardi and Pesole (2010)

parameters will outperform any software trained on another, even related, organism. Recycling models or parameter-settings is certainly useful when a gene set for training needs to be created, as, even if not perfect, sketches of gene models can be used to further curate gene structures manually that eventually will be used to fine tune parameters.

The way nowadays software evolve is the degree to which they are able to integrate divers information related to genes, from sources like protein similarity, RNAseq, expression (=coverage), etc., other than the statistical models used for ab initio gene prediction. From the onset of gene prediction a lot of effort was made to model gene features in statistical entities scoring sequences, like coding potential and splice site detection. Different flavors, e.g., of Markov Models (MMs) that captures states of the sequence, were developed from Interpolated MMs combining different orders of MMs to Generalized MMs that capture sequence strings rather than individual nucleotides. Typically MMs are broadly used to distinguish protein-coding regions from non-coding regions (intron, intergenic, UTR; Krogh et al. 1994; Kulp et al. 1996). These efforts were linked to the fact that, back then, not many other genomes were sequenced, and that programs therefore needed to rely mostly on good statistical models. The earliest extrinsic information that was made available was sanger-ESTs (expressed sequence tags) to help structure genes on an anonymous genome sequence (Table 9.2).

Besides MMs, some software rely on third party software (like for splice sites: SpliceMachine (de Groeve et al.), NetGene2 (Brunak et al. 1991; Hebsgaard et al. 1996) or GeneSplicer (Pertea et al. 2001) within Eugene (Foissac et al. 2008; Schiex et al. 2001), mSplicer (Rätsch et al. 2007) for mCode (Schweikert et al. 2009) that also need to be trained prior to be included into the gene-modeler. These signal sensors most of the time use divers methods to extract and score motifs across the genome sequence (those named above use neural networks or Support Vector Machines, e.g., Li and Jiang 2005) (Fig. 9.1).

The flexibility of Eugene is part related to its internal way to score the genomic sequence that needs to be annotated. The data structure, a direct acyclic graph (DAG) collects the scores for every transition between each nucleotide. The content sensors (IMM, BLASTX, ESTs) typically modify the score horizontally along the tracks while the signal sensors allow switching between certain tracks. Currently, Eugene's DAG has over 40 tracks that score the sequence. These tracks can store, e.g., protein similarity to indicate which track relative to the reading frame should be used, while RNAseq will indicate both which parts of the genomic sequence are part of a transcribed gene, regardless of the reading frame, and which splice sites, derived from the junctions, should be followed to switch from an exon/transcript track to an intron track. The combination of the different tracks needs to be trained to assign proper weights that will eventually guide the prediction. The final prediction will be the best scoring path through the different tracks and switches, incorporating as much as possible the provided information. The approach followed leads to predictions that, in theory, should correspond with the current data and knowledge. The danger of using such a method is the amount of noise in the provided data, and the difficulty to filter the better data from the rest.

Nowadays, with the ever-increasing availability of plant genomes and even more animal and prokaryotic genomes in databases, we dispose of a wealth of genes that can be used to hint gene prediction software toward regions that show (high) similarity, and thus should be included in the gene models. Sanger-ESTs, that were generated for the organism being annotated, were used early on with gene finders, and are now replaced by RNAseq (e.g., Tisserant et al. 2011; Mizrachi et al. 2010; Coleman et al. 2010) with the advantage of producing evidence at a much broader scale than ESTs ever did. Besides, ESTs were more expensive and labor intensive to produce, they were also limited, due to the cloning protocol, to the most occurring form of the genes or transcripts, even if alternatives existed. Because of this limitation, evidence for alternative transcripts was hardly seen as reliable unless the number of transcripts that reproducibly could be sequenced was high enough.

The side effect of this growing wealth of information, certainly from the RNAseq side, is that we see much more transcribed regions, and that earlier concepts need to be revised. These

Table 9.2 List of software for gene prediction (not exhaustive)

Program	Web page (http://)	Ab initio + Evidence	References
Genscan	http://genes.mit.edu/GENSCANinfo.html	No	Burge and Karlin (1997)
GeneID	http://genome.crg.es/software/geneid/index.html	EST, proteins, (RNAseq)	Guigó et al. (1992)
SNAP	http://korflab.ucdavis.edu/software.html	No	Korf (2004)
GlimmerHMM	http://ccb.jhu.edu/software/glimmerhmm/	No	Delcher et al. (1999)
GeneMark	http://exon.gatech.edu/GeneMark/	No	Besemer and Borodovsky (2005)
AUGUSTUS	http://bioinf.uni-greifswald.de/augustus/	ESTs, cDNAs, and proteins	Stanke et al. (2006)
SGP2	http://genome.crg.es/software/sgp2/index.html	TBLASTX hits (+ GeneID)	Parra et al. (2003)
GENOMESCAN	http://genes.mit.edu/genomescan.html	BLASTX hits	Yeh et al. (2001)
TWINSCAN	http://mblab.wustl.edu/nscan/submit/	BLASTN hits and ESTs	Gross and Brent (2006a, b)
Eugene	http://eugene.toulouse.inra.fr	ESTs, cDNAs, and proteins, RNAseq + external ab initio	Schiex et al. (2001)
N-SCAN	http://mblab.wustl.edu/nscan/submit/	ESTs, complete genomes	Gross and Brent (2006a, b)
EXOGEAN	www.biologie.ens.fr/dyogen/spip.php?rubrique4&lang=en	ESTs, cDNAs, and proteins	Djebali et al. (2006)
ASPIC	http://150.145.82.212/aspic/aspicgeneid.tar.gz	ESTs and cDNAs (+ GeneID)	Bonizzoni et al. (2005)
FGENESH	www.softberry.com	proteins or cDNA	Salamov and Solovyev (2000)
CONRAD	www.broadinstitute.org/annotation/conrad/	cDNA, other related genome	DeCaprio et al. (2007)
Mappers			
GenomeThreader	http://www.genomethreader.org	cDNA & proteins	Gremme et al. (2005)
GeneWise	http://www.ebi.ac.uk/Tools/psa/genewise/	cDNA & proteins	Birney et al. (2004)
EXONERATE	http://www.ebi.ac.uk/~guy/exonerate/beginner.html	cDNAs	Slater and Birney (2005a, b)
PASA	http://pasa.sourceforge.net	cDNAs	Haas et al. (2003a)
Combiners			
GAZE	www.sanger.ac.uk/Software/analysis/GAZE/	All available + external ab initio	Howe et al. (2002)
JIGSAW	www.cbcb.umd.edu/software/jigsaw/	All available + external ab initio	Allen and Salzberg (2005)
MAKER	www.yandell-lab.org/software/maker.html	All available + external ab initio (in an automated pipeline)	Cantarel et al. (2008)

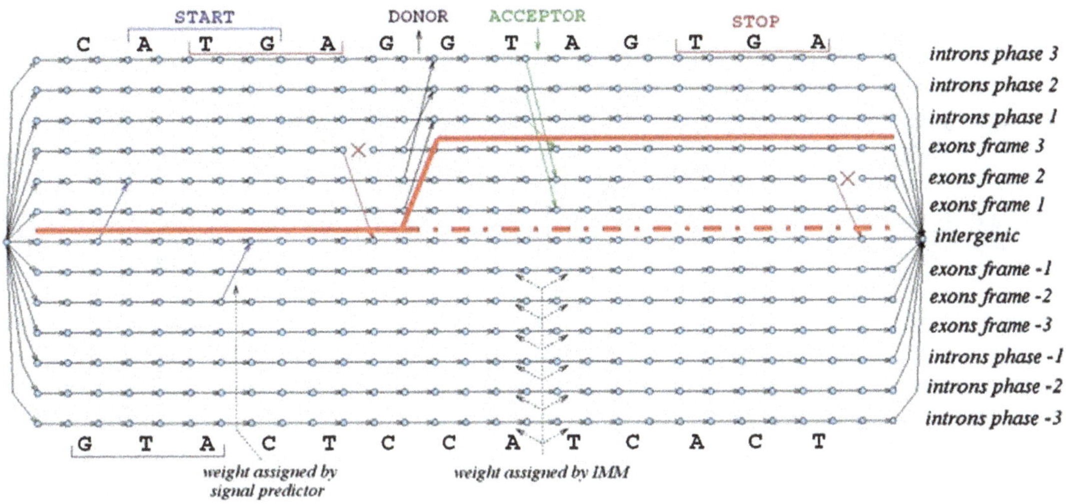

Fig. 9.1 Simplified direct acyclic graph (DAG) structure depicting internal scoring scheme

new concepts have consequences on the gene-callers that need to be able to model our increasing knowledge of objects coded on the chromosomes. Indeed we now start to see the extend of alternatively spliced genes, also in plants, genes in introns or protein-coding genes overlapping with non-coding genes as well as to which extend UTRs overlap neighboring genes. Furthermore, besides protein coding genes, that we were used to predict, we now find more and more evidence for other 'objects' (like non-coding genes) being transcribed. This flood of information becomes an increasing issue when reaching high depth with potential artifacts that reach levels that confuse man and machine. Indeed, with the increased depth, more transcribed regions are appearing, making the assimilation of the data cumbersome. One aspect of the higher coverage (more depth), are reads that cannot be incorporated in any known gene structure. As these reads have the same weight as any other read, part of genuine (protein-) coding gene, they become a noise that needs to be dealt with. But as coverage increases with the technology getting better at sequencing smaller amounts, this noise becomes increasingly difficult to distinguish from genes with low expression, misleading prediction systems. To incorporate RNAseq, coverage needs to be taken into account dynamically, such that local low amounts of reads can still be used for low-expressed genes and not with a general cut-off over the whole dataset.

In the frame of the tomato project, stranded RNAseq (as presented in Passalacqua et al. 2012) was not available yet and the coverage still manageable if only a limited set of well-chosen libraries were used. Using everything leads to an increase of noise, while picking libraries allowed for a broader coverage rather than a deeper coverage. The software used for the iTAG annotation, Eugene, was at the time of the project not suitable to take coverage into account, but still, was the only software available that was flexible enough to incorporate the new type of data easily. The RNAseq that was shared from different tomato projects (not part of the genome project), was all un-stranded (stranded RNAseq became available later Passalacqua et al. 2012) and unpaired for most libraries. The filter applied on the reads, aiming at preserving a maximum of information, while reducing the noise level, was by limiting the input to the junctions spanning introns: knowing where the introns are makes the prediction of exons 'easy'. This filter would include low expressed genes (as long as they had a spliced gene structure), while still avoiding reads coming from transposable elements (TE) and other spuriously mapped reads. This filter also had as consequences that the minimum read length usable was 100 nt (75 nt reads yielded very low amounts of useful data). Furthermore, these reads, besides indicating where the introns are, became strand specific, imposed by the splice sites that can only occur on one or the other strand. Also, as most transposable elements carry intron-less genes we avoid signals that could trigger a gene prediction on transposable element related loci. The only remaining difficulty for which we still had no elegant solution was the occurrence of alternative splicing. Indeed, all possible junctions reported all registered cases of potential splicing events genes were having in the given conditions linked to the RNA sampling. Filtering on coverage was then done using a general threshold keeping the most occurring splice events.

With the RNAseq, the problem of cost, while increasing coverage or depth got solved, and with it, also the access to rare transcription events that was not described before. Among these, we see transcription from regions were no proteins could be decoded, at least as we currently know them, but also splice sites that were rarely seen before suddenly became more abundant. Although with the latter, we need to be cautious as we rely here on mapping software that are limited to known splice sites (an assumption to enhance speed and quality of the mapping, at to some extend a trade-off on the exhaustivity). These assumptions can cause software to mistakenly force mapping toward some splice sites while others should have been reported. The ENCODE project (ENCODE consortium 2004), aiming at an in-depth annotation of a portion of the human genome reported many observations, showing that a larger part of the genome was transcribed. These observations pushed them also to develop new software that would allow more flexibility (e.g., STAR, Dobin et al. 2013) in detecting rare events. Different software exists like GSNAP (Wu and Nacu 2010), Tophat2, built on top of bowtie2 (Kim et al. 2013), or CRAC (Philippe et al. 2013) and GEM (Marco-Sola et al. 2012) all published recently, and advertise their capacity to find rare (previously unreported) splice events by being more versatile in their mapping. But important is to realize that for prediction we rely on the mapping quality of the reads.

When predicting genes, transposable elements (TE) are seen as undesirable side objects that should be kept apart from the (protein coding) gene sets. But, when running RNAseq experiments, TE produce reads too, and these mislead the prediction. Transposable elements, on their own should be therefore carefully annotated as they are part of the features that can be described on a genome sequence. In fact TE should be taken care of prior to the prediction of the protein coding genes. These repeated elements do code for their own proteins, like reverse transcriptases, integrates, etc, necessary for their proliferation, and the presence and order of their proteins defines their type or family. The reason why these elements should be handled before predicting protein coding genes is that the protein-genes from the TE have largely the same statistical features, like codon usage, to code their gene set. And these TE, when occurring in the vicinity of protein coding genes that are not related to any TE, might get fused during the prediction process. But one needs to be careful when masking. The typical characteristic of TE is that they occur in multiple copies, and pipelines like ANGELA (Nussbaumer et al. 2013) or RepeatModeler will use this feature to identify candidate repeats. Unfortunately, recently duplicated genes occurring in multiple copies, showing high similarity, share this characteristic while they should not be counted among the TE. Also some TE, due to multiple insertions can carry a copy of a gene that by no means is involved in the subsistence of the TE. This process, potentially beneficial for the organisms by enabling the enlargement of gene families, should not result in a representative TE that would also mask members of a gene family. It is therefore advisable to hard-mask consensus sequences of transposable elements using genuine genes in order to ensure no hitchhiking genes would be even partially masked.

Working on genomes that result from Next Generation Sequencing (NGS: 454, Ilumina) short reads also have an influence on gene prediction. Indeed, these technologies have some biases like homo-polynucleotides, or difficulties with GC rich areas. Besides these biases related to the technologies, simple errors occur, that translated to gene prediction cause frame shifts and early stop-codons. These errors (indels), are difficult to correct, and in a number of cases should not be corrected as they are real and alter indeed genes (truncated gene structures). An approach to improve a genome assembly and resolve sequencing and assembly errors is by mapping all of the available RNAseq on the genome. Based on simple majority-rules the genome can be transformed, incorporating nucleotides, changing others according to the most occurring nucleotide from the RNAseq reads. Including this cleaning, prior to run gene-callers enhances the gene models resolving issues at the level of genomic sequence.

To functionally annotate the predicted genes, appropriate human readable description need to be generated. Therefore, top-scoring BLAST results (filtered on e-values) are scored combining alignment scores and quality of the hit descriptions. The "quality" of the descriptions results from a lexical scoring of individual "words" based on their frequency in the bulk of collected descriptions. Furthermore filters are included to weight words in function of blacklisted uninformative words, e.g., "hypothetical protein" or "similarity to". To declare a protein as "unknown" a cut-off on the score is used. Descriptions from BLAST that fit predicted GO terms are preferred as these use the standard terminology already. To extend the readable descriptions, InterPro results, if available, are appended. In the case multiple GO terms matches are found, terms most close to the branch-ends in the GO-three were reported as the most informative term, excluding parents.

Procedure

The annotation of the tomato genome performed by the iTAG consortium (international Tomato Annotation Group) relied on a pipeline operating as a distributed, worldwide network of resources and experts. It used SGN (http://solgenomics.net/) as a central data repository and exchange node. For the iTAG pipeline, used for tomato and potato, we relied on Eugene (Schiex et al. 2001; Foissac et al. 2008) as it has besides its own ab initio prediction capabilities, also an extended flexibility to integrate and combine a high diversity of extrinsic data, and other prediction results from other software. The software was trained using ~200 genes, manually curated genes in their genomic context (with their intron-exon structure, interspersed with gene-free intergenic regions). This training set was further shared with other groups to train their respective software. At the time, neither of these other software integrated extrinsic data, and thus all produced pure ab initio results. Besides the other gene-callers that were integrated, protein homologies, EST mapping as well as RNAseq junctions were added to the prediction scheme.

Prior to running the pipeline, genomic sequences need to be masked to avoid that the coding characteristics from genes part of transposable elements would lead to gene models. The de novo repeat collection was build using the ANGELA (Automated Nested Genetic Element Annotation) pipeline (Nussbaumer et al. 2013). This pipeline combines different repeat mining programs to finally output a nonredundant but representative data set that can be used as a library for RepeatMasker. (Smit et al.). The soft-masked (masked sequence in lower case) is then further used for the gene prediction pipeline.

Prior to run the final prediction, extrinsic data needs to be mapped and formatted for the Eugene software.

Proteins can be given to Eugene via 2 plugins:

1. BLASTX for more remote homology that would lead to mismatches or gapped alignments with unreliable ends. This information points mainly to the reading frame that should be used, not so much to delineate exon borders, translation starts, etc.
2. Proteins from more closely related organisms can be mapped more stringently, with proper intron borders and give information on translation starts and/or stops. Here we use software like Genomethreader (Gremme et al.

2005), Genewise (Birney et al. 2004) or others. The higher order information can be given to Eugene via the AnnotaStruct plugin using the GFF-like format and the proper keywords or Sequence Ontology terms (SO). For the tomato genome, the *Arabidopsis thaliana* TAIR10 protein set was mapped on the tomato chromosomes. For potato, TAIR10 as well as iTAG tomato was mapped to inform the prediction system.

Transcript data, whether sanger-ESTs or longer NGS reads (454) was mapped using Genomethreader (Gremme et al. 2005), exonerate (Slater and Birney 2005a, b), PASA (Haas et al. 2003) or other software that is able to map longer transcript sequences, spanning more that one intron, taking splicing donors and acceptors into account.

For the shorter Illumina reads (minimum 100 nt) fast, but reliable mappers are needed. Here we rely much on the correctness of the spliced junctions that will inform the software on the introns and their borders. In our case at the time of the tomato genome annotation we used TopHat (Trapnell et al. 2009).

In any case, whether it is protein or transcript data, quality primes. This means that the software used for mapping should be most reliable and deliver quality above quantity of mapped reads.

Eugene (Schiex et al. 2001; Foissac et al. 2008) has a great flexibility to integrate extrinsic information including results from other gene-callers. These used in conjunction with Eugene were GenelD (Parra et al. 2000), AUGUSTUS (Stanke et al. 2006), TwinScan (Gross and Brent 2006a, b), GeneMark (Lukashin and Borodovsky 1998; Besemer and Borodovsky 2005) and Glimmer (Delcher et al. 1999). Each of these softwares were trained for tomato, and run independently (Fig. 9.2).

The data properly formatted for the specific plugins or in GFF3 format including SO-labels was made available for Eugene. The result of the prediction delivered the coordinates of the protein-coding genes that in theory should represent the consensus genes models, including most of the extrinsic data. EuGene's prediction, followed by manual expert curation, produced a consensus annotation of 34,727 and 35,004 protein encoding genes for the tomato (iTAG v2.3) and the potato nuclear genomes, respectively. The gene coordinates, translated into protein sequence, were further processed for functional annotation. To initiate functional characterization of the predicted protein set OrthoMCL (Li et al. 2003) clustering including several dicots as well as rice, was run, resulting in 8615 gene families shared among all species, 562 tomato-specific gene families and 8886 non-clustered tomato genes (singletons).

The aim of the functional annotation was at delivering reliable rather than exhaustive functional descriptions. Automatic human readable descriptions (AHRD) were assigned to 78 % of tomato proteins, while 22 % were designated as "unknown protein". 42 % or 14,565 annotations fulfilled all quality criteria, being

- BLASTP,
  a. Bit score of the BLASTP result is ≥ 50 and *e*-value is $\leq$e−10.
  b. Overlap of the BLAST result is ≥ 60 %.
  c. Top token score from lexical analysis is ≥ 0.5 (see below).
  Results were obtained from Swissprot, TAIR, and TrEMBL databases
- Domain search results from InterProScan and
- Gene ontology (GO) terms predicted by PhyloFUN.

InterProScan identified 240,027 protein domains of 13,752 distinct domain types. 87 % of the genes (30,148 out of 34,727 genes in total) have been assigned with at least one domain.

Using PhyloFUN (Schoof et al.) and Interpro2GO we could assign in total 39,192 GO terms to 19,662 or 57 % of the 34,727 tomato proteins. The overlap between both methods reached 24 % of the assigned GO terms. While Interpro2GO solely, could retrieve 9082 or 26 % of the GO terms; PhyloFUN added only 6 %. Both tools showed an overlap of 106 unique GO terms. Interpro2GO has been reported as being more sensitive and can annotate more proteins.

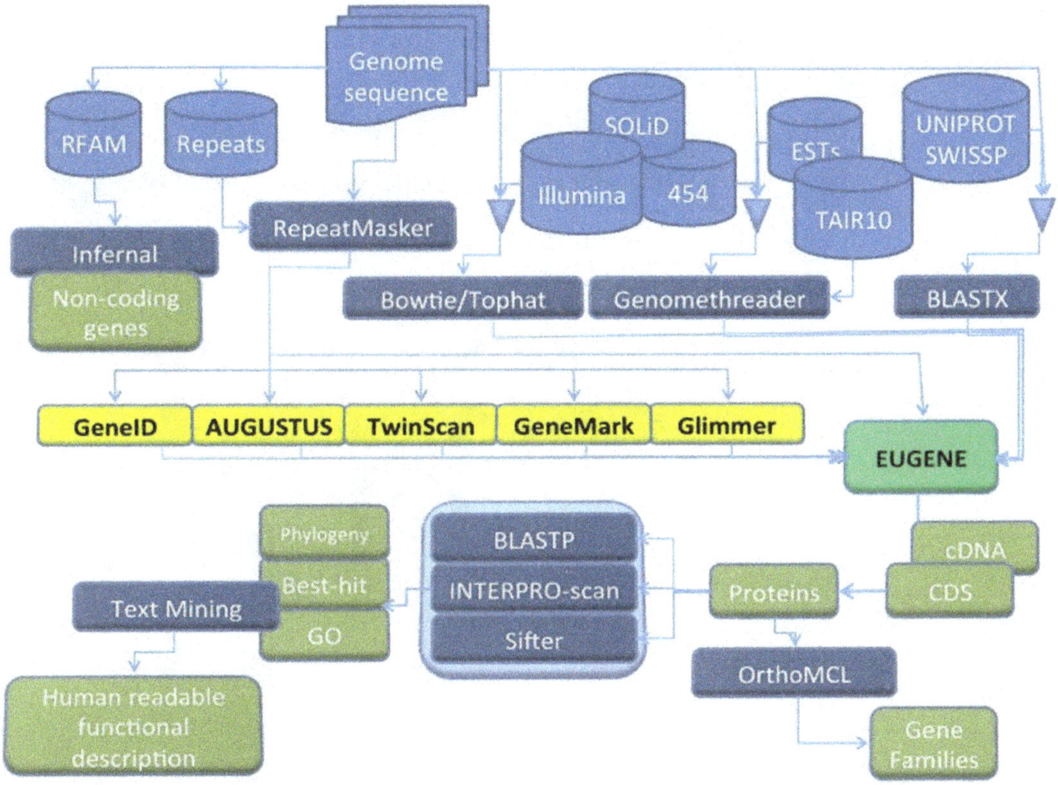

Fig. 9.2 iTAG annotation pipeline scheme

The PhyloFUN pipeline is more specific and can annotate more specific GO terms from further down the GO hierarchy.

As a result from the 34,727 genes predicted in tomato, nearly 10 % or 3371 annotations coincide with the most reliable GO term assigned by PhyloFUN. The presented number seems low due to the more stringent criteria used. Running Blast2GO (Götz et al. 2008) or interpro2GO (Camon et al. 2005), each on their own, returned more genes with GO labels. Often though, the additionally assigned labels remained quite general and less informative to what the function of a gene was.

To propagate and homogenize the better descriptions from the automatically assigned human readable descriptions to members of gene families OrthoMCL clustering was used (Li et al. 2003). As OrthoMCL is prone to errors due to faulty gene-predictions, obtained gene families were combined with selected sets of whole proteomes of largely experimentally verified GO annotations. From these extended gene families, phylogenetic trees were computed to establish proper relations between genes in each OrthoMCL cluster. To establish cases of sound relationships the software SIFTER was used and GO terms transferred according to the phylogenetic tree.

As multiple source of information are combined to produce the gene-description, a quality-tag was introduced to trace back how the description was built by the automated procedure, allowing evaluating the reliability of the description.

These data, the gene models with their functional description, were loaded in an online user interface for community manual curation called ORCAE (bioinformatics.psb.ugent.be/orcae/; Sterck et al. 2012).

The system allows registered users to directly work and modify the data via a web interface. The system keeps a history of all changes made to the data and has a build-in alert system to

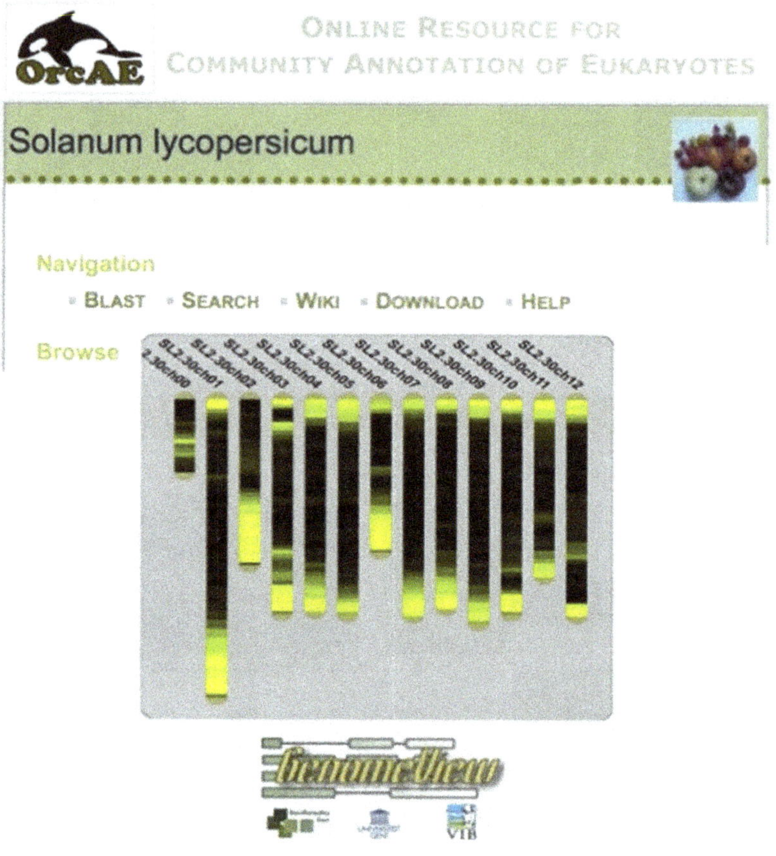

Fig. 9.3 ORCAE interface for manual curation

follow specific genes. The platform displays multiple pre-computed analyses to help expert annotators to evaluate the quality of the gene models and validate the functional description. Some fields can be updated textually, while gene structures can be manipulated through the graphical interface provided by GenomeView (http://genomeview.org; Abeel et al. 2012). Having a graphical interface, showing the underlying sequence and the 3-frame translations avoid many mistakes that could be made using input fields requesting numbers. Also, as frames "jump" with exon borders being modified, a user can easily see whether introduced modifications are possible. Upon modification applied to gene models, the system automatically updates all the pre-computed data allowing further curation (Fig. 9.3).

From the originally 36,287 predicted genes (raw results), still ~1000 genes escaped the masking and were reclassified as transposable elements. A remaining ~500 genes were reclassified as pseudo-genes or discarded in favor of better gene models built by the experts. The current status reports 34,727 predicted genes, from which 1346 genes underwent an expert intervention (3.8 %) whether it was to complement/correct the functional description or correct erratic/incomplete gene models.

Conclusion

Since the first release in 2012, the assembly and the annotation of the tomato genome data was further improved. More sequence data has been

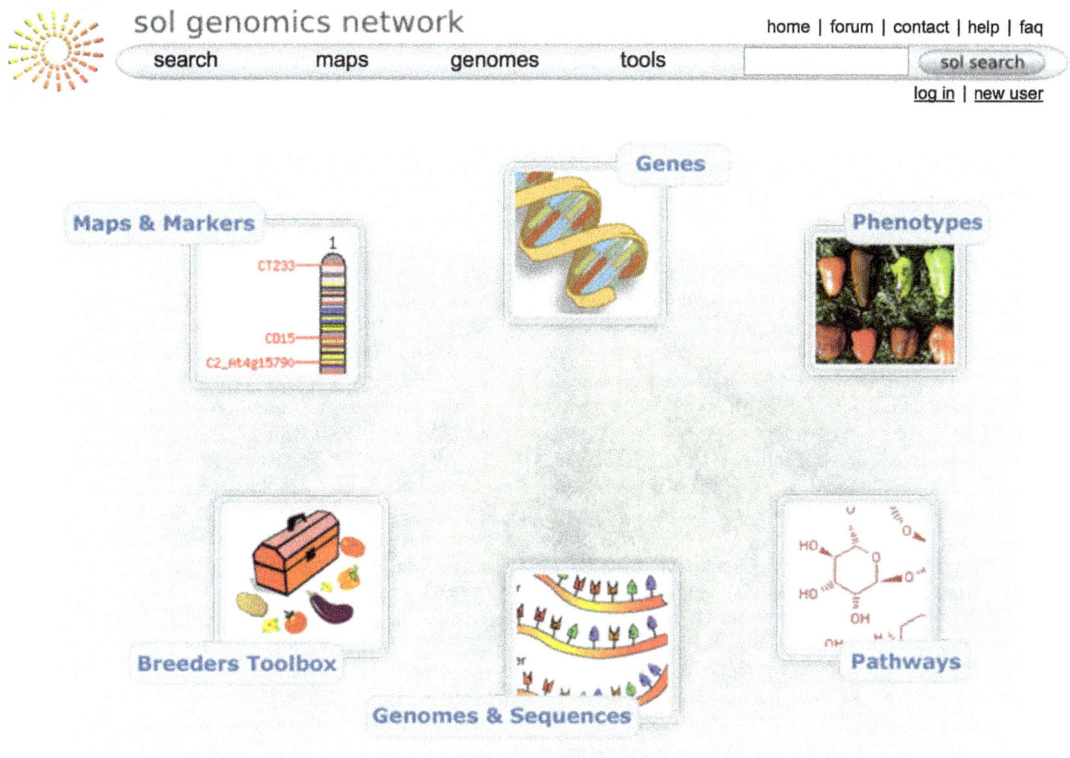

Fig. 9.4 http://solgenomics.net

generated and more FISH has been done to locate BAC sequences on chromosomes. This has lead to the release in 2014 of an assembly 2.5. The iTAG annotation at this point has been transferred to that new assembly.

The central node from iTAG still remains as the main access point for solanaceae data, centralized around Tomato and it's relatives (Fig. 9.4).

References

Abeel T, Van Parys T, Saeys Y, Galagan J, Van de Peer Y (2012) Genomeview: a next-generation genome browser. Nucleic Acids Res 40(2):e12

Allen JE, Salzberg SL (2005) JIGSAW: integration of multiple sources of evidence for gene prediction. Bioinformatics 21(18):3596–3603

Altschul SF et al (1990) Basic local alignment search tool. J Mol Biol 215:403–410

Besemer J, Borodovsky M (2005) GeneMark: web software for gene finding in prokaryotes, eukaryotes and viruses. Nucleic Acids Res 33:W451–W454

Besemer J, Lomsadze A, Borodovsky M (2001) GeneMarkS: a self-training method for prediction of gene starts in microbial genomes. Implications for finding sequence motifs in regulatory regions. Nucleic Acids Res 29(12):2607–2618

Birney E, Clamp M, Durbin R (2004) GeneWise and genomewise. Genome Res 14(5):988–995

Bonizzoni P, Rizzi R, Pesole G (2005) ASPIC: a novel method to predict the exon–intron structure of a gene that is optimally compatible to a set of transcript sequences. BMC Bioinform 6:244

Brunak S, Engelbrecht J, Knudsen S (1991) Prediction of human mRNA donor and acceptor sites from the DNA sequence. J Mol Biol 220:49–65

Burge C, Karlin S (1997) Prediction of complete gene structures in human genomic DNA. J Mol Biol 268:78–94

Camon EB, Barrell DG, Dimmer EC, Lee V, Magrane M, Maslen J, Binns D, Apweiler R (2005) An evaluation of GO annotation retrieval for BioCreAtIvE and GOA. BMC Bioinform 6(Suppl 1):S17

Cantarel BL, Korf I, Robb SMC, Parra G, Ross E, Moore B et al (2008) MAKER: an easy-to-use annotation pipeline designed for emerging model organism genomes. Genome Res 18(1):188–196

Coghlan et al (2008) nGASP—the nematode genome annotation assessment project. BMC Bioinform 19 (9):549

Coleman SJ, Zeng Z et al (2010) Structural annotation of equine protein-coding genes determined by mRNA sequencing. Anim Genet 41:121–130

DeCaprio D, Vinson JP, Pearson MD, Montgomery P, Doherty M, Galagan JE (2007) Conrad: Gene prediction using conditional random fields. Genome Res 17 (9):1389–1398

Delcher AL, Harmon D et al (1999) Improved microbial gene identification with GLIMMER. Nucleic Acids Res 27:4636–4641

Djebali S, Delaplace F, Roest Crollius H (2006) Exogean: a framework for annotating protein-coding genes in eukaryotic genomic DNA. Genome Biology 7(Suppl 1):S7.1–S7.10

Dobin A, Davis CA, Schlesinger F, Drenkow J, Zaleski C, Jha S, Batut P, Chaisson M, Gingeras TR (2013) STAR: ultrafast universal RNA-seq aligner. Bioinformatics 29(1):15–21

ENCODE Project Consortium (2004) The ENCODE (Encyclopedia of DNA elements) project. Science 306(5696):636–640

Foissac S et al (2008) Genome annotation in plants and fungi: EuGene as a model platform. Curr Bioinform 3:87–97

Götz S, García-Gómez JM, Terol J, Williams TD, Nagaraj SH, Nueda MJ, Robles M, Talón M, Dopazo J, Conesa A (2008) High-throughput functional annotation and data mining with the Blast2GO suite. Nucleic Acids Res 36(10):3420–3435

Gremme G, Brendel V, Sparks ME, Kurtz S (2005) Engineering a software tool for gene structure prediction in higher organisms. Inf Softw Technol 47 (15):965–978

Gross SS, Brent MR (2006a) Using multiple alignments to improve gene prediction. J Comput Biol 13: 379–393

Gross SS, Brent MR (2006b) Using multiple alignments to improve gene prediction. J Comput Biol 13(2):379–393

Guigó R, Knudsen S, Drake N, Smith T (1992) Prediction of gene structure. J Mol Biol 226(1):141–157

Haas BJ, Delcher AL, Mount SM, Wortman JR, Smith RK Jr, Hannick LI, Maiti R, Ronning CM, Rusch DB, Town CD, Salzberg SL (2003) Improving the Arabidopsis genome annotation using maximal transcript alignment assemblies. Nucleic Acids Res 31 (19):5654–5666

Hebsgaard SM, Korning PG, Tolstrup N, Engelbrecht J, Rouze P, Brunak S (1996) Splice site prediction in Arabidopsis thaliana DNA by combining local and global sequence information. Nucleic Acids Res 24 (17):3439–3452

Howe KL, Chothia T, Durbin R (2002) GAZE: a generic framework for the integration of gene-prediction data by dynamic programming. Genome Res 12(9): 1418–1427

Kim D, Pertea G, Trapnell C, Pimentel H, Kelley R, Salzberg SL (2013) TopHat2: accurate alignment of transcriptomes in the presence of insertions, deletions and gene fusions. Genome Biol 14(4):R36

Korf I (2004) Gene finding in novel genomes. BMC Bioinform 5:59

Krogh A, Mian IS et al (1994) A hidden Markov model that finds genes in *E. coli* DNA. Nucleic Acids Res 22:4768–4778

Kulp D, Haussler D et al (1996) A generalized hidden Markov model for the recognition of human genes in DNA. Proc Int Conf Intell Syst Mol Biol 4:134–142

Li H, Jiang T (2005) A class of edit kernels for SVMs to predict translation initiation sites in eukaryotic mRNAs. J Comput Biol 12:702–718

Li L, Stoeckert CJ Jr, Roos DS (2003) OrthoMCL: identification of ortholog groups for eukaryotic genomes. Genome Res 13(9):2178–2189

Lukashin AV, Borodovsky M (1998) GeneMark.hmm: new solutions for gene finding. Nucleic Acids Res 26 (4):1107–1115

Marco-Sola S, Sammeth M, Guigó R, Ribeca P (2012) The GEM mapper: fast, accurate and versatile alignment by filtration. Nat Methods 9(12):1185–1188

Mizrachi E, Hefer CA et al (2010) De novo assembled expressed gene catalog of a fast-growing Eucalyptus tree produced by Illumina mRNA-Seq. BMC Genomics 11:681

Nussbaumer T et al (2013) MIPS PlantsDB: a database framework for comparative plant genome research. Nucleic Acids Res. 41:D1144–D1151

Parra G, Blanco E, Guigó R (2000) GeneID in Drosophila. Genome Res 10(4):511–515

Parra G, Agarwal P et al (2003) Comparative gene prediction in human and mouse. Genome Res 13: 108–117

Passalacqua KD, Varadarajan A et al (2012) Strand-specific RNA-seq reveals ordered patterns of sense and antisense transcription in Bacillus anthracis. PLoS One 7:e43350

Pertea M, Lin X, Salzberg SL (2001) GeneSplicer: a new computational method for splice site prediction. Nucleic Acids Res 29(5):1185–1190

Philippe N, Salson M, Commes T, Rivals E (2013) CRAC: an integrated approach to the analysis of RNA-seq reads. Genome Biol 14(3):R30

Picardi E, Pesole G (2010) Computational methods for ab initio and comparative gene finding. Methods Mol Biol 609:269–284

Rätsch Gunnar, Sonnenburg S, Srinivasan J, Witte H, Müller KR, Sommer RJ, Schölkopf B (2007) Improving the C. elegans genome annotation using machine learning. PLoS Comput Biol 3(2):e20

Salamov A, Solovyev V (2000) Ab initio gene finding in Drosophila genomic DNA. Genome Res 10:516–522

Schiex T, Moisan A, Rouzé P (2001) EuGène: An eucaryotic gene finder that combines several sources of evidence. Lect. Notes Comput Sci 2066:111–125

Schoof et al. (2012) https://github.com/groupschoof/PhyloFun

Schweikert G, Behr J, Zien A et al (2009) mGene.web: a web service for accurate computational gene finding. Nucleic Acids Res 37:W312–W316

Slater GStC*, Birney E (2005a) Automated generation of heuristics for biological sequence comparison. BMC Bioinform 2005(6):31

Slater GS, Birney E (2005b) Automated generation of heuristics for biological sequence comparison. BMC Bioinform 6:31

Smit AFA, Hubley R, Green P (1996) RepeatMasker at http://repeatmasker.org

Stanke M, Schoffmann O et al (2006) Gene prediction in eukaryotes with a generalized hidden Markov model that uses hints from external sources. BMC Bioinform 7:62

Sterck L, Billiau K et al (2012) ORCAE: online resource for community annotation of eukaryotes. Nat Methods 9(11):1041

Tisserant E, Da Silva C et al (2011) Deep RNA sequencing improved the structural annotation of the Tuber melanosporum transcriptome. New Phytol 189:883–891

Trapnell C*, Pachter L, Salzberg SL (2009) TopHat: discovering splice junctions with RNA-Seq. Bioinformatics 25(9):1105–1111

Wu TD, Nacu S (2010) Fast and SNP-tolerant detection of complex variants and splicing in short reads. Bioinformatics 26(7):873–881

Yandell M, Ence D (2012) A beginner's guide to eukaryotic genome annotation. Nat Rev Genet 13:329–342

Yeh RF, Lim LP et al (2001) Computational inference of homologous gene structures in the human genome. Genome Res 11:803–816

Zhang MQ (2002) Computational prediction of eukaryotic protein-coding genes. Nat Rev Genet 3:698–709

Repeat Sequences in the Tomato Genome

10

Maria Luisa Chiusano and Chiara Colantuono

Abstract

The sequencing of the tomato genome revealed that, though the moderated size when compared to most of the Solanaceae and other plant species, it comprises more than the 60 % of DNA repeats. This is in contrast with initial estimations assessing that the total genome comprised only about the 10–22 % of repetitive sequences. These preliminary hypotheses were probably biased by the presence of single-copy DNA within the repetitive portion of the genome and by the high sequence divergence of the repeat content. Though the release of the first version of the genome sequences in 2012, the complete view of the repeated regions in tomato at sequence level is still partial, because of difficulties due mainly to DNA repeat sequencing and assembling. However, deeper knowledge on the repeat content of the genome and its distribution was consistently supported by cytogenetics, molecular markers and reassociation kinetics, accompanied by advanced approaches such as *Fluorescence* In Situ *Hybridization* (FISH) and more recently *Optical Mapping*. These techniques helped to clarify many of the principal aspects related to the distribution and the organization of the major repeat classes in tomato, contributing to a consistent overview of this essential part of the genome. The main focus of this chapter is to describe the repeat content of the tomato genome as revealed from the sequencing effort and associated bioinformatics, mainly considering the distribution of highly and moderately repeated DNA sequences. We provide a general overview on plant genome complexity and repeat content, presenting the main repeat categories and their organization. Then we describe the bioinformatics for DNA repeats sequence analysis, focusing on most common approaches for investigations in large genomic sequences, as well as on major repeated sequence collections available to support plant genome annotations. Details on the

M.L. Chiusano (✉) · C. Colantuono
Department of Agraria, University Federico II of Naples, Naples, Italy
e-mail: chiusano@unina.it

M. Causse et al. (eds.), *The Tomato Genome*, Compendium of Plant Genomes,
DOI 10.1007/978-3-662-53389-5_10

methods employed to analyze the tomato genome sequences (assembly v. 2.40) published in 2012 will be presented. The description of what is known from tomato concerning the major DNA repeat classes is therefore overviewed highlighting the major results or confirmations obtained thanks to the genome sequencing effort. The discussion is mainly focused on the general description of repeat occurrence in the tomato genome, though questions on the specific role and evolution of these extended regions in tomato and in plant genomes, as well as in other eukaryotes, still remain open.

Keywords
Tomato · Repeat · Bioinformatics · Duplication · Cytogenetics

Introduction

The exploitation of evolving experimental techniques, starting from early cytological approaches, molecular markers, *Fluorescence* In Situ *Hybridization (FISH)* and *Optical Mapping,* till the nucleotide sequencing of entire genomes, contributed relevant discoveries on genome organization, also determining relationships among chromosomal peculiarities, in phylogeny, in evolution.

Comparative approaches highlighted that many structure features of plant genomes are remarkably similar among different species, and are also shared with other eukaryotes, animals and fungi (Heslop-Harrison 2000). All eukaryotes have their genomic DNA organized in chromosomes, associated with proteins, showing almost the same organization. Centromeric regions are located in regions that are almost conserved along the chromosome structure, and the terminal regions are organized in telomeres.

Comparative approaches also highlighted the relevance of polyploidy in plants, with chromosome number which varies widely among plant species, such that 2*n* ranges in value from 4 to more than 1000, although the number within any given species is usually constant. Occurrence of polyploidy may be also associated to diploidization events, with rearrangements also implying genome reshuffling, translocations, fusion and fission of chromosomes. These events have been discussed to be some of the consequences why plant genomes are highly duplicated (Lysak et al. 2005; Cui et al. 2006; Tang et al. 2008a, b; Jiao et al. 2011, 2012; Sangiovanni et al. 2013). Beyond the interesting issue of investigating on the mechanisms implied in the occurrence of polyploidy and diploidization events in plants, even in a relatively short time span, tracing plant genome evolution and diversification (Jaillon et al. 2007; Tomato Genome Consortium 2012; Denoeud et al. 2014), it would also be rather intriguing to understand what enabled angiosperms to efficiently manage the presence of homologous chromosomes in comparison to all other eukaryotes, where polyploids are rare. However, in the context of this chapter, it is remarkable to focus on the effects that whole-genome and segmental duplications had on the redundancy of genome regions and of gene copies, with the definition of novel gene families. Though it is not the aim of this chapter to discuss repeats in DNA due to polyploidization events or to retaining of duplicated regions, it is noteworthy, indeed, to underline also here that one of the main outcomes of the tomato genome sequencing effort was the tracing of two consecutive genome triplications in the *Solanum* lineage. The more ancient event was shared with rosids, while, a more recent one appeared specific to the *Solanum* lineage (Tomato Genome Consortium 2012; Denoeud et al. 2014). These events had a relevant impact on diversification and evolution of novel functionalities in these clade of plants. However, it is discussed that the repeated regions tracing these possible events in the tomato genome were mainly detected only at sequence level (Tomato Genome Consortium 2012), presumably because of the high divergence determined by gene loss or mutations since the last hypothesized polyploidization event (Shearer et al. 2014).

The dynamics of genome evolution in plants offers striking opportunities to have multiple copies of the genome content, i.e. to repeat it, and to keep it duplicated even when diploidization occurred. Furthermore, the transfer of genes or of entire parts of the DNA from organelles to nucleus is now well documented both in plants and animals (Martin and Herrmann 1998; Vaughan et al. 1999).

Worthy to note, though the different occurrences of genome rearrangements in plants, the gene numbers as well as their order are almost conserved over substantial evolutionary distances in plants (Gebhardt et al. 1991; Ahn et al. 1993; Devos and Gale 1993, 1997, 2000).

The tomato genome, as an example, is highly syntenic with those of other economically important Solanaceae (Potato Genome Sequencing Consortium 2011; Tomato Genome Consortium 2012; Hirakawa et al. 2014; Kim et al. 2014; Sierro et al. 2014) as well as other plants (Jaillon et al. 2007). However, plant genome size can strongly vary among different species. Indeed, repetitive sequences contribute significantly to genome size in plants. Understanding the mechanisms and inferring on possible functional reasons favouring these variability and plasticity is still an open challenge.

DNA Content in the Cell

The amount of DNA (in picograms) in an unreplicated haploid cell, which corresponds to the constant value or C-value (Swift 1950; Greilhuber et al. 2005), is relatively homogeneous within a species. However, it is evident that the C-value is particularly variable between species. This variability is not related to the complexity of the organisms in terms of size or developmental mechanisms. The DNA content of the unicellular amoeba was 200 times higher than in human cells, though mammals have evident higher developmental complexity. This initially "unexpected" phenomenon represents the so-called "C-value paradox". The paradox is today explained knowing that the DNA content in a species can be abundant in repetitive sequences, though the numbers of coding genes are of the same order of magnitude in all eukaryotes, which ranges from about 6000 in the unicellular *Saccharomyces cerevisiae* to approximately 20,000 to 25,000 in the human genome (which is 200 times bigger than the genome of the yeast) (Richard et al. 2008).

In general, the term "repetitive sequences" refers to highly similar DNA fragments that are present in multiple copies in a genome. In particular the major contribution to the haploid genome size in eukaryotes is due to highly and moderately repeated sequences, i.e. DNA motifs, ranging in length from a single couple of nucleotides to thousands of nucleotides, repeated many hundreds or thousands of times. These repeated motifs are ubiquitous in eukaryotic genomes (Charlesworth et al. 1994; Kumar and Bennetzen 1999; Bowen and Jordan 2002) and represent a large portion of the chromosome structure (von Sternberg 2002), ranging between 50 and 90 % or more of all the nuclear DNA content. As an example, more than the 50 % of the human genome is composed by repeats (Richard et al. 2008).

In higher plants, the amount of DNA is particularly variable between species (Flavell et al. 1974; Bennett and Smith 1976; Ouyang and Buell 2004; Hawkins et al. 2009). The lowest content reported for *A. thaliana* is one of the main reasons why this genome was the first one to be sequenced among plant species (NSF 1990; Arabidopsis Genome Initiative 2000). Accordingly, mainly thanks to its "modest" genome size, poplar was the first tree to be sequenced (Brunner et al. 2004). Also in the case of plant genomes, the proportion of protein-coding regions is rather similar among the species (Table 10.1). Indeed, the structural and developmental complexity of plant species with very different amounts of DNA per cell is not fundamentally different from those with the highest amounts (Smyth 1991). It is also evident (Table 10.1) that the contribution of repeats to each genome has a wide range of variability starting from very low percentages, like in *Arabidopsis thaliana*, reaching a very high relative content like in *Capsicum annum* (~82 %) and in several monocots (~85 %).

DNA Repeat Classes

Repetitive DNA was first detected because of its rapid reassociation kinetics when denatured, since the rate at which a particular sequence reassociates is proportional to the number of times it is found in the genome. Based on the renaturation rates, in denaturation–renaturation experiments of genomic DNA after heat exposure, it is possible to identify three major classes of DNA sequence types: the highly repetitive sequences,

Table 10.1 List of plants with sequenced genomes

Scientific name	Monocot/dicot	#Chr (*n*)	Size (Mb)	#Gene	%Repeat	References
Arabidopsis lyrata	Dicot	8	207	32.670	30	Hu et al. (2011)
Arabidopsis thaliana	Dicot	5	125	25.498	14	The Arabidopsis Genome Initiative (2000)
Brassica rapa	Dicot	10	485	41.174	40	The Brassica rapa Genome Sequencing Project Consortium (2011)
Capsicum annum cultivate/wild	Dicot	12	3349/3480	35.336/34.476	81/82	Qin et al. (2014)
Carica papaya	Dicot	9	372	28.629	43	Ming et al. (2008)
Coffee canephora	Dicot	11	710	25.574	50	Denoeud et al. (2014)
Cucumis sativus	Dicot	7	367	26.682	24	Huang et al. (2009)
Fragaria vesca	Dicot	7	240	34.809	23	Shulaev et al. (2011)
Glycine max	Dicot	20	1115	46.430	57	Schmutz et al. (2010)
Hordeum vulgare	Monocot	7	5100	30.400	84	The International Barley Genome Sequencing Consortium (2012)
Lotus japonicus	Dicot	6	472	30.799	56	Sato et al. (2008)
Musa acuminata	Monocot	11	523	36.542	44	D'Hont et al. (2012)

(continued)

Table 10.1 (continued)

Scientific name	Monocot/dicot	#Chr (n)	Size (Mb)	#Gene	%Repeat	References
Nelumbo nucifera	Dicot	8	929	26.685	57	Ming et al. (2013)
Nicotiana tabacum K326/TN90/BX	Dicot	24 (2n)	4600/4410/4570	91.870/81.404/93.303	73/79/73	Sierro et al. (2014)
Oryza brachyantha	Monocot	12	300	32.038	29	Chen et al. (2013)
Oryza sativa	Monocot	12	389	37.544	26	International Rice Genome Sequencing Project (2005)
Phoenix dactylifera	Monocot	18	658	28.890	40	Al-Mssallem et al. (2013)
Solanum lycopersicum	Dicot	12	900	34.727	63	The Tomato Genome Consortium (2012)
Solanum melongena	Dicot	12	1100	85.446	71	Hirakawa et al. (2014)
Solanum tuberosum	Dicot	12	844	39.031	62	The Potato Genome Sequencing Consortium (2011)
Sorghum bicolor	Monocot	10	818	34.496	62	Paterson et al. (2009)
Theobroma cacao	Dicot	10	430	28.798	24	Argout et al. (2011)
Triticum aestivum	Monocot	42 (6n)	17,000	124.201	80	IWGSC (2014)
Triticum urartu	Monocot	7	4940	34.879	67	Ling et al. (2013)
Vitis vinifera	Dicot	19	475	30.434	41	Jaillon et al. (2007)
Zea mays	Monocot	10	2300	32.540	85	Schnable et al. (2009)

Type (monocot or dicot), number of chromosomes (#Chr), size (Mb) and haploid number (n), number of annotated genes (#Gene), percentage of repeats and related bibliographic references (author, year) are also reported

representing DNA fragments that reassociate very rapidly; the moderately repetitive ones, i.e. DNA fragments that reassociate at an intermediate rate, the single copy (or very low copy number class) representing fragments that do not repeat at a consistent frequency in DNA sequences. Such approaches to estimate the repetitive content of genomic DNAs in different organisms, though possible underestimations due to diverging repetitive elements, are remarkable since they give out a global accurate picture of genome composition in the absence of sequence information. In parallel to

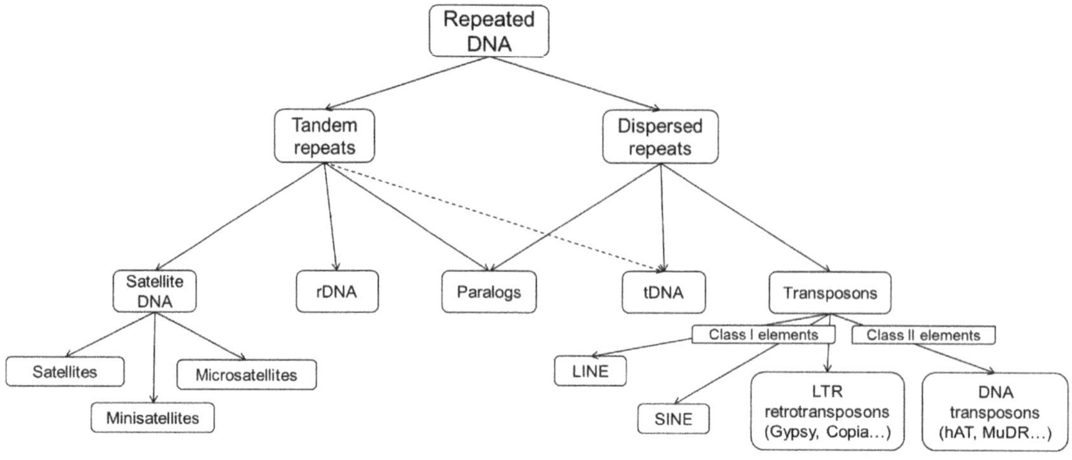

Fig. 10.1 Repeated DNA sequences in eukaryotic genomes. The two main categories of repeated elements (tandem and dispersed repeats) are shown, along with their subcategories

the reassociation kinetics properties, repeated sequences can be also divided in two major categories based on their organization or distribution in a genome: "tandem repeats" and "dispersed repeats" (Fig. 10.1). Tandem repeats are generally corresponding to the highly repetitive sequences. They mostly localize on large conspicuous heterochromatic DNA blocks at the distal ends and interstitial parts of the chromosome (Schmidt and Heslop-Harrison 1998) and include sequences that are repeated in tandem along the genome sequences such as ribosomal DNA repeat arrays (rDNA) and satellite DNA. Among tandem repeats, duplicated protein-coding genes (paralogs) can also be included. Dispersed repeats are usually corresponding to moderately repeated sequences, and include transposons and dispersed gene paralogs. Transfer RNA genes (tDNA) are often distributed in tandem, but they are usually included among the dispersed repeats (Richard et al. 2008).

Tandem Repeats

rDNA

rDNAs represent non protein-coding multigene families usually classified as tandem repeats. rDNAs (Fig. 10.1) are usually head-to-tail arrays of genes encoding the precursor (45S) of the three largest ribosomal RNAs (18S, 5.8S and 25S in plants). The corresponding DNA region generally contains several tandem copies, including active rRNA genes and silent rRNA genes, which are often highly compacted in dense heterochromatin. The rDNA region gives rise to secondary constrictions in metaphase chromosomes that are called the nucleolus organizer regions (NOR), around which the nucleolus forms. rRNA coding genes are usually transcribed by RNA polymerase I. The 5S rRNA genes, highly conserved genes of around 120nts in length, are distributed independently from the 45S rDNA, in multiple copies arranged as tandem arrays separated by a high variable spacer in sequence and in length. The number of copies of the core unit, from 200 to 900 nucleotides, can vary from 1000 to 50,000 copies. The sequences can be adjacent or not to the 45S rDNA region and are usually transcribed by the RNA polymerase III.

Satellite DNA

The name "satellite DNA" refers to a "satellite" band different in density from bulk DNA in a density gradient, due to repetitions of short DNA sequences. It consists of almost large number of repeat units, distributed as tandem arrays of DNA. Satellite DNA is in itself also distinguished in minisatellites or microsatellites. Both subcategories are variable in number of repeats

(Variable Number of Tandem Repeats or VNTR). Minisatellites consist of a core repeat units of 10 to 60–90 nucleotides. Microsatellites (also known as "Simple Sequence Repeats" or SSRs, or "Short Tandem Repeats" or STRs) consist of a core of around 2–6–10 nucleotides. In general satellite DNA can be distributed throughout the chromosomes (King et al. 1997; Richard et al. 2008), both in heterochromatin and euchromatin regions (Cuadrado and Schwarzacher 1998; Cuadrado and Jouve 2007a, b; Chang et al. 2008), in genes, both in the protein-coding regions, in introns, or in their regulatory regions, and within transposable elements.

The tandem satellite DNA sequences exhibit in general characteristic chromosomal locations, with roles depending on their locations. They can be at telomeric, subtelomeric and centromeric regions, with repetitive families that can be shared within a taxonomic family or a genus, or may be specific to the species, genome or even a chromosome (Sharma and Raina 2005). These features have formed the basis of extensive utilization of repetitive sequences for taxonomic and phylogenetic studies. Satellite DNA is the main component of centromeres, with a core units from 9 to 64 bp long, and of telomeric regions, with a conserved core units of around 6 bp, and repetition numbers that can range from hundreds to thousands, depending on the species (Podlevsky et al. 2008), forming the main structural constituent of heterochromatin. Centromeres are essential for chromosome segregation, yet their DNA sequences evolve rapidly in contrast with the high conservation of the core units of telomeres (Henikoff et al. 2001). Centromeres differ greatly in their sequence organization among different species. In *Saccharomyces cerevisiae* a "point centromere" of 125-bp sequence is sufficient to confer centromere function (Meraldi et al. 2006). In most animals and plants, centromeres contain megabase-scale arrays of simple tandem repeats, sometimes interspersed with long terminal repeat transposons (Heslop-Harrison et al. 2003) and, despite their relevant role, very little is known about the degree to which centromere tandem repeats share common properties between different species (Melters et al. 2013). However, the key kinetochore proteins are conserved in both plants and animals, particularly the centromere-specific histone H3-like protein (CENH3) highlighting the importance of epigenetic mechanisms in the establishment and maintenance of centromere identity (Houben and Schubert 2003). Telomere repeats occur predominantly at the ends of eukaryotic chromosomes, arranged in tandem to form large uninterrupted blocks often associated to subtelomeric satellite repeats (Ganal et al. 1991). They appear to protect chromosome ends from degradation and shortening during replication (Mason and Biessmann 1995).

Microsatellites may have high variability in length, due to unequal crossing over, rolling circle amplification and replication slippage, even before meiosis (Tautz and Schlotterer 1994), making these regions endowed of a high rate of mutation per locus per generation (Jarne and Lagoda 1996; Kruglyak et al. 1998). This is why these sequences are important for different approaches (Buschiazzo and Gemmell 2006). Indeed microsatellites can be amplified using unique sequences at the flanking regions to define primers for amplifications, producing variable patterns of fragments lengths which are useful for population studies, fingerprinting, marker assisted selection, and study of breeding patterns of wild or domesticated species (Martinez-Zapater et al. 1986; Maluszynska and Heslop-Harrison 1991; Michelmore et al. 1991; Martin et al. 1992; Maughan et al. 1995; Liu et al. 1996; McCouch et al. 1997; Milbourne et al. 1997; Livingstone et al. 1999).

Dispersed Repeats

tDNA

Genes coding for transfer RNAs represent a non protein-coding multigene family, as rRNA coding genes. Though often distributed in tandem, they are usually classified as dispersed repeats.

In addition to its essential function in protein synthesis, recent studies have shown that tRNAs are multifunctional molecules involved in many processes of cellular metabolism (Minajigi and Francklyn 2010). Furthermore, tRNA-derived RNAs appear to be used in the RNA silencing pathway, and are a major source of short interspersed nuclear elements (Bermudez-Santana et al. 2010; Phizicky and Hopper 2010).

It is postulated that all tRNA genes (tDNAs) derive from an ancestral molecule (Eigen et al. 1989) that during evolution gave rise to a full set of tRNA genes generated as the result of numerous mutation, duplication and reorganization events. The number of tRNA pseudogenes and organellar-like tRNA genes present in nuclear genomes varies greatly from one plant species to another. Generally, there is no correlation between genome size and tDNA copy number in the nuclear genome (Richard et al. 2008). However, Michaud et al. (2011), in their analysis of tRNA gene distribution in plant genomes, revealed that the tRNA gene content in plants is rather homogenous, and is mostly correlated with genome size.

Transposable Elements

Among dispersed repeats, transposable elements (TEs) are DNA sequences that are capable of "moving" in the cell, integrating into a new site within the genome where they originated from (Craig et al. 2002), creating changes and amplifying and altering the cell's genome size. This is why they were also termed "jumping elements". They were discovered in plants by Barbara McClintock who earned her Nobel Prize for this scientific contribution in 1983 (McClintock 1953). She not only found that genes could move, but also that they could be turned on or off according to the environmental conditions or during different stages of cell development. Transposons consist of two major classes: retrotransposons (class I elements) and DNA transposons (class II elements) (Fig. 10.1), depending on the mechanisms that determine their excision and insertion in the genome.

Retrotransposons replicate by forming RNA intermediates, which are then reverse transcribed to DNA sequences and inserted into new genomic locations. Therefore, retrotransposons need transcription and a reverse transcriptase to move, while DNA transposons are excised from the genome, and the "cut-and-paste" mechanisms for transposition require transposases (Craig et al. 2002). Retrotransposons are commonly grouped in LTR or non-LTR retrotransposons according to the presence or not of long terminal repeats (LTR). In LTR retrotransposons, the terminal repeats range from ~100 bp to over 5 kb in size. They are the most high representative class in plant genomes (Kumar and Bennetzen 1999; Bennetzen 2000) and may be further subclassified into different classes, differing by the degree of sequence similarity and by the order of encoded gene products along their structure. Among these, Ty1-copia-like and Ty3-gypsy-like are commonly found in high copy number in plants genomes, but also in animals, fungi and protista. Retroviruses are often classified separately from the LTR retrotransposons though they share many features with them. A major difference with Ty1-copia and Ty3-gypsy retrotransposons is that Retroviruses have an Envelope protein (ENV) and have domains that enable extracellular mobility (Cotton 2001).

Non-LTR retrotransposons include long interspersed elements (LINEs) and short interspersed elements (SINEs). LINEs encodes for functionalities that are essential for retrotransposition, such as reverse transcriptase and endonucleases activities, and are transcribed by the RNA polymerase II, like mRNAs. Their mechanisms of transposition, however, differ from that of other LTR elements (Bibillo and Eickbush 2004). SINEs are nonautonomous retroelements, with length ranging from 100 to 900 bp, and copy not identical in the genome (Kramerov and Vassetzky 2005). They do not encode reverse transcriptase, and presumably co-opt the LINE machinery to be retrotransposed (Jurka 1997). They are transcribed by RNA polymerase III, being organized at their 5' end like a typical tRNA promoter (Defraia and Slotkin 2014).

Bioinformatics for Repeat Detection

Repeat Sequence Databases

Due to the presence of different types of repeats, there are different dedicated databases that organize repeats, such as *Repbase* (Jurka et al. 2005), the *Tandem Repeats Database* (Gelfand et al. 2007), *RepeatsDB* (Di Domenico et al. 2014). In particular, *RepBase* is a comprehensive repeat collection including prototypes of repetitive DNA sequences derived from the consensus of each of the repeat families from each eukaryotic species. The *Tandem Repeats Database* is specific for repeated regions in tandem, while *RepeatsDB* specifically contains tandem repeats found in protein sequences. In parallel to these resources, *Rfam* (Burge et al. 2013) contains families of non protein-coding RNAs, and is useful to support annotation of the corresponding genes in a genome, rRNA and tRNA coding genes included.

Some available databases are specific for plants, *PGSB Repeat Database* (Nussbaumer et al. 2013) and the *Plant Repeat Database* organized starting from the *TIGR Plant repeat database* (Ouyang and Buell 2004), this last updated till 2008, both designed as comprehensive repeat collections. *PlantSat* (Macas et al. 2002) and *Plant rDNA database* (Garcia et al. 2012) are dedicated to satellite repeats and rDNAs, respectively. Some of these databases have the possibility to allow search for repeated region in specific genera or species, such as the *Plant Repeat Database*, that is made of subsections dedicated to Solanaceae, Gramineae or other plants, or *Plant rDNA database*.

Methodologies

Bioinformatics strategy to identify and annotate repeats in genome sequences is almost similar even in different species. In general, the currently available methods can be based on comparative approaches, which aim to identify and therefore classify the repeated regions aligning a query sequence, the one to be analyzed, with sequences representing repeat classes organized in dedicated databases. Other approaches are based on de novo detections of repeats along a sequence, these methods supporting the identification of novel repeat sequences, i.e. sequences not available in dedicated collections since not yet discovered and classified.

RepeatMasker (Smit et al. 1996) or *Censor* (Kohany et al. 2006) are some of the well-known similarity-based search tools, useful to support the annotation of the repeats detected along a sequence and to provide its masked version, i.e. a sequence in which all the regions identical to repeats are changed to X or Ns, to be ignored in subsequent analyses, like those necessary to detect coding genes.

Similarity methods also may consider comparisons with established genome sequence references find occurrence of similar repeat regions.

Tandem Repeats Finder (Benson 1999) and *mreps* (Kolpakov 2003) are other specific tools helpful to find and annotate tandem repeats in DNA sequences. Like *LTR_STRUC* (McCarthy and McDonald 2003), *Recon* (Bao and Eddy 2002) and *RepeatScout* (Price et al. 2005), they detect repeated DNA sequences by de novo approaches. These approaches are generally based on self-comparisons of repeated similar regions. The exploitation of associated clustering approaches usually permits also to group-related sequences, to classify them into families and or subfamilies.

The identification and the annotation of repeated gene loci, such as those coding for non protein-coding genes (tRNA, rRNA), can be performed by dedicated tools like *Infernal* (Nawrocki et al. 2009), also useful for the identification of other non protein-coding RNAs. Specifically, *Infernal* is used to search RNA families dedicated databases for similar sequences such as *Rfam*. Infernal builds a profile from a structurally annotated multiple sequence alignments of RNA families with a position-specific scoring system. The scoring approach also takes into consideration secondary structure organization of the family being modelledQuery, such as base pairing, combining different levels of structure information to get to the most appropriate result. Other tools, such as *tRNAscan-SE* (Schattner et al. 2005) and *ARAGORN* (Laslett and Canback 2004) or *SnoReport* (Hertel et al. 2008) are specific for some classes of RNAs, like tRNAs and snoRNAs, respectively.

Repeats in the Tomato Genome

Protein-coding Gene Paralogs

Though the description of protein-coding paralog genes is not the main topic of this chapter, preferred to briefly reported on their distribution in the tomato genome since they represent repeat sequences in a genome and their occurrence contributed to reveal the two consecutive triplications events of the *Solanum* lineage, that moulded the gene set controlling fruit characteristics (Tomato Genome Consortium 2012). The total number of genes with at least one paralog in tomato is 25,992, about 75 % of the total gene content. In Fig. 10.2 we report the distribution of paralog gene numbers per chromosome. This reflects the high duplication level of mRNA coding genes reported in the tomato genome (Tomato Genome Consortium 2012).

Non Protein-coding Repeated Genes

Among paralogs we may also consider large multigene families such as ribosomal RNAs (rDNA) and tRNAs (tDNA) genes.

Non protein-coding RNAs in the tomato genome sequences were annotated by *Infernal* using the *Rfam* database (version 9.1) (specifically, the collection available at ftp://ftp.sanger.ac.uk/pub/databases/Rfam/9.1/infernal-latest.tar.gz and compatible with *Infernal* 1.0) (Tomato Genome Consortium 2012).

Long rDNAs were excluded from the analyses of the tomato assembly released by the consortium, because of a specific option used by the authors when running the software *Infernal,* that excluded the annotation of these specific regions (Tomato Genome Consortium 2012, supplementary materials 2.3.2). Therefore the analysis resulted to be limited to the identification of 1853 non protein-coding RNAs of 90 distinct *Rfam* families in which almost 48 % of all the targets represented tRNA coding genes (RF00005) (Tomato Genome Consortium 2012).

Table 10.2 summarizes the results included in the *iTAG2.4_infernal.gff3* file made available by the tomato genome sequencing consortium at the ftp section of the Sol Genomics Network (http://solgenomics.net/). Moreover, in order to complete the annotation of the non protein-coding rDNAs, we performed a *BLASTn* of the tomato chromosomes versus the Large Subunit sequences (LSU, RF02543), which include the 25S RNA, and the Small Subunit (SSU, RF01960) sequences, corresponding to 18S, both collections available in the *Rfam* database (release 12.0). We considered only locus that corresponded to matches with identity and coverage ≥98 %.

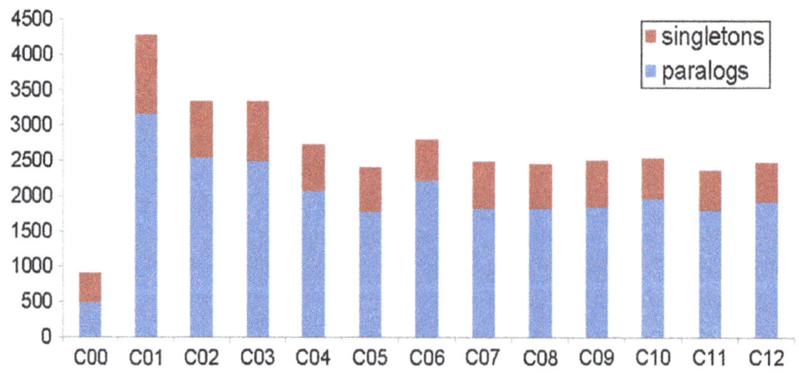

Fig. 10.2 Paralog gene distribution per chromosome. The data source from which we report this summary is obtained from BioMart section of *EnsemblPlants* (http://plants.ensembl.org/)

Table 10.2 Number of 5.8S rRNA, 5S rRNA, tRNA as reported by the Tomato Genome Consortium (2012)

	iTAG v. 2.4			Udated	
	5.8S rRNA	5S rRNA	tRNA	18S rRNA	25S rRNA
chr 00	11	3	16	4	20
chr 01	2	38	109	4	9
chr 02	0	1	76	1	6
chr 03	2	3	83	5	6
chr 04	0	1	71	1	4
chr 05	3	0	60	2	1
chr 06	7	0	102	5	11
chr 07	1	2	52	2	4
chr 08	0	0	70	1	8
chr 09	0	2	44	2	3
chr 10	0	0	90	1	8
chr 11	13	4	48	12	21
chr 12	1	0	64	2	6
Sum	40	54	885	42	107

Updated contents of 25S rRNA and 18S rRNA gene are also shown

5.8S rRNA genes defined by the consortium are listed mainly on chromosomes 11 and 6, while higher figures are reported by our updating corresponding to regions similar to 25S sequences (Table 10.2). It is also evident that there are still matches on the unassigned sequences collected as *unassembled* on "chromosome 0", probably because the difficulties in assigning repeated sequences during the assembly of large and complex genomes.

The table also shows a high number of 5S coding regions on chromosome 1 (Fig. 10.3a), confirming the loci identified as repeated in tandem by FISH on pachythene chromosomes on the short arm of chromosome 1 (1S), close to the centromeric region (Vallejos et al. 1986; Lapitan et al. 1991; Xu and Earle 1996a, b). Though, as explained, the information on the long rDNA regions (45S or at least 18S and 25S families) was not available from the sequencing and annotation effort, we reviewed the information collected from analyses preceding the tomato genome sequencing and exploited our updating based on the *BLASTn* analysis. Indeed, it was known that ribosomal DNA represents the most abundant repetitive DNA family in tomato, comprising approximately 3 % of the genome. From experimental analysis, 5S and 45S rRNA genes were detected as tandemly repeated with 1000 and 2300 copies. Karyotyping in combination with fluorescence in situ hybridization (FISH) on tomato pachytene chromosomes allowed the identification and mapping of the 45S rDNA on the satellite of the short arm of chromosome 2 (2S) and a minor locus on 2L, though these evidence are not confirmed by the tomato genome sequencing, from which no match, neither with the only considered marker 5,8S, was detected (Vallejos et al. 1986; Tanksley et al. 1988; Lapitan et al. 1991; Xu and Earle 1996a, b). However, these results find some confirmation from our updated analysis, with few matches from the 25S confirmed on chromosome 2. Other minor loci were also revealed at 6S, 9S and 11S (Xu and Earle 1996a, b), the first and the last also finding some confirmation by the annotation from the consortium, with stronger support by our update. Indeed, the updated analysis shows regions similar to the 25S (LSU) in all the

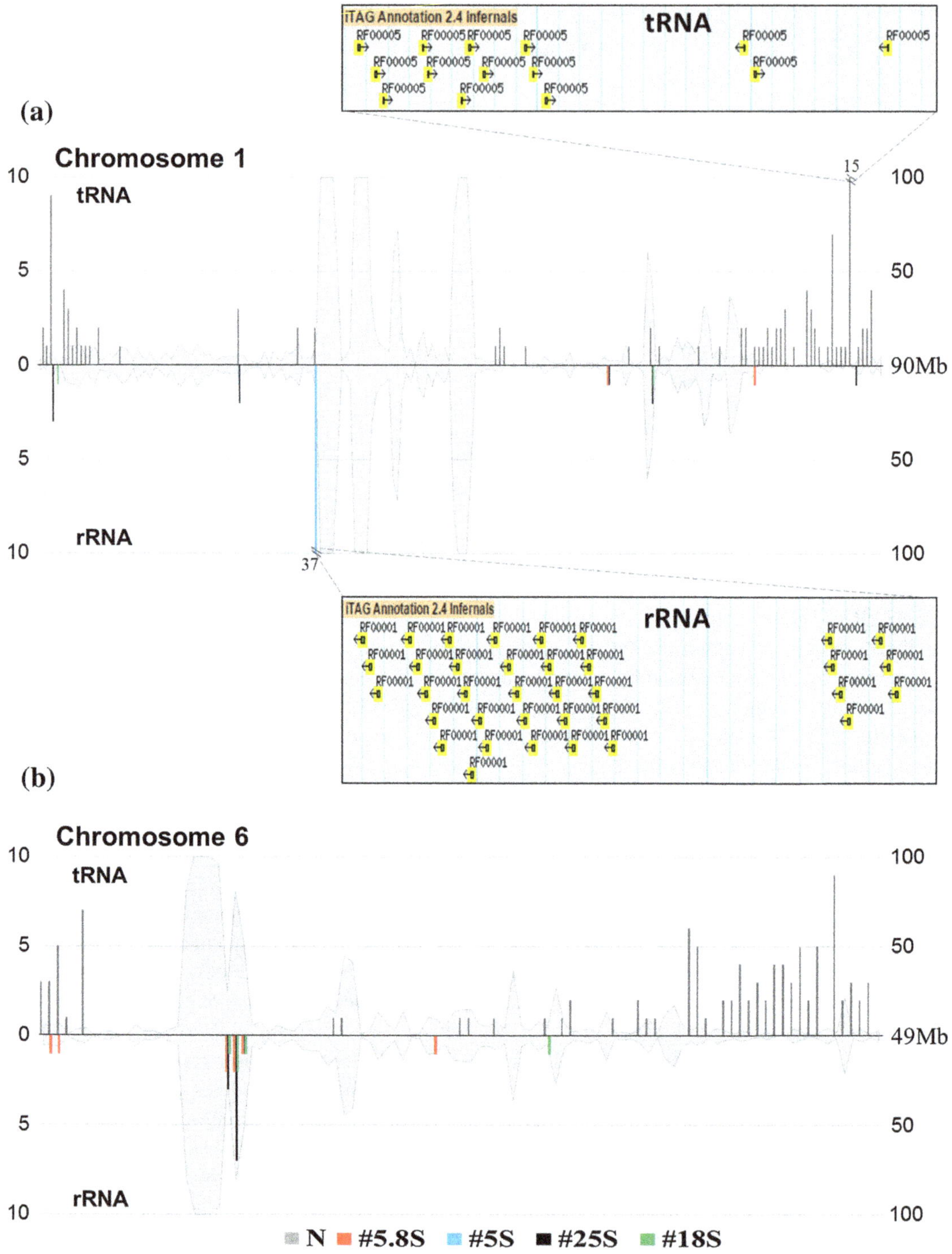

Fig. 10.3 Distribution per chromosome 1 (**a**) and chromosome 6 (**b**) of repeated non protein-coding genes. Percentage of *N* is also reported by a nonoverlapping window analysis of chromosomes divided per 500 Kb, with a total of 197 windows for chromosome 1 and 100 windows for chromosomes 6. Details of regions with 5S rRNA and tRNA in tandem on chromosome 1 are shown

chromosomes, accompanied by a similar distribution by the 18S, though with lower numbers, in contrast with what expected from previous analysis.

In Fig. 10.3a, b the distribution of non protein-coding genes on chromosomes 1 and 6 are shown, respectively. Data are from the *iTAG2.4_infernal.gff* file made available by the tomato genome consortium at ftp://ftp.solgenomics.net/genomes/Solanum_lycopersicum/annotation/ITAG2.4_release/. Moreover, the results from the updated analysis here provided are also shown in the figure.

Our updated analysis also permitted the clear identification of an rDNA locus associated to the occurrence of 45S loci on chromosome 6, since 18S 5.8S and 25S are all located in the region (Fig. 10.3b).

tDNA distribution is shown both in Table 10.2 and in Fig. 10.3. Interestingly to notice, their occurrence is reported in all the chromosomes.

Noncoding Tandem Repeats

Noncoding tandem repeat sequences in tomato chromosomes were detected using the de novo approach of *Tandem Repeats Finder* (Benson 1999), with default parameters. This permitted to classify the sequences by length into microsatellites (2–9 bp), minisatellites (10–99) and satellites (≥100-bp), while overlapping annotations of more than one of the three classes were classified as hybrid type.

The whole collections of tandem repeats resulted to cover 3.2 % of the genome, with the major contribution from minisatellites (1.7 of the entire genome and 53.7 % of the tandem repeats). Microsatellite repeats in tomato genome were also analyzed by Suresh et al. (2014), who detected a total of 68,641 microsatellite repeat motifs. Dinucleotide repeats (60.18 %) resulted much more abundant than tri (19.56 %) and other repeats, of which ~82.90 and ~17.10 % were simple and compound repeats, respectively. A total of 5841 and 4773 SSRs were present in the assigned genes and their 5′-upstream sequences, with average frequencies of 0.172 SSRs/gene and 0.14 SSRs/5′-upstream sequences, respectively. Data are accessible at the *Tomato Genomic Resources Database* (http://59.163.192.91/tomato2/).

Telomere

Beyond rDNAs, telomeres are the most ubiquitous tandem repeated arrays in the genome of eukaryotes.

The telomere repeats have been studied extensively in species of the Solanaceae family, which show mostly the Arabidopsis-type telomere (TTTAGGG). The typical tomato telomeric repeat (TR) (TT(T/A)AGGG) is arranged in tandem to form large uninterrupted blocks (Ganal et al. 1991). A block of 162-bp subtelomeric repeats (TGRI) is localized a few hundred kb from the terminal telomere repeats in 20 of the 24 homologous chromosomes (Ganal et al. 1988, 1991; Schweizer et al. 1988; Lapitan et al. 1989). These repeated blocks together accounts for around the 2 % of the total chromosomal DNA and, though the TR repeat is highly conserved, the long range physical structure of these arrays has been shown to be highly variable in different varieties (Broun et al. 1992) and within the genome (Zhong et al. 1998). Zhong et al. (1998) investigated on the relative length and distribution of the TR the spacer and the TGRI blocks in tomato chromosomes. The major evidence from Zhong et al. work was to highlight differences in TR-spacer-TGRI organization in most if not all the chromosome ends in tomato. Concerning the role of the spacer and the TGRI repeats it is assumed that they could represent buffering blocks separating chromosome ends from unique sequences or alternatively, playing a role in favouring or preventing chromosome degradation, fusions and fissions (Meyne et al. 1990). However, they have also been speculated to be regions susceptible to unequal crossing over between homologous and even nonhomologous chromosomes, yielding to high polymorphisms even in conserved genomes (Broun et al. 1992).

Interestingly, interstitial telomeric repeats (ITRs) were also revealed hybridizing the TR repeat on lambda clones of tomato, showing unexpected telomere homologous sequences on 8 of the 12 tomato centromeres (Ganal et al. 1991; Presting et al. 1996).

ITRs are organized as short tandem arrays and are expected to be evolutionary relics derived from chromosomal rearrangements and DNA repairs (He et al. 2013). However, megabase-sized ITR arrays were reported in *Solanum* species (Tek and Jiang 2004). These results showed that some ITR subfamilies were amplified and invaded the functional centromeres of Solanaceae chromosomes revealing possible other roles than simply being relics of chromosomal rearrangements. The epigenetic landscape and transcription of telomeres and ITRs were also investigated. As an example, in *Nicotiana tabacum* (with no detectable ITRs), and in *Ballantinia antipoda*, (with large blocks of pericentromeric ITRs and relatively short telomeres) Majerová et al. (2014) revealed that genuine telomeres displayed heterochromatic as well as euchromatic marks, while ITRs were just heterochromatic. Methylated cytosines were present at telomeres and ITRs, but showed a bias with more methylation towards distal telomere positions and different blocks of ITRs methylated to different levels (Majerová et al. 2014). Interestingly, the authors also showed that telomeres and ITRs are transcribed, and that the level of telomerase transcripts is tissue dependent, contributing novel insights for the understanding of the specific role and regulation activity of the associated transcripts.

Centromere

The tomato genome sequencing confirmed the presence of a high DNA repeat content in the heterochromatin pericentromeric regions, however no value added information was provided by the sequencing effort to characterize centromeric tandem repeated regions. It is known, however, that both the centromeric satellites and the retroelements are essential for centromere recognition by kinetochore proteins (Zhong et al. 2002; Nagaki and Murata 2005; Nagaki et al. 2011), and previous efforts also revealed the mosaic structure of centromeres in plant species (Nagaki et al. 2012). Interestingly, though it was evident that centromeric repeats evolve rapidly (Melters et al. 2013), Gong et al. (2012) recently reported that six of the 12 potato centromeres contain megabase-sized arrays of satellite repeats different in each centromere. By contrast, five potato centromeres are shown to be composed of single- and low-copy DNA sequences, with no satellite repeats detected. These five potato centromeres structurally resemble neocentromeres. Moreover, they also showed that most of the centromeric satellite repeats in potato were amplified recently from retrotransposon-related sequences and are not present in wild *Solanum* species closely related to potato.

A deeper comparative analysis revealed that different centromeric haplotypes were found to be associated with three potato centromeres, including haplotypes containing megabase-sized satellite repeats and haplotypes that do not contain the same repeats (Wang et al. 2014).

To further understand the evolution of centromeric DNA in *Solanum* species, (Zhang et al. 2014) conducted a genome-wide analysis of DNA sequences associated with the cenH3 nucleosomes in *Solanum verrucosum* ($2n = 2x = 24$), a wild species closely related to potato. They demonstrated a rapid divergence of the centromeric sequences between these two closely related species. Therefore, they hypothesized that centromeric satellite repeats may undergo boom–bust cycles of evolution from which a structurally favourable repeat lengths, maybe favouring the structure ideal for cenH3 nucleosome organization, could take place.

Many existing centromeres are believed to have originated as neocentromeres that activated de novo from noncentromeric regions by acquiring specific histones in the nucleosome (for example, the canonical histone H3 is replaced by cenH3 histone in plants or by CENP-A in animals (Kalitsis and Choo 2012; Rocchi et al. 2012). Newly formed neocentromeres are associated with gene "desert" regions and initially do not contain satellite repeats (Marshall et al. 2008; Wang et al. 2014). The evolutionarily new centromeres presumably accumulate satellite repeats and/or retrotransposons during evolution and eventually evolve rapidly to become repeat-based centromeres (Yan et al. 2006; Kalitsis and Choo 2012; Sharma et al. 2013).

Transposons

Considering the dispersed repeats, we already reported on tDNA distribution in the tomato genome in the paragraph on non protein-coding repeated gene families.

The other relevant class among dispersed repeats includes the transposons. In Table 10.3, we report the nucleotide coverage in terms of transposon classes of all the chromosomes, as derived from the annotation reported in the *iTAG2.4_repeat.gff3* file released by the tomato genome consortium (Tomato Genome Consortium 2012) and available at http://solgenomics.net.

While the pseudomolecules images in the Nature paper report the general behaviour of repeat content along tomato and potato pseudomolecules, in this chapter we provide, as an example, a more detailed view with a similar approach showing the distribution of all single class of repeats along tomato chromosomes 1 and 6 (Fig. 10.4a, b).

As reported from Nature 2012, full length LTR retrotransposons in the tomato genome sequence, were detected by a curated analysis starting from a de novo approach based on *LTR-STRUC* (McCarthy and McDonald 2003). 1647 intact LTR retrotransposons were detected. These sequences were assigned to the gypsy or copia subgroups which were identified thanks to the order of their inner protein domains.

Additional full length LTR elements were found by sequence similarity, leading to a total of 4052 still intact elements. Moreover, a cluster analyses of these sequences highlighted that tomato and potato (Potato Genome Sequencing Consortium 2011) genome sequences shared common LTR retrotransposons (Tomato Genome Consortium 2012).

The insertion events of LTR retrotransposons were also dated by the sequence divergence between left and right LTRs (Wiley et al. 2009). Interestingly, this analysis showed fewer copies in tomato and potato when compared to sorghum and older insertion age. This appears to be a peculiarity of tomato, and apparently also of potato, among angiosperms (Tomato Genome Consortium 2012).

Transposons along tomato chromosomes were annotated by the *wublast* version of *RepeatMasker* (http://www.repeatmasker.org) against the dicots section of *mipsREdat* (REdat_v8.9_Eudico). This transposon library is connected to a repeat classification scheme (mips_REcat) and contains a collection of known transposons as well as de novo detected LTR retrotransposons from tomato (1647) and potato (1309). The *RepeatMasker* output was subjected to two post-processing filter steps: (a) removal of low confidence hits (length <50 bp, score ≥255) and (b) cleaning of overlapping annotations, considering higher score hits first, and overlapping lower scored hits either shortened or, if the overlap exceeded 80 % of their length, removed.

In Table 10.3 we redefined the nucleotide coverage in terms of repeat classes for all the tomato chromosomes, starting from the available annotation from the consortium (Tomato Genome Consortium 2012).

Moreover, while the pseudomolecule images in the Nature 2012 paper (Tomato Genome Consortium 2012) reports the general behaviour or the global repeat content along tomato pseudomolecules, in this chapter we provide a more detailed view with a similar approach showing the distribution of all single classes of repeats along chromosomes 1 and 6 (Fig. 10.4a, b).

Moreover, in Fig. 10.5 we report the distribution of the transposons by the delta repeat minus gene content in a 500 kb window in chromosome 6. The plots confirmed the high content of LTR retrotransposon in repeat-rich regions, that should correspond to heterochromatin regions (Di Filippo et al. 2012) with higher content of the gypsy-like class and much lower content of the copia-like one. The plots also show that, among non-LTR retrotransposon, the SINE are more frequent in gene richer regions, as also demonstrated at BAC level (Di Filippo et al. 2012), with a similar trend also from LINE.

Table 10.3 Number of nucleotides covered by transposons per chromosomes

	Length	*N*	Retro transposon	LTR	LTR copia	LTR gypsy	LINE	SINE	DNA transposon	DNA En-Spm	DNA Harbinger	DNA hAT	DNA MuDR	Other
C00	21,805,821	3,139,100	5	11,455	1762	4741	10	58	135	18	7	29	26	14
C01	98,543,444	12,423,381	3962	32,077,426	5,750,265	19,161,095	72,439	129,011	290,884	43,507	14,911	96,182	16,837	14,521
C02	55,340,444	8,083,432	1605	14,898,384	2,445,478	8,553,327	66,702	75,569	189,419	18,401	11,143	33,888	9377	11,392
C03	70,787,664	9,925,850	3167	22,389,432	3,411,165	13,988,572	66,267	99,920	220,311	29,141	10,310	85,887	9375	5984
C04	66,470,942	6,021,919	4998	23,441,720	4,051,988	13,641,676	93,403	113,876	234,454	30,573	9137	50883	14,996	12,163
C05	65,875,088	4,634,458	4760	24,682,363	4,329,908	16,365,009	66,477	96,407	179,533	27,364	6125	84,475	11,744	9455
C06	49,751,636	6,178,685	3216	14,940,362	2,694,369	8,510,709	64,352	79,036	176,359	15,585	9356	49,379	6312	7342
C07	68,045,021	6,084,209	3232	25,568,639	4,245,047	15,897,705	60,272	103,023	172,764	27,979	7901	62,597	10,107	6780
C08	65,866,657	6,081,969	3236	24,947,566	3,749,174	15,354,025	55,707	90,328	180,081	38,447	7729	72,172	10,124	8276
C09	72,482,091	7,614,308	4151	26,933,831	4,358,334	16,786,921	41,483	109,364	179,824	31,557	8050	36,901	12,588	4313
C10	65,527,505	4,736,321	3387	25,077,433	4,499,404	15,169,358	40,894	94,067	196,755	64,172	5481	99,554	12,679	6774
C11	56,302,525	6,045,240	4510	19,645,202	3,058,258	11,949,620	75,219	93,800	163,561	22,893	8539	44,329	8894	10,355
C12	67,145,203	5,338,821	5274	26,165,887	4,402,233	16,376,894	49,614	119,045	210,762	18,246	7514	44,773	9091	8398
Sum	823,944,041	86,307,693	45,503	280,779,700	46,997,385	171,759,652	752,839	1,203,504	2,394,842	367,883	106,203	761,049	132,150	105,767

The LTR column includes annotations without further classification. Chromosome lengths (length) and number of Ns per chromosome are also specified

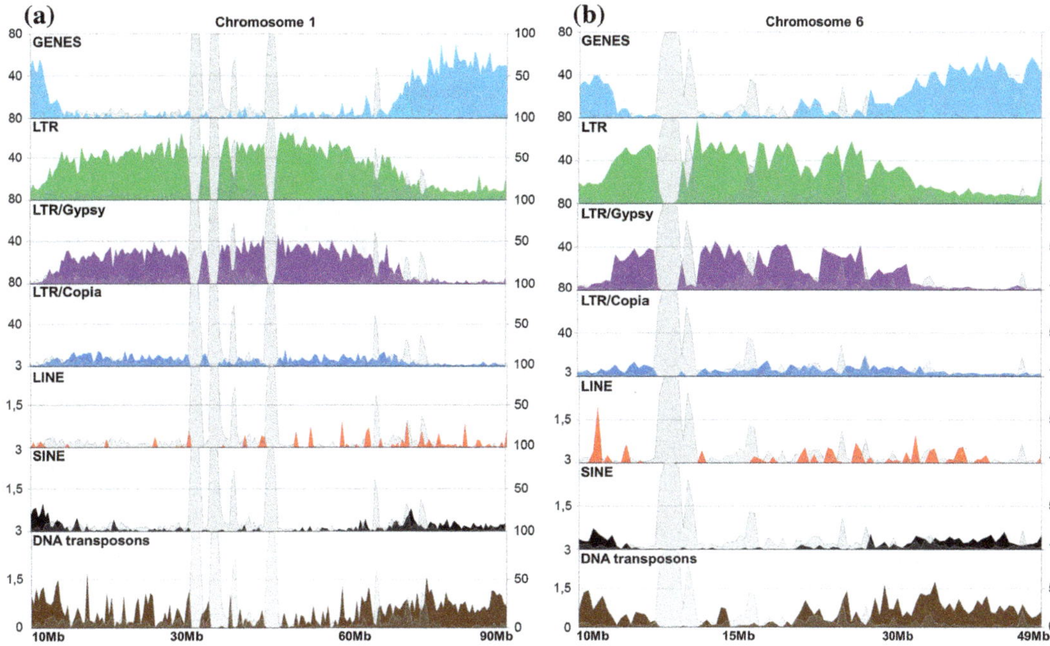

Fig. 10.4 Distribution of gene and repeat content along chromosomes 1 and 6. *Annotation of line*, LTR, Gypsy, Copia, Sine and DNA transposons were obtained from *ITAG2.4_repeats.gff3*; gene annotations were from *ITAG2.4_gene_models.gff3*, both available at http://solgenomics.net/. Data are reported by a 500 Kb nonoverlapping window. *Left* and *right* y-axes represent different percentages. The *right* y-axes represent the number of undefined nucleotide (*N*) per window

The *iTAG2.4_repeats.gff3* file used to perform this analysis was downloaded from the ftp section at http://solgenomics.net/.

Discussion

Solanaceae is an unusually divergent family consisting of approximately 90 genera and 3000–4000 species (Knapp et al. 2004) and almost all members share the same chromosome number ($x = 12$) (Wikstrom et al. 2001). Though the genomes appeared to have undergone relatively small numbers of chromosomal rearrangements (Park et al. 2011), they maintained a conserved gene content and order (Bonierbale et al. 1988; Tanksley et al. 1988; Prince et al. 1993; Livingstone et al. 1999; Wang et al. 2008; Wu et al. 2009). Though, the sequencing of different genotypes of the same species revealed microscale heterogeneity between cultivated and wild species (Traini et al. 2013; Ercolano et al. 2014; Qin et al. 2014), the overall conservation of the Solanaceae gene regions was generally described as conserved, even at the level of syntenic segments (Wang et al. 2011). The level of conservation revealed at gene level, however, is not confirmed when considering genome size, repetitive sequence content and composition. Within the Solanaceae family, *Solanum lycopersicum* (tomato) has a genome size of ~950 Mb, the genome size of *Solanum tuberosum* (potato) is 840 Mb and *Capsicum annuum* (pepper) genomes is of 3349 Mb, though the estimated gene content is comparable, suggesting a possible significant role of repeats in the speciation of these clade of plants (Zhu et al. 2008).

The 12 tomato chromosomes consist of an extended heterochromatic region (>60 % genome), mostly representing the telomeres and extended pericentromeric regions. The euchromatin regions locate in the distal part of the chromosome (Peterson et al. 1996, 1998), composed of most single-copy sequences with fewer

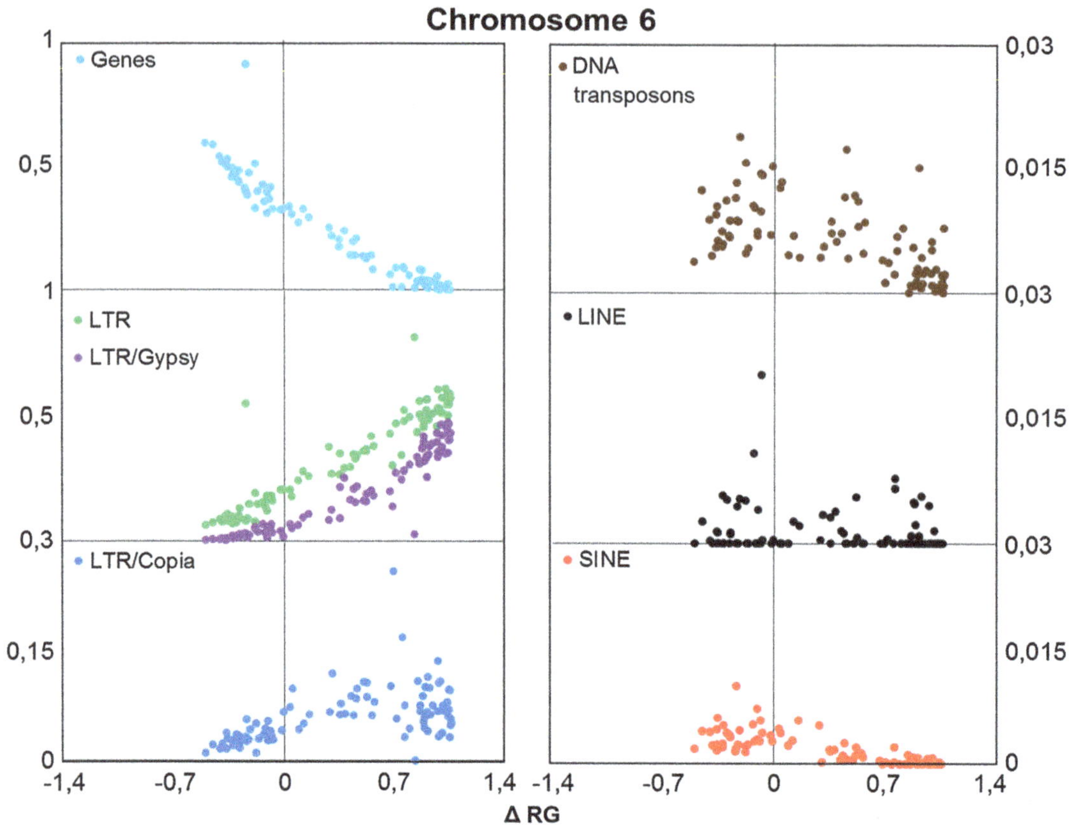

Fig. 10.5 Distribution of main repeat classes by windows of 500 kb along chromosome 6. The data are reported as frequency in the window versus the difference between repeat and gene content frequency (ΔRG). Annotations were obtained as for Fig. 10.3

retrotransposon and the 90 % of the genes (Chang et al. 2008).

Pericentromeric heterochromatin is generally assumed to be gene poor and repeat-rich, where crossing over is severely repressed (Sherman and Stack 1995). The pericentromeric heterochromatic segments contain a large portion of retrotransposons, other types of repeated sequences and some single-copy sequences, which also include a lower but representative gene content (Di Filippo et al. 2012).

Among tandem repeats, ribosomal DNA represents one of the most abundant repetitive DNA family. The repeat unit, estimated to be 9.1 Kb, was expected of 2300 copies and at the end of chromosome 2 by Ganal et al. (1988). rDNA should represent the 3 % of the tomato genome and its distribution was described also by several other efforts (Vallejos et al. 1986; Lapitan et al. 1991). As reported in this chapter, the rDNA regions appear not to be exhaustively covered by the tomato genome sequencing and by the associated annotation, and this is presumably the reason why they are not broadly discussed in the effort (Tomato Genome Consortium 2012). However, the presence of satellite DNA joint to the intergenic spacer of rDNA units also reveals the strong association of these two types of repeats and a possible initiation of satellite repeats from these loci (Jo et al. 2009).

Previous analysis also confirmed a 162 bp satellite repeat, named TGRI, with 77,000 copies in the genome as localized within a few hundred kb of the terminal 7 bp telomeric repeat TT(T/A)AGGG in tomato, at 20 of 24 chromosome ends (Ganal et al. 1988). In addition, internal telomeric repeats (ITR) were also found at a few centromeric and interstitial sites (Lapitan et al. 1989; Ganal et al. 1992; Presting et al. 1996), opening interesting questions on the reasons of this organization, as also highlighted in this chapter.

Two other tomato genomic repeats, TGRII and TGRIII, are less abundant, and were estimated with 4200 and 2100 copies, respectively. TGRII is apparently randomly distributed with quite a regular spacing of 133 kb (Ganal et al. 1988), while TGRIII is predominantly clustered in the pericentromeric region. The TGRIV repeat was discovered later and it was found mainly associated to satellite repeats in the centromere (Chang et al. 2008).

Microsatellite polymorphism and genomic distribution were studied in tomato by fingerprinting using labelled oligonucleotide probes complementary to GATA or GACA microsatellites (Vosman et al. 1992; Grandillo and Tanksley 1996). The mapping of individual fingerprint bands showed main association to centromeres (Arens et al. 1995). The copy number and the size of microsatellite containing restriction fragments were proved to be highly variable between tomato cultivars (Arens et al. 1995). Structure, abundance, variability and location were also evaluated (Broun and Tanksley 1996) and successfully used for genotyping tomato cultivars and accessions (Smulders et al. 1997; Bredemeijer et al. 2002). Interestingly, what is evident in tomato is the presence of compound satellite repeats, highly variable in length and strongly specific to the species. Ganal et al. (1988), underlined that the distribution of the major classes of tandem repeats described in tomato is limited to this species. This is probably due to high evolving rate of these regions. Zamir and Tanksley (1988) also reported a positive correlation between copy number and rate of diver[illegible] of repeats among DNA sequences from [illegible]olanaceae species. This means that [illegible]d regions are less conserved when [illegible]-copy regions, coherently also [illegible]tive pressure on the two [illegible]alyses revealed rapid [illegible]imal sequences (Presting et al. 1996) which is also confirmed from analysis in other Solanaceae (Gong et al. 2012; Melters et al. 2013; Wang et al. 2014; Zhang et al. 2014).

Among all classes of repeats, transposons comprise a large proportion of the tomato genome. In general, the highest contribution to dispersed repeats in plant genomes is mainly due to LTR retrotransposons (Piegu et al. 2006; Richard et al. 2008; Lee and Kim 2014). Plants show more C-value variation than other taxa (http://data.kew.org) (Bennett and Leitch 2005), which appears to be correlated with LTR retrotransposon abundance (Michael 2014). In animals non-LTR elements appear to be more abundant (Sakowicz et al. 2009). DNA transposons have minor impact on genome size because of the way they expand (Lee and Kim 2014). In particular, repeat-rich regions of the tomato genome revealed abundance of the LTR retroelements Ty3–gypsy and Ty1–copia (Yasuhara and Wakimoto 2006; Chang et al. 2008; Szinay et al. 2008; Tang et al. 2008a, b; Peters et al. 2009; Di Filippo et al. 2012), though the second class is present at a less extent, as also confirmed by the tomato genome annotation (Table 10.3; Fig. 10.5).

In Di Filippo et al. (2012), tomato genome sequences obtained by the preliminary BAC sequencing that preceded the whole-genome shotgun approach were analyzed to correlate heterochromatin and euchromatin regions with the relative gene and repeat content. Moreover, in the same effort, molecular markers, available to define the eu/heterochromatin boundaries along each tomato chromosome (data from the Solanaceae Genome Network website), and all the BACs associated to the chromosome structure by fluorescence in situ hybridization (FISH) (de Jong 1998; de Jong et al. 2000; Wang et al. 2006; Szinay et al. 2008; Tang et al. 2008a, b; Peters et al. 2009) were used to analyze the associated sequences. This gave out a preliminary confirmation based on sequence analysis that BACs associated to euchromatin in the tomato genome were indeed richer in gene and lower in repeat content when compared to BACs associated to heterocromatin regions. The analyses presented in Di Filippo et al. (2012), while confirming the initial assumption that genes were predominantly located in repeat-poor euchromatin regions, proved that the repeat-rich heterochromatic BACs were not completely depleted of genes (Yasuhara and Wakimoto 2006; Mueller et al. 2009). Interestingly, Di Filippo et al. (2012) also proposed an immediate approach to show the specific content of repeat classes in tomato gene or repeat richer BACs, corresponding to euchromatic and heterochromatic BACs, respectively. We also exploited the same approach here to confirm, at chromosome level, the distribution of different repeat classes in compositionally different genome regions (Fig. 10.5).

Today it is well known that transposons play various relevant roles in genome evolution, gene expression regulation and genetic instability. They can change position within the genome, contributing to genome reorganizations and altering the genome size, since transposition often results in duplication of the transposable elements, contributing with their movement to changes in cell function and organisms development (Nowacki et al. 2009) as well as to genome reorganization. Interestingly, in most cases transposable elements are silenced through epigenetics mechanism like methylation and chromatin remodelling. As a consequence, no phenotypic effects nor the movement of transposons occur when, in the wild type plant, they are silenced (Martienssen and Colot 2001; Reik et al. 2001). It is important to note, however, that DNA methylation is not conceived as a factor provoking heterochromatin formation (some species may lack methylation) but rather as a factor stabilizing heterochromatin structures (for review, see Wolffe and Matzke 1999).

Type, number and size of repeat domains in a genome can vary among species, but even differ between close genotypes or accessions, being useful as genome markers in karyotype analysis and chromosome markers in a segregating population. However, based on the assumption that a portion that comprises such a large extent of higher eukaryotes genome sequence cannot be without specific reasons, more interesting could be the understanding of the role and, possibly, advantages, if any, in repeat expansion or reduction, as well as association of these phenomena with heterochromatin formation. A prerequisite for heterochromatin formation appears to be the structural organization of the repeats rather than the nature of the particular sequences, or their repetitive character. It is evident that DNA repeats have specific structure role in constitutive heterochromatin, essential in multicellular organisms at chromosomal and nuclear level. At the chromosomal level, constitutive heterochromatin is present around vital areas such as telomeres and centromeres. The centromeric satellite DNA and retrotransposons are known to be essential in the recognition of the kinetochore (Zhong et al. 2002; Nagaki et al. 2003). The pericentromeric repeats are considered important in the recruitment of histone modification enzymes promoting the formation and maintenance of heterochromatin (Hall et al. 2002; Volpe et al. 2002; Zhong et al. 2002; Bender 2004; Lippman et al. 2004) and conferring protection and strength to the centromere. Around secondary constrictions, heterochromatic blocks may ensure against evolutionary change of ribosomal DNA by decreasing the frequency of crossing over in these regions during meiosis, also absorbing the effects of mutagenesis. Indeed, repetitive sequences in the form of constitutive heterochromatin appeared concomitant with the localization of the portion of the genome that was concerned with synthesis of ribosomal RNA, and with the need to protect chromosome structure and function by telomeres and centromeres, when the mitotic spindle developed in evolution. During meiosis heterochromatin may also aid in the initial alignment of chromosomes, facilitating speciation by allowing chromosomal rearrangement but also providing, through the species specificity of its DNA, barriers against cross-fertilization. At the nuclear level, constitutive heterochromatin may help to maintain the spatial relationships through all the steps of cell cycle. The repetitive DNA was therefore kept through natural selection and, because of its innate attitude to amplify and expand, it favoured eukaryotes genome expansion and evolution (Yunis and Yasmineh 1971; Bennetzen and Kellogg 1997). This occurred in the limit of a efficient management of other cellular activities (Knight et al. 2005). In principle, repeats are prone to expand but there exist also mechanisms to decrease dramatically their content, if necessary, including illegitimate or unequal recombination and other type of deletions (Grover and Wendel 2010). However, beyond the relevance here discussed, and the impact DNA repeats can have on genome evolution and expansion, it would also be rather important to investigate on further possible roles of species specific repeats in structuring and protecting the genome though the energy requirements that genome expansion can take from cell functionality.

References

Ahn S, Anderson JA, Sorrells ME, Tanksley SD (1993) Homoeologous relationships of rice, wheat and maize chromosomes. Mol Gen Genet 241(5–6):483–490

Al-Mssallem IS, Hu S, Zhang X, Lin Q, Liu W et al (2013) Genome sequence of the date palm Phoenix dactylifera L. Nat Commun 4:2274

Arabidopsis Genome Initiative T (2000) Analysis of the genome sequence of the flowering plant *Arabidopsis thaliana*. Nature 408(6814):796–815

Arens P, Odinot P, Heusden AV, Lindhout P, Vosman B (1995) GATA-and GACA-repeats are not evenly distributed throughout the tomato genome. Genome 38(1):84–90

Argout X, Salse J, Aury JM, Guiltinan MJ, Droc G et al (2011) The genome of *Theobroma cacao*. Nat Genet 43(2):101–108

Bao Z, Eddy SR (2002) Automated de novo identification of repeat sequence families in sequenced genomes. Genome Res 12(8):1269–1276

Bender J (2004) Chromatin-based silencing mechanisms. Curr Opin Plant Biol 7(5):521–526

Bennett M, Leitch I (2005) Plant DNA C-values database. Royal Botanic Gardens, Kew

Bennett MD, Smith JB (1976) Nuclear dna amounts in angiosperms. Philos Trans R Soc Lond B Biol Sci 274 (933):227–274

Bennetzen JL (2000) Comparative sequence analysis of plant nuclear genomes: microcolinearity and its many exceptions. Plant Cell 12(7):1021–1030

Bennetzen JL, Kellogg EA (1997) Do plants have a one-way ticket to genomic obesity? Plant Cell 9 (9):1509–1514

Benson G (1999) Tandem repeats finder: a program to analyze DNA sequences. Nucleic Acids Res 27 (2):573–580

Bermudez-Santana C, Attolini CS, Kirsten T, Engelhardt J, Prohaska SJ et al (2010) Genomic organization of eukaryotic tRNAs. BMC Genomics 11(1):270

Bibillo A, Eickbush TH (2004) End-to-end template jumping by the reverse transcriptase encoded by the R2 retrotransposon. J Biol Chem 279(15):14945–14953

Bonierbale MW, Plaisted RL, Tanksley SD (1988) RFLP maps based on a common set of clones reveal modes of chromosomal evolution in potato and tomato. Genetics 120(4):1095–1103

Bowen NJ, Jordan IK (2002) Transposable elements and the evolution of eukaryotic complexity. Curr Issues Mol Biol 4(3):65–76

Bredemeijer G, Cooke R, Ganal M, Peeters R, Isaac P et al (2002) Construction and testing of a microsatellite database containing more than 500 tomato varieties. Theor Appl Genet 105(6–7):1019–1026

Broun P, Ganal MW, Tanksley SD (1992) Telomeric arrays display high levels of heritable polymorphism among closely related plant varieties. Proc Natl Acad Sci 89(4):1354–1357

Broun P, Tanksley S (1996) Characterization and genetic mapping of simple repeat sequences in the tomato genome. Mol Gen Genet MGG 250(1):39–49

Brunner AM, Busov VB, Strauss SH (2004) Poplar genome sequence: functional genomics in an ecologically dominant plant species. Trends Plant Sci 9 (1):49–56

Burge SW, Daub J, Eberhardt R, Tate J, Barquist L et al (2013) Rfam 11.0: 10 years of RNA families. Nucleic Acids Res 41:D226–D232

Buschiazzo E, Gemmell NJ (2006) The rise, fall and renaissance of microsatellites in eukaryotic genomes. BioEssays 28(10):1040–1050

Chang SB, Yang TJ, Datema E, van Vugt J, Vosman B et al (2008) FISH mapping and molecular organization of the major repetitive sequences of tomato. Chromosome Res 16(7):919–933

Charlesworth B, Sniegowski P, Stephan W (1994) The evolutionary dynamics of repetitive DNA in eukaryotes. Nature 371(6494):215–220

Chen J, Huang Q, Gao D, Wang J, Lang Y et al (2013) Whole-genome sequencing of Oryza brachyantha reveals mechanisms underlying Oryza genome evolution. Nat Commun 4:1595

Cotton J (2001) Retroviruses from retrotransposons. Genome Biol 2(2):1

Craig NL, Craigie R, Gellert M, Lambowitz AM (2002) Mobile DNA II. ASM Press, Washington, DC

Cuadrado A, Jouve N (2007a) The nonrandom distribution of long clusters of all possible classes of trinucleotide repeats in barley chromosomes. Chromosome Res 15(6):711–720

Cuadrado A, Jouve N (2007b) Similarities in the chromosomal distribution of AG and AC repeats within and between Drosophila, human and barley chromosomes. Cytogenet Genome Res 119(1–2):91–99

Cuadrado A, Schwarzacher T (1998) The chromosomal organization of simple sequence repeats in wheat and rye genomes. Chromosoma 107(8):587–594

Cui L, Wall PK, Leebens-Mack JH, Lindsay BG, Soltis DE et al (2006) Widespread genome duplications throughout the history of flowering plants. Genome Res 16(6):738–749

D'Hont A, Denoeud F, Aury J-M, Baurens F-C, Carreel F et al (2012) The banana (*Musa acuminata*) genome and the evolution of monocotyledonous plants. Nature 488(7410):213–217

de Jong JH (1998) High resolution FISH reveals the molecular and chromosomal organization of repetitive sequences in tomato. Cytogenet Cell Genet 81:104

de Jong JH, Zhong XB, Fransz PF, Wennekes-van Eden J, Jacobsen E et al (2000) High resolution FISH reveals the molecular and chromosomal organisation of repetitive sequences of individual tomato chromosomes. In: Olmo E, Redi C (eds) Chromosomes today. Birkhäuser, Basel, pp 267–275

Defraia C, Slotkin RK (2014) Analysis of retrotransposon activity in plants. Methods Mol Biol 1112:195–210

Denoeud F, Carretero-Paulet L, Dereeper A, Droc G, Guyot R et al (2014) The coffee genome provides insight into the convergent evolution of caffeine biosynthesis. Science 345(6201):1181–1184

Devos KM, Gale MD (1993) Extended genetic maps of the homoeologous group 3 chromosomes of wheat, rye and barley. Theor Appl Genet 85(6–7):649–652

Devos KM, Gale MD (1997) Comparative genetics in the grasses. Plant Mol Biol 35(1–2):3–15

Devos KM, Gale MD (2000) Genome relationships: the grass model in current research. Plant Cell 12 (5):637–646

Di Domenico T, Potenza E, Walsh I, Parra RG, Giollo M et al (2014) RepeatsDB: a database of tandem repeat protein structures. Nucleic Acids Res 42:D352–D357

Di Filippo M, Traini A, D'Agostino N, Frusciante L, Chiusano ML (2012) Euchromatic and heterochromatic compositional properties emerging from the analysis of *Solanum lycopersicum* BAC sequences. Gene 499(1):176–181

Eigen M, Lindemann BF, Tietze M, Winkler-Oswatitsch R, Dress A et al (1989) How old is the genetic code? Statistical geometry of tRNA provides an answer. Science 244(4905):673–679

Ercolano MR, Sacco A, Ferriello F, D'Alessandro R, Tononi P et al (2014) Patchwork sequencing of tomato San Marzano and Vesuviano varieties highlights genome-wide variations. BMC Genomics 15:138

Flavell RB, Bennett MD, Smith JB, Smith DB (1974) Genome size and the proportion of repeated nucleotide sequence DNA in plants. Biochem Genet 12(4):257–269

Ganal MW, Broun P, Tanksley SD (1992) Genetic mapping of tandemly repeated telomeric DNA sequences in tomato (Lycopersicon esculentum). Genomics 14(2):444–448

Ganal MW, Lapitan NL, Tanksley SD (1988) A molecular and cytogenetic survey of major repeated DNA sequences in tomato (Lycopersicon esculentum). Mol Gen Genet MGG 213(2–3):262–268

Ganal MW, Lapitan NL, Tanksley SD (1991) Macrostructure of the tomato telomeres. Plant Cell 3(1):87–94

Garcia S, Garnatje T, Kovarik A (2012) Plant rDNA database: ribosomal DNA loci information goes online. Chromosoma 121(4):389–394

Gebhardt C, Ritter E, Barone A, Debener T, Walkemeier B et al (1991) RFLP maps of potato and their alignment with the homoeologous tomato genome. Theor Appl Genet 83(1):49–57

Gelfand Y, Rodriguez A, Benson G (2007) TRDB—the tandem repeats database. Nucleic Acids Res 35(suppl 1):D80–D87

Gong Z, Wu Y, Koblížková A, Torres GA, Wang K et al (2012) Repeatless and repeat-based centromeres in potato: implications for centromere evolution. Plant Cell Online 24(9):3559–3574

Grandillo S, Tanksley S (1996) QTL analysis of horticultural traits differentiating the cultivated tomato from the closely related species *Lycopersicon pimpinellifolium*. Theor Appl Genet 92(8):935–951

Greilhuber J, Doležel J, Lysák MA, Bennett MD (2005) The origin, evolution and proposed stabilization of the terms 'genome size' and 'C-value' to describe nuclear DNA contents. Ann Bot 95(1):255–260

Grover CE, Wendel JF (2010) Recent insights into mechanisms of genome size change in plants. J Bot 2010:8

Hall IM, Shankaranarayana GD, Noma K-I, Ayoub N, Cohen A et al (2002) Establishment and maintenance of a heterochromatin domain. Science 297 (5590):2232–2237

Hawkins JS, Proulx SR, Rapp RA, Wendel JF (2009) Rapid DNA loss as a counterbalance to genome expansion through retrotransposon proliferation in plants. Proc Natl Acad Sci U S A 106 (42):17811–17816

He L, Liu J, Torres GA, Zhang H, Jiang J et al (2013) Interstitial telomeric repeats are enriched in the centromeres of chromosomes in Solanum species. Chromosome Res 21(1):5–13

Henikoff S, Ahmad K, Malik HS (2001) The centromere paradox: stable inheritance with rapidly evolving DNA. Science 293(5532):1098–1102

Hertel J, Hofacker IL, Stadler PF (2008) SnoReport: computational identification of snoRNAs with unknown targets. Bioinformatics 24(2):158–164

Heslop-Harrison JS (2000) Comparative genome organization in plants: from sequence and markers to chromatin and chromosomes. Plant Cell 12 (5):617–636

Heslop-Harrison JS, Brandes A, Schwarzacher T (2003) Tandemly repeated DNA sequences and centromeric chromosomal regions of Arabidopsis species. Chromosome Res 11(3):241–253

Hirakawa H, Shirasawa K, Miyatake K, Nunome T, Negoro S et al (2014) Draft genome sequence of eggplant (Solanum melongena L.): the representative solanum species indigenous to the old world. DNA Res 21(6):649–660

Houben A, Schubert I (2003) DNA and proteins of plant centromeres. Curr Opin Plant Biol 6(6):554–560

Hu TT, Pattyn P, Bakker EG, Cao J, Cheng JF et al (2011) The Arabidopsis lyrata genome sequence and the basis of rapid genome size change. Nat Genet 43(5):476–481

Huang S, Li R, Zhang Z, Li L, Gu X et al (2009) The genome of the cucumber, Cucumis sativus L. Nat Genet 41(12):1275–1281

International Barley Genome Sequencing, Mayer CKF, Waugh R, Brown JW, Schulman A et al (2012) A physical, genetic and functional sequence assembly of the barley genome. Nature 491(7426):711–716

International Rice Genome Sequencing, P (2005) The map-based sequence of the rice genome. Nature 436 (7052):793–800

IWGSC (2014) A chromosome-based draft sequence of the hexaploid bread wheat (Triticum aestivum) genome. Science 345(6194):1251788

Jaillon O, Aury JM, Noel B, Policriti A, Clepet C et al (2007) The grapevine genome sequence suggests ancestral hexaploidization in major angiosperm phyla. Nature 449(7161):463–467

Jarne P, Lagoda PJ (1996) Microsatellites, from molecules to populations and back. Trends Ecol Evol 11 (10):424–429

Jiao Y, Leebens-Mack J, Ayyampalayam S, Bowers JE, McKain MR et al (2012) A genome triplication associated with early diversification of the core eudicots. Genome Biol 13(1):R3

Jiao Y, Wickett NJ, Ayyampalayam S, Chanderbali AS, Landherr L et al (2011) Ancestral polyploidy in seed plants and angiosperms. Nature 473(7345):97–100

Jo S-H, Koo D-H, Kim J, Hur C-G, Lee S et al (2009) Evolution of ribosomal DNA-derived satellite repeat in tomato genome. BMC Plant Biol 9(1):42

Jurka J (1997) Sequence patterns indicate an enzymatic involvement in integration of mammalian retroposons. Proc Natl Acad Sci 94(5):1872–1877

Jurka J, Kapitonov VV, Pavlicek A, Klonowski P, Kohany O et al (2005) Repbase Update, a database of eukaryotic repetitive elements. Cytogenet Genome Res 110(1–4):462–467

Kalitsis P, Choo KH (2012) The evolutionary life cycle of the resilient centromere. Chromosoma 121(4):327–340

Kim S, Park M, Yeom SI, Kim YM, Lee JM et al (2014) Genome sequence of the hot pepper provides insights into the evolution of pungency in Capsicum species. Nat Genet 46(3):270–278

King DG, Soller M, Kashi Y (1997) Evolutionary tuning knobs. Endeavour 21(1):36–40

Knapp S et al (2004) Solanaceae—a model for linking genomics with biodiversity. Comp Funct Genomics 5(3):285–291

Knight CA, Molinari NA, Petrov DA (2005) The large genome constraint hypothesis: evolution, ecology and phenotype. Ann Bot 95(1):177–190

Kohany O, Gentles AJ, Hankus L, Jurka J (2006) Annotation, submission and screening of repetitive elements in repbase: repbasesubmitter and censor. BMC Bioinform 7:474

Kolpakov R (2003) mreps: efficient and flexible detection of tandem repeats in DNA. Nucleic Acids Res 31 (13):3672–3678

Kramerov DA, Vassetzky NS (2005) Short retroposons in eukaryotic genomes. Int Rev Cytol 247:165–221

Kruglyak S, Durrett RT, Schug MD, Aquadro CF (1998) Equilibrium distributions of microsatellite repeat length resulting from a balance between slippage events and point mutations. Proc Natl Acad Sci USA 95(18):10774–10778

Kumar A, Bennetzen JL (1999) Plant retrotransposons. Annu Rev Genet 33:479–532

Lapitan NL, Ganal MW, Tanksley SD (1989) Somatic chromosome karyotype of tomato based on in situ hybridization of the TGRI satellite repeat. Genome 32 (6):992–998

Lapitan NLV, Ganal MW, Tanksley SD (1991) Organization of the 5S ribosomal RNA genes in the genome of tomato. Genome 34(4):509–514

Laslett D, Canback B (2004) ARAGORN, a program to detect tRNA genes and tmRNA genes in nucleotide sequences. Nucleic Acids Res 32(1):11–16

Lee S-I, Kim N-S (2014) Transposable elements and genome size variations in plants. Genom Inform 12 (3):87–97

Ling HQ, Zhao S, Liu D, Wang J, Sun H et al (2013) Draft genome of the wheat A-genome progenitor Triticum urartu. Nature 496(7443):87–90

Lippman Z, Gendrel A-V, Black M, Vaughn MW, Dedhia N et al (2004) Role of transposable elements in heterochromatin and epigenetic control. Nature 430 (6998):471–476

Liu Z-W, Biyashev R, Maroof MS (1996) Development of simple sequence repeat DNA markers and their integration into a barley linkage map. Theor Appl Genet 93(5–6):869–876

Livingstone KD, Lackney VK, Blauth JR, Van Wijk R, Jahn MK (1999) Genome mapping in Capsicum and the evolution of genome structure in the Solanaceae. Genetics 152(3):1183–1202

Lysak MA, Koch MA, Pecinka A, Schubert I (2005) Chromosome triplication found across the tribe Brassiceae. Genome Res 15(4):516–525

Macas J, Meszaros T, Nouzova M (2002) PlantSat: a specialized database for plant satellite repeats. Bioinformatics 18(1):28–35

Majerová E, Mandáková T, Vu GTH, Fajkus J, Lysak MA et al (2014) Chromatin features of plant telomeric sequences at terminal vs. internal positions. Front Plant Sci 5:593

Maluszynska J, Heslop-Harrison J (1991) Localization of tandemly repeated DMA sequences in Arabidopsis thaliana. Plant J 1(2):159–166

Marshall OJ et al (2008) Neocentromeres: new insights into centromere structure, disease development, and karyotype evolution. Am J Hum Genet 82(2):261–282

Martienssen RA, Colot V (2001) DNA methylation and epigenetic inheritance in plants and filamentous fungi. Science 293(5532):1070–1074

Martin GB, Ganal MW, Tanksley SD (1992) Construction of a yeast artificial chromosome library of tomato and identification of cloned segments linked to two disease resistance loci. Mol Gen Genet MGG 233(1–2):25–32

Martin W, Herrmann RG (1998) Gene transfer from organelles to the nucleus: how much, what happens, and why? Plant Physiol 118(1):9–17

Martinez-Zapater JM, Estelle MA, Somerville CR (1986) A highly repeated DNA sequence in *Arabidopsis thaliana*. Mol Gen Genet MGG 204(3):417–423

Mason JM, Biessmann H (1995) The unusual telomeres of Drosophila. Trends Genet 11(2):58–62

Maughan P, Maroof MS, Buss G (1995) Microsatellite and amplified sequence length polymorphisms in cultivated and wild soybean. Genome 38(4):715–723

McCarthy EM, McDonald JF (2003) LTR_STRUC: a novel search and identification program for LTR retrotransposons. Bioinformatics 19(3):362–367

McClintock B (1953) Induction of instability at selected loci in maize. Genetics 38(6):579–599

McCouch SR, Chen X, Panaud O, Temnykh S, Xu Y et al (1997) Microsatellite marker development, mapping and applications in rice genetics and breeding. Plant Mol Biol 35(1–2):89–99

Melters DP, Bradnam KR, Young HA, Telis N, May MR et al (2013) Comparative analysis of tandem repeats from hundreds of species reveals unique insights into centromere evolution. Genome Biol 14(1):R10

Meraldi P, McAinsh AD, Rheinbay E, Sorger PK (2006) Phylogenetic and structural analysis of centromeric DNA and kinetochore proteins. Genome Biol 7(3):R23

Meyne J, Baker RJ, Hobart HH, Hsu T, Ryder OA et al (1990) Distribution of non-telomeric sites of the (TTAGGG) n telomeric sequence in vertebrate chromosomes. Chromosoma 99(1):3–10

Michael TP (2014) Plant genome size variation: bloating and purging DNA. Brief Funct Genomics 13(4):308–317

Michaud M, Cognat V, Duchêne A-M, Maréchal-Drouard L (2011) A global picture of tRNA genes in plant genomes. Plant J 66(1):80–93

Michelmore RW, Paran I, Kesseli R (1991) Identification of markers linked to disease-resistance genes by bulked segregant analysis: a rapid method to detect markers in specific genomic regions by using segregating populations. Proc Natl Acad Sci 88(21):9828–9832

Milbourne D, Meyer R, Bradshaw JE, Baird E, Bonar N et al (1997) Comparison of PCR-based marker systems for the analysis of genetic relationships in cultivated potato. Mol Breed 3(2):127–136

Minajigi A, Francklyn CS (2010) Aminoacyl transfer rate dictates choice of editing pathway in threonyl-tRNA synthetase. J Biol Chem 285(31):23810–23817

Ming R, Hou S, Feng Y, Yu Q, Dionne-Laporte A et al (2008) The draft genome of the transgenic tropical fruit tree papaya (*Carica papaya* Linnaeus). Nature 452(7190):991–996

Ming R, VanBuren R, Liu Y, Yang M, Han Y et al (2013) Genome of the long-living sacred lotus (*Nelumbo nucifera* Gaertn.). Genome Biol 14(5):R41

Mueller LA, Lankhorst RK, Tanksley SD, Giovannoni JJ, White R et al (2009) A snapshot of the emerging tomato genome sequence. Plant Gen 2(1):78–92

Nagaki K, Murata M (2005) Characterization of CENH3 and centromere-associated DNA sequences in sugarcane. Chromosome Res 13(2):195–203

Nagaki K, Shibata F, Kanatani A, Kashihara K, Murata M (2012) Isolation of centromeric-tandem repetitive DNA sequences by chromatin affinity purification using a HaloTag7-fused centromere-specific histone H3 in tobacco. Plant Cell Rep 31(4):771–779

Nagaki K, Shibata F, Suzuki G, Kanatani A, Ozaki S et al (2011) Coexistence of NtCENH3 and two retrotransposons in tobacco centromeres. Chromosome Res 19 (5):591–605

Nagaki K, Song J, Stupar RM, Parokonny AS, Yuan Q et al (2003) Molecular and cytological analyses of large tracks of centromeric DNA reveal the structure and evolutionary dynamics of maize centromeres. Genetics 163(2):759–770

Nawrocki EP, Kolbe DL, Eddy SR (2009) Infernal 1.0: inference of RNA alignments. Bioinformatics 25(10):1335–1337

Nowacki M, Higgins BP, Maquilan GM, Swart EC, Doak TG et al (2009) A functional role for transposases in a large eukaryotic genome. Science 324(5929):935–938

NSF (1990) Document 90–80. A long-range plan for the multinational coordinated Arabidopsis thaliana genome research project. National Science Foundation, Washington, DC

Nussbaumer T, Martis MM, Roessner SK, Pfeifer M, Bader KC et al (2013) MIPS PlantsDB: a database framework for comparative plant genome research. Nucleic Acids Res 41:D1144–D1151

Ouyang S, Buell CR (2004) The TIGR plant repeat databases: a collective resource for the identification of repetitive sequences in plants. Nucleic Acids Res 32:D360–D363

Park M, Jo S, Kwon J-K, Park J, Ahn JH et al (2011) Comparative analysis of pepper and tomato reveals euchromatin expansion of pepper genome caused by differential accumulation of Ty3/Gypsy-like elements. BMC Genomics 12(1):85

Paterson AH, Bowers JE, Bruggmann R, Dubchak I, Grimwood J et al (2009) The Sorghum bicolor genome and the diversification of grasses. Nature 457(7229):551–556

Peters SA, Datema E, Szinay D, van Staveren MJ, Schijlen EG et al (2009) *Solanum lycopersicum* cv. Heinz 1706 chromosome 6: distribution and abundance of genes and retrotransposable elements. Plant J 58(5):857–869

Peterson DG, Pearson WR, Stack SM (1998) Characterization of the tomato (*Lycopersicon esculentum*)

genome using in vitro and in situ DNA reassociation. Genome 41(3):346–356

Peterson DG, Stack SM, Price HJ, Johnston JS (1996) DNA content of heterochromatin and euchromatin in tomato (*Lycopersicon esculentum*) pachytene chromosomes. Genome 39(1):77–82

Phizicky EM, Hopper AK (2010) tRNA biology charges to the front. Genes Dev 24(17):1832–1860

Piegu B, Guyot R, Picault N, Roulin A, Saniyal A et al (2006) Doubling genome size without polyploidization: dynamics of retrotransposition-driven genomic expansions in Oryza australiensis, a wild relative of rice. Genome Res 16(10):1262–1269

Podlevsky JD, Bley CJ, Omana RV, Qi X, Chen JJL (2008) The telomerase database. Nucleic Acids Res 36:D339–D343

Potato Genome Sequencing Consortium, T (2011) Genome sequence and analysis of the tuber crop potato. Nature 475(7355):189–195

Presting GG, Frary A, Pillen K, Tanksley SD (1996) Telomere-homologous sequences occur near the centromeres of many tomato chromosomes. Mol Gen Genet MGG 251(5):526–531

Price AL, Jones NC, Pevzner PA (2005) De novo identification of repeat families in large genomes. Bioinformatics 21(Suppl 1):i351–i358

Prince JP, Pochard E, Tanksley SD (1993) Construction of a molecular linkage map of pepper and a comparison of synteny with tomato. Genome 36 (3):404–417

Qin C, Yu C, Shen Y, Fang X, Chen L et al (2014) Whole-genome sequencing of cultivated and wild peppers provides insights into Capsicum domestication and specialization. Proc Natl Acad Sci USA 111 (14):5135–5140

Reik W, Dean W, Walter J (2001) Epigenetic reprogramming in mammalian development. Science 293 (5532):1089–1093

Richard G-F, Kerrest A, Dujon B (2008) Comparative genomics and molecular dynamics of DNA repeats in eukaryotes. Microbiol Mol Biol Rev 72(4):686–727

Rocchi M, Archidiacono N, Schempp W, Capozzi O, Stanyon R (2012) Centromere repositioning in mammals. Heredity 108(1):59–67

Sakowicz T, Gadzalski M, Pszczółkowski W (2009) Short interspersed elements (SINEs) in plant genomes. Adv Cell Biol 1:1–12

Sangiovanni M, Vigilante A, Chiusano ML (2013) Exploiting a reference genome in terms of duplications: the network of paralogs and single copy genes in *Arabidopsis thaliana*. Biol (Basel) 2(4):1465–1487

Sato S, Nakamura Y, Kaneko T, Asamizu E, Kato T et al (2008) Genome structure of the legume, Lotus japonicus. DNA Res 15(4):227–239

Schattner P, Brooks AN, Lowe TM (2005) The tRNAscan-SE, snoscan and snoGPS web servers for the detection of tRNAs and snoRNAs. Nucleic Acids Res 33:W686–W689

Schmidt T, Heslop-Harrison J (1998) Genomes, genes and junk: the large-scale organization of plant chromosomes. Trends Plant Sci 3(5):195–199

Schmutz J, Cannon SB, Schlueter J, Ma J, Mitros T et al (2010) Genome sequence of the palaeopolyploid soybean. Nature 463(7278):178–183

Schnable PS, Ware D, Fulton RS, Stein JC, Wei F et al (2009) The B73 maize genome: complexity, diversity, and dynamics. Science 326(5956):1112–1115

Schweizer G, Ganal M, Ninnemann H, Hemleben V (1988) Species-specific DNA sequences for identification of somatic hybrids between Lycopersicon esculentum and Solanum acaule. Theor Appl Genet 75(5):679–684

Sharma S, Raina SN (2005) Organization and evolution of highly repeated satellite DNA sequences in plant chromosomes. Cytogenet Genome Res 109(1–3):15–26

Sharma SK, Bolser D, de Boer J, Sønderkær M, Amoros W et al (2013) Construction of reference chromosome-scale pseudomolecules for potato: integrating the potato genome with genetic and physical maps. G3 Genes Genomes Genet 3(11):2031–2047

Shearer LA, Anderson LK, de Jong H, Smit S, Goicoechea JL, et al (2014) Fluorescence in situ hybridization and optical mapping to correct scaffold arrangement in the tomato genome. G3 Genes Genomes Genet 4(8):1395–1405

Sherman JD, Stack SM (1995) Two-dimensional spreads of synaptonemal complexes from solanaceous plants. VI. High-resolution recombination nodule map for tomato (*Lycopersicon esculentum*). Genetics 141 (2):683–708

Shulaev V, Sargent DJ, Crowhurst RN, Mockler TC, Folkerts O et al (2011) The genome of woodland strawberry (*Fragaria vesca*). Nat Genet 43 (2):109–116

Sierro N, Battey JN, Ouadi S, Bakaher N, Bovet L et al (2014) The tobacco genome sequence and its comparison with those of tomato and potato. Nat Commun 5:3833

Smit AFA, Hubley R, Green P (1996) RepeatMasker Open-3.0. http://www.repeatmasker.org

Smulders MJM, Bredemeijer G, Rus-Kortekaas W, Arens P, Vosman B (1997) Use of short microsatellites from database sequences to generate polymorphisms among *Lycopersicon esculentum* cultivars and accessions of other Lycopersicon species. Theor Appl Genet 94(2):264–272

Smyth DR (1991) Dispersed repeats in plant genomes. Chromosoma 100(6):355–359

Suresh BV, Roy R, Sahu K, Misra G, Chattopadhyay D (2014) Tomato genomic resources database: an integrated repository of useful tomato genomic information for basic and applied research. PLoS One 9(1):e86387

Swift H (1950) The constancy of desoxyribose nucleic acid in plant nuclei. Proc Natl Acad Sci USA 36 (11):643–654

Szinay D, Chang SB, Khrustaleva L, Peters S, Schijlen E et al (2008) High-resolution chromosome mapping of BACs using multi-colour FISH and pooled-BAC FISH as a backbone for sequencing tomato chromosome 6. Plant J 56(4):627–637

Tang H, Bowers JE, Wang X, Ming R, Alam M et al (2008a) Synteny and collinearity in plant genomes. Science 320(5875):486–488

Tang X, Szinay D, Lang C, Ramanna MS, van der Vossen EA et al (2008b) Cross-species bacterial artificial chromosome-fluorescence in situ hybridization painting of the tomato and potato chromosome 6 reveals undescribed chromosomal rearrangements. Genetics 180(3):1319–1328

Tanksley SD, Bernatzky R, Lapitan NL, Prince JP (1988) Conservation of gene repertoire but not gene order in pepper and tomato. Proc Natl Acad Sci USA 85 (17):6419–6423

Tautz D, Schlotterer (1994) Simple sequences. Curr Opin Genet Dev 4(6):832–837

Tek AL, Jiang J (2004) The centromeric regions of potato chromosomes contain megabase-sized tandem arrays of telomere-similar sequence. Chromosoma 113 (2):77–83

Tomato Genome Consortium, T (2012) The tomato genome sequence provides insights into fleshy fruit evolution. Nature 485(7400):635–641

Traini A, Iorizzo M, Mann H, Bradeen JM, Carputo D, Frusciante L, Chiusano ML (2013) Genome microscale heterogeneity among wild potatoes revealed by diversity arrays technology marker sequences. Int Journal Genomics 2013:9

Vallejos CE, Tanksley SD, Bernatzky R (1986) Localization in the tomato genome of DNA restriction fragments containing sequences homologous to the rRNA (45s), the major chlorophyll a/b binding polypeptide and the ribulose bisphosphate carboxylase genes. Genetics 112(1):93–105

Vaughan H, Heslop-Harrison J, Hewitt G (1999) The localization of mitochondrial sequences to chromosomal DNA in orthopterans. Genome 42(5):874–880

Volpe TA, Kidner C, Hall IM, Teng G, Grewal SI et al (2002) Regulation of heterochromatic silencing and histone H3 lysine-9 methylation by RNAi. Science 297(5588):1833–1837

von Sternberg R (2002) On the roles of repetitive DNA elements in the context of a unified genomic-epigenetic system. Ann NY Acad Sci 981:154–188

Vosman B, Arens P, Rus-Kortekaas W, Smulders M (1992) Identification of highly polymorphic DNA regions in tomato. Theor Appl Genet 85(2–3):239–244

Wang L, Zeng Z, Zhang W, Jiang J (2014) Three potato centromeres are associated with distinct haplotypes with or without megabase-sized satellite repeat arrays. Genetics 196(2):397–401

Wang X, Wang H, Wang J, Sun R, Wu J et al (2011) The genome of the mesopolyploid crop species *Brassica rapa*. Nat Genet 43(10):1035–1039

Wang Y, Diehl A, Wu F, Vrebalov J, Giovannoni J et al (2008) Sequencing and comparative analysis of a conserved syntenic segment in the Solanaceae. Genetics 180(1):391–408

Wang Y, Tang X, Cheng Z, Mueller L, Giovannoni J et al (2006) Euchromatin and pericentromeric heterochromatin: comparative composition in the tomato genome. Genetics 172(4):2529–2540

Wikstrom N, Savolainen V, Chase MW (2001) Evolution of the angiosperms: calibrating the family tree. Proc Biol Sci 268(1482):2211–2220

Wiley G, Macmil S, Qu C, Wang P, Xing Y et al (2009) Methods for generating shotgun and mixed shotgun/paired-end libraries for the 454 DNA sequencer. Curr Prot Hum Genet 18.11:11-18–11-21

Wolffe AP, Matzke MA (1999) Epigenetics: regulation through repression. Science 286(5439):481–486

Wu F, Eannetta NT, Xu Y, Durrett R, Mazourek M et al (2009) A COSII genetic map of the pepper genome provides a detailed picture of synteny with tomato and new insights into recent chromosome evolution in the genus Capsicum. Theor Appl Genet 118(7):1279–1293

Xu J, Earle ED (1996a) Direct FISH of 5S rDNA on tomato pachytene chromosomes places the gene at the heterochromatic knob immediately adjacent to the centromere of chromosome 1. Genome 39(1):216–221

Xu J, Earle ED (1996b) High resolution physical mapping of 45S (5.8S, 18S and 25S) rDNA gene loci in the tomato genome using a combination of karyotyping and FISH of pachytene chromosomes. Chromosoma 104(8):545–550

Yan H, Ito H, Nobuta K, Ouyang S, Jin W et al (2006) Genomic and genetic characterization of rice Cen3 reveals extensive transcription and evolutionary implications of a complex centromere. Plant Cell 18 (9):2123–2133

Yasuhara JC, Wakimoto BT (2006) Oxymoron no more: the expanding world of heterochromatic genes. Trends Genet 22(6):330–338

Yunis JJ, Yasmineh WG (1971) Heterochromatin, satellite DNA, and cell function. Science 174(4015):1200–1209

Zamir D, Tanksley S (1988) Tomato genome is comprised largely of fast-evolving, low copy-number sequences. Mol Gen Genet MGG 213(2–3):254–261

Zhang H, Koblizkova A, Wang K, Gong Z, Oliveira L et al (2014) Boom-bust turnovers of megabase-sized centromeric DNA in solanum species: rapid evolution of DNA sequences associated with centromeres. Plant Cell 26(4):1436–1447

Zhong CX, Marshall JB, Topp C, Mroczek R, Kato A et al (2002) Centromeric retroelements and satellites interact with maize kinetochore protein CENH3. Plant Cell Online 14(11):2825–2836

Zhong XB, Fransz PF, Wennekes-van Eden J, Kammen AV, Zabel P et al (1998) FISH studies reveal the molecular and chromosomal organization of individual telomere domains in tomato. Plant J 13(4):507–517

Zhu W, Ouyang S, Iovene M, O'Brien K, Vuong H et al (2008) Analysis of 90 Mb of the potato genome reveals conservation of gene structures and order with tomato but divergence in repetitive sequence composition. BMC Genomics 9:286

Two Paleo-Hexaploidies Underlie Formation of Modern Solanaceae Genome Structure

11

Jingping Li, Haibao Tang, Xiyin Wang and Andrew H. Paterson

Abstract

Polyploidy, multiplication of whole genome content, is an important evolutionary force. Paleo-polyploidies (ancient genome duplications) have been identified in early lineages of animals, yeasts, and ciliates, but are particularly widespread in plants, with more than 32 events described. Deep impacts of paleo-polyploidies on plant evolution and diversity are a research focus in recent years. There are three unequivocally known paleo-hexaploidy (ancient genome triplication) events: one predated divergence of core eudicots ("γ"), one predated divergence of Solanaceae lineages ("T"), and one predated divergence of *Brassica* species. Two of the three events, γ and T, have affected the ancestors of all modern Solanaceae species, which includes tomato (*Solanum lycopersicum*).

J. Li · X. Wang · A.H. Paterson
Plant Genome Mapping Laboratory, University of Georgia, Athens, GA 30602, USA
e-mail: jli4@uga.edu

X. Wang
e-mail: wangxy@uga.edu

J. Li · A.H. Paterson (✉)
Institute of Bioinformatics, University of Georgia, Athens, GA 30602, USA
e-mail: Paterson@plantbio.uga.edu

H. Tang
J. Craig Venter Institute, Rockville, MD 20850, USA
e-mail: htang@jcvi.org

H. Tang
Center for Genomics and Biotechnology, Fujian Agriculture and Forestry University, Fuzhou 350002, Fujian Province, China

X. Wang
Center for Genomics and Computational Biology, School of Life Sciences, School of Sciences, Hebei United University, Tangshan 063000, Hebei, China

M. Causse et al. (eds.), *The Tomato Genome*, Compendium of Plant Genomes,
DOI 10.1007/978-3-662-53389-5_11

Signatures of the paleo-hexaploidy T were first described in the tomato genome, and confirmed in the potato (*Solanum tuberosum*) genome. Comparison among several asterid genomes revealed that T likely occurred in the Solanaceae lineage, and may have been chronologically close to the Solanaceae–Rubiaceae divergence. The successive γ and T paleo-hexaploidies produced nine theoretical copies of each ancestral locus in a modern Solanaceae haploid genome, although only a fraction of these were retained. Following triplication, the paleo-genomes underwent massive nonrandom gene loss and extensive structural rearrangement, resulting in adaptive genetic changes and evolutionary novelties. In this chapter we will review recent research on the timing and formation of the γ and T paleo-hexaploidies, and their evolutionary effects on the shaping of modern Solanaceae genomes.

Keywords
Paleo-hexaploidy · Paleo-polyploidy · Synteny · Genome evolution · Tomato · Solanaceae

Introduction

The first two asterid plant genomes, those of tomato and potato from the Solanaceae (nightshade) family, were sequenced about a decade after the first plant genome was published, that of *Arabidopsis thaliana* (a rosid) (Arabidopsis Genome Initiative 2000). They greatly expanded our knowledge of angiosperms (flowering plants), the Earth's dominant vegetation, which contains about 80 % of known plant species. Today's angiosperms consist of about 250,000 recorded species in about 450 families, of which about 75 % or 198,000 species in about 336 families are eudicots (The Angiosperm Phylogeny Group 2009; Stevens 2012; Hedges and Kumar 2009). Eudicots, characterized by two embryonic cotyledons and tricolpate pollen grains, contain two major crown clades of taxa, the rosids (~70,000 species) and the asterids (~80,000 species), which diverged about 125–93 million years ago (MYA) in early- to mid-Cretaceous (Bell et al. 2010; Moore et al. 2010; Wang et al. 2009; Bremer et al. 2004). The asterid plants consist of ~102 families, many of which are very closely associated with humans, such as tomatoes, potatoes, blueberries (Ericaceae family), coffee (Rubiaceae family), lavender (Lamiaceae family), olives (Oleaceae family), elderberries (Adoxaceae family), dogwoods (Cornaceae family), and sunflower (Asteraceae family).

One question that benefits greatly from whole genome sequencing is the effects of paleo-polyploidies, or ancient whole genome duplications (WGDs), on the evolution of plant genome structure (see Sect. 13.2). Paleo-polyploidy refers to ancient polyploidy (whole genome multiplication) events that have subsequently been diploidized (returning to disomic inheritance), resulting in the present-day haploid genome content containing more than one set of the ancestral genome. For example, a paleo-tetraploid genome has two sets of haploid genomes each containing two sets of the pre-duplication ancestral haploid genomes. Paleo-polyploidies have been reported in the eukaryotic kingdoms of Animalia (Dehal and Boore 2005; Ohno 1970), Fungi (Kellis et al. 2004; Wolfe and Shields 1997), and Chromalveolata (Aury et al. 2006), but are most widespread in Plantae. All angiosperms are paleo-polyploids, having experienced at least one, and usually more, WGDs in their lineage histories (Jiao et al. 2011; Soltis et al. 2009; Tang et al. 2008a; Cui et al. 2006;

Stebbins 1966; Masterson 1994; Blanc and Wolfe 2004). More than 32 paleo-polyploidy events have been identified in sequenced angiosperm genomes.

Even before any plant genome was sequenced, comparative mapping of molecular markers suggested that the small genome of *Arabidopsis thaliana* actually contains many paralogous regions, which may be descended from paleo-polyploidy events (Kowalski et al. 1994; Paterson et al. 1996). This inference was supported by later studies using sequence from the first plant genome of *A. thaliana* (Grant et al. 2000; Ku et al. 2000; Vision et al. 2000; Simillion et al. 2002; Bowers et al. 2003; Paterson et al. 2000). One of the key findings from the first sequenced plant genomes was the pan-core eudicot paleo-hexaploidy ($2n = 6x$) "γ" (discussed in Sect. 13. 5). Paleo-hexaploidy (ancient genome triplication) occurs or survives much less frequent than paleo-tetraploidy (ancient genome duplication, or doubling). Before the sequencing of the tomato genome, the only two other paleo-hexaploidies identified were one in the *Brassica* lineage estimated to have occurred 13–17 MYA (Wang et al. 2011), and γ. The tomato genome revealed the third case of paleo-hexaploidy (also the first case in asterids), the T event (Tomato Genome Consortium 2012), discussed in detail in Sects. 13.3 and 13.4 of this chapter.

This chapter focuses on the two paleo-hexaploidies experienced by Solanaceae ancestors. We will start by a very brief methodological overview. Then we will first discuss the pan-Solanaceae T event because it was the terminal WGD event in this lineage and therefore easier to study than the more ancient γ event that was nested inside T. After that we will discuss the pan-core eudicot γ event by first profiling it using the grape (rosids) genome where γ is a terminal WGD (grape genome experienced no reduplication following γ), and then prove that it was also shared by ancestral asterids. In the end we will discuss the evolutionary effects of γ and T on the tomato genome structure, and raise a few questions for future studies on these two and more paleo-hexaploidy events.

Methods to Identify Paleo-Polyploidy

Paleo-polyploidy events are difficult to identify because they occurred in the ancient past, during which time conservation of sequence and synteny between paralogous regions has been severely eroded. Typically more than 70–80 % of the genes duplicated in a paleo-polyploidy are subsequently lost. The remaining loci are further shuffled by post-WGD genome rearrangements. Therefore it is necessary to collect genome-wide signals for detection of WGDs. Because a paleo-polyploidy event duplicates all loci in the progenitor genome at the same time, the histogram of their paralogous genes Ks (nucleotide substitutions per synonymous site) values forms a peak corresponding to the event (Lynch and Conery 2000). Those distributions can therefore be used to identify paleo-polyploidies, with the limitations that Ks divergence cannot be resolved when it is either too small or too large, and that the rate of accumulation of mutations varies among gene families.

When genome sequence is available, the most sensitive and accurate paleo-polyploidy detection methods are synteny-based, which have been used in studies in yeasts (Kellis et al. 2004), vertebrates (Dehal and Boore 2005; Smith et al. 2013) and plants (Bowers et al. 2003; Tang et al. 2008a). In addition, synteny conservation is preserved across very long evolutionary distances, for example across eudicot-monocot comparison, and is unaffected by DNA substitution rate variation. Two synteny detection programs that are capable of aligning multiple genomes are MCscan (Tang et al. 2008a, b; Wang et al. 2012) and ADHoRe (Simillion et al. 2008; Proost et al. 2012). On the other hand, because paralogous regions from a paleo-polyploidy event usually undergo reciprocal gene loss, having a reference genome that did not experience the paleo-polyploidy (and subsequent gene loss) under study is very helpful in recovering maximum syntenic mapping between the regions. For example, in rosids some genomes have not experienced additional WGDs after γ, such as grape (Jaillon et al. 2007), papaya (Ming et al. 2008), and peach (Verde et al. 2013). These often serve as outgroups when studying more recent WGDs in other rosid lineages. For more comprehensive reviews of the methods used in paleo-polyploidy identification, readers are referred to Paterson et al. (2010) and Chap. 8 in Paterson (2014).

The Paleo-Hexaploidy T: Triplication of the Solanaceae Ancestral Genome

Paleo-polyploidy in Solanaceae was first detected from studies of genetic map data, and supported by EST data. Early comparison of a 293 loci potato genetic map with the *A. thaliana* genome suggested possible ancient segmental duplications (Gebhardt et al. 2003). Based on patterns of paralogous genes synonymous (third codon position) substitutions (Ks) in tomato and potato EST sequences, this event was inferred to be a genome-wide duplication, and estimated to predate tomato–potato divergence (Blanc and Wolfe 2004; Schlueter et al. 2004; Cui et al. 2006). Using 1,392 duplicated gene families shared by 8 plant species, Schlueter et al. (2004) modeled a log normal Ks component (median 0.632) in tomato corresponding to an inferred WGD ~52 MYA. Independent study by Blanc and Wolfe (2004) analyzed 7963 tomato and 6597 potato paralogs, and estimated a modal Ks peak of ~0.60. Using constant-rate birth–death process as a null model (Cui et al. 2006) identified a significant Ks peak (median ~0.79) in tomato paralogous genes from 10,028 EST and 5303 Unigene sequences, further supporting this paleo-polyploidy event.

Analysis of the tomato genome sequence revealed this WGD event to be a paleo-hexaploidy (triplication) (Tomato Genome Consortium 2012), which was called "T" for easy reference. Distribution of Ks values between syntenic tomato paralogs confirmed previous inferences of the paleo-polyploidy. To dissect the patterns of homeology, syntenic regions, i.e., with matching gene content and order, were aligned between the genomes of tomato and the rosid plant grape (*Vitis vinifera*) that has been free of additional WGDs after the pan-core eudicot γ event (Tang et al. 2008a, b; Jaillon et al. 2007), and is therefore a valuable reference in plant genome comparisons. This comparison clearly showed the shared γ event between the two lineages and the unshared T event in tomato (Tomato Genome Consortium 2012). Because of massive gene loss following paleo-polyploidy, most (~95.8 %) T triplicates in tomato have lost 1–2 homeologs. However across the entire genome signals of synteny are strong enough to allow detection of the triplication patterns. Genome-wide, 73 % of tomato gene loci are in blocks that are each orthologous to one grape region, collectively covering 84 % of the grape gene space. Among those grape regions, 26.8 % map to one orthologous region in tomato, 47.4 % to two, and 25.7 % to three, a pattern most parsimoniously explained by a historical triplication in tomato. By aligning against single orthologous grape genomic regions, the present-day tomato genome can be partitioned into three nearly nonoverlapping T "subgenomes" (Fig. 2 in Tomato Genome Consortium 2012). Each of the three subgenomes now spans all 12 tomato chromosomes, indicating extensive genome rearrangement since the triplication. After polyploidization, there is sometimes noticeable difference in the evolution of the subgenomes, known as biased fractionation or subgenome dominance (Schnable et al. 2011; Sankoff and Zheng 2012; Tang et al. 2012; Thomas et al. 2006). The three paleo-subgenomes in the present-day tomato genome cover 45.5, 21.5, and 9.9 % of gene loci, respectively, possibly reflecting this phenomenon.

The potato, another species in the genus *Solanum* that diverged from tomato ~7.3 MYA, was sequenced at about the same time (Potato Genome Sequencing et al. 2011), and shared the T event. The potato and tomato genomes are highly colinear (Fig. 11.6). There is relatively small ~8.7 % nucleotide divergence and 9 major inversions between the two genomes (Tomato Genome Consortium 2012). Comparison of potato and grape genomes showed single grape regions corresponding to 1–3 potato regions. Overall 27.8 % of grape genes are in regions orthologous to one region in potato,

38.1 % to two regions, and 14.5 % to three regions, collectively spanning 68 % of the gene space in potato and 80 % in grape, consistent with the results between tomato and grape. Patterns of Ks distribution among triplicated potato paralogs closely resemble those of tomato as well, and are clearly distinct from those of γ paralogs (Tomato Genome Consortium 2012). The only discrepancy lied in that the potato genome paper (Potato Genome Sequencing et al. 2011) reported this event as a duplication instead of a triplication. However, careful reexamination of Supplementary Fig. 6b of the paper, which aligned syntenic regions between grape, *Arabidopsis*, poplar, and potato, revealed that the figure missed the third T region on potato chromosome 8. Therefore, both independent analyses of the potato genome and reexamination of previous results support that T was a triplication that predated potato–tomato divergence.

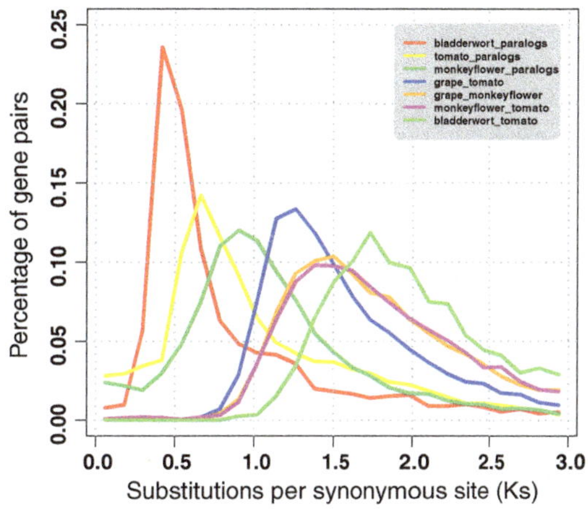

Fig. 11.1 Histograms of Ks (nucleotide substitutions per synonymous site) between paralogous and orthologous gene pairs in tomato, monkey flower, bladderwort, and grape. The x-axis is Ks values filtered as [0, 3] since Ks < 0 reflects invalid calculation in PAML and Ks > 3 exceeds empirical threshold for saturation of nucleotide divergence. The y-axis is percentage of gene pairs. The curves are plotted with different colors, but also labeled by their peak order from left to right. Comparison among the Ks distributions indicated that tomato has average nucleotide substitution rate slower than that of monkey flower, while bladderwort has the highest rate

Further Circumscribing the T Event Using Additional Asterid Genomes

Based on Ks distributions of paralogous tomato genes the triplication T was estimated to have occurred 90.4–51.6 MYA Fig. 11.1; (Tomato Genome Consortium 2012). The divergence of ancestral Euasterid I and II lineages is around 123–85 MYA (Hedges et al. 2006), making it possible that T was shared by those lineages. In order to evaluate these possibilities, newly published genomes of asterid species monkey flower (*Mimulus guttatus*, Scrophulariaceae family), bladderwort (*Utricularia gibba*, Lentibulariaceae family), kiwifruit (*Actinidia chinensis*, Actinidiaceae family), and 6 BACs from coffee (*Coffea Arabica*, Rubiaceae family) were analyzed and compared to the tomato genome. The circumscription of WGD events in these lineages is summarized in Fig. 11.5.

Kiwifruit (*Actinidia chinensis*) belongs to the basal asterid order Ericales. The kiwifruit genome experienced the γ triplication, after which it experienced two lineage-specific WGDs that were not shared with the Euasterid I and II lineages (Huang et al. 2013). Comparing the kiwifruit genome to the tomato genome revealed a synteny pattern of 4-to-3 correspondence (Fig. 11.2), indicating that the T triplication event was not shared by kiwifruit, as otherwise a 1-to-4 synteny correspondence would be observed. This inference is consistent with dating of the relative WGD and speciation events on the two lineages based on molecular data (not shown), and inferences from the kiwifruit genome paper (Huang et al. 2013).

The recently published genomes of monkey flower (*Mimulus guttatus*) and bladderwort (*Utricularia gibba*) helped confine the timing of T within the Euasterids. Bladderwort has one of the smallest genomes among flowering plants (~82 Mb). However it has experienced the γ triplication as well as three more WGDs in its lineage (Ibarra-Laclette et al. 2013) that were close in time (Fig. 11.1). Detailed synteny analysis revealed that the first of these three WGDs was shared with its sister lineage *Mimulus* of the Lamiales (Ibarra-Laclette et al. 2013), which is also the only lineage-specific

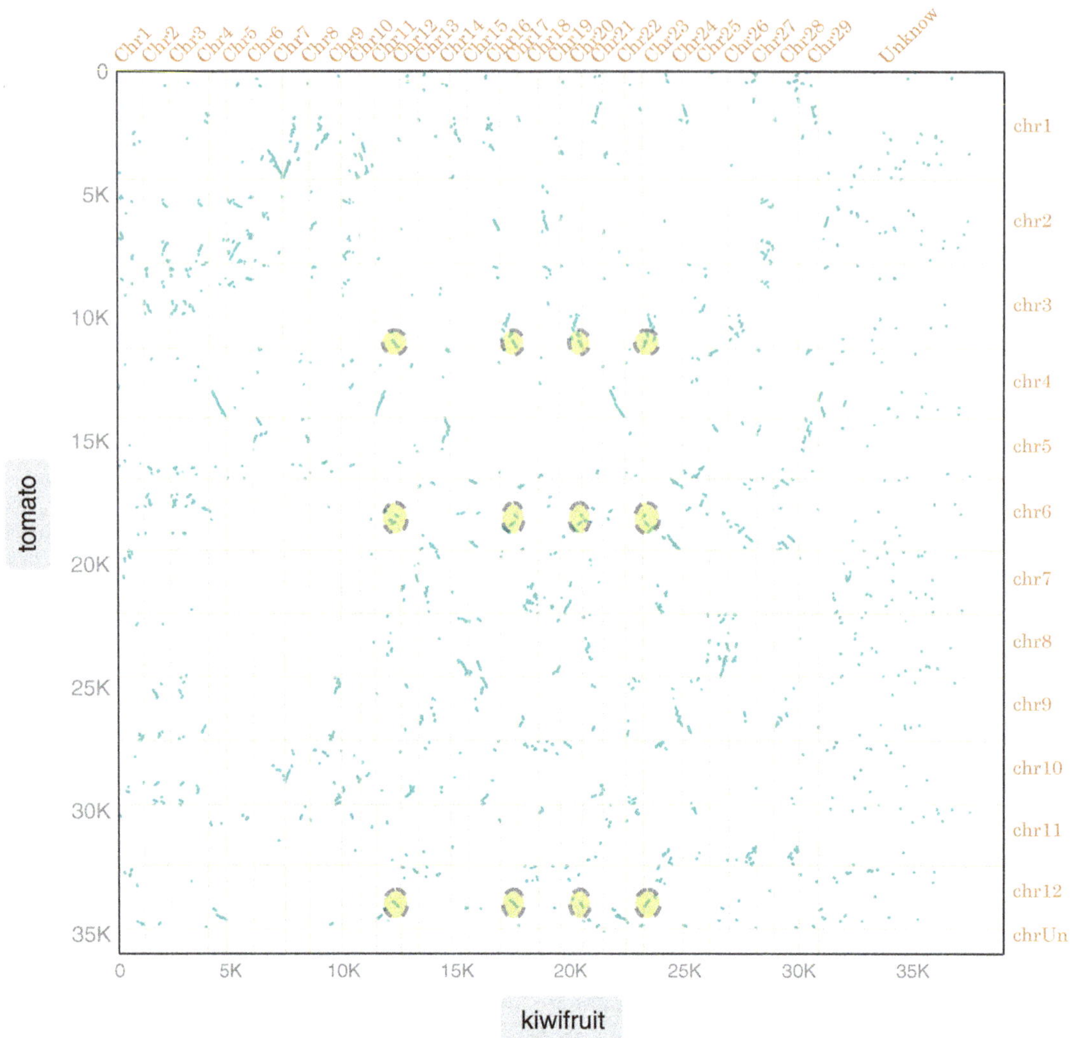

Fig. 11.2 Alignment of tomato and kiwifruit genomes. The segments labeled as "Unknown" and "chrUn" are unanchored scaffolds in the genome assemblies. *Each dot* represents a pair of syntenic genes. Continuous stretches of synteny matching are broken down by gene loss and rearrangement. *Yellow circles* with dashed borders highlight an exemplary set of syntenic regions with multiple-to-multiple (in this case 3 tomato–4 kiwifruit) correspondences, reflecting lineage-specific triplication T (3×) in tomato and 2 duplications (4×) in kiwifruit

WGD in *Mimulus* (Fig. 11.5). Since ancestral linkages are preserved better in the monkey flower genome which experienced fewer WGDs than bladderwort, the former is compared to the tomato genome (Fig. 11.3). Each set of T paralogous regions in tomato (up to 3 regions retained in the present-day genome) corresponds to up to two paralogous regions in monkey flower (Fig. 11.3), collectively spanning 88.4 % of the monkey flower genome and 82.0 % of the tomato genome. Distribution of the synteny blocks' anchor gene pairs median Ks values (an approximation of evolutionary distance between the syntenic regions) forms a single population, again suggesting that the tomato–monkey flower split predated their lineage-specific WGDs. Therefore T is likely not shared with the Lamiales, an inference also supported in the bladderwort genome paper (Ibarra-Laclette et al. 2013).

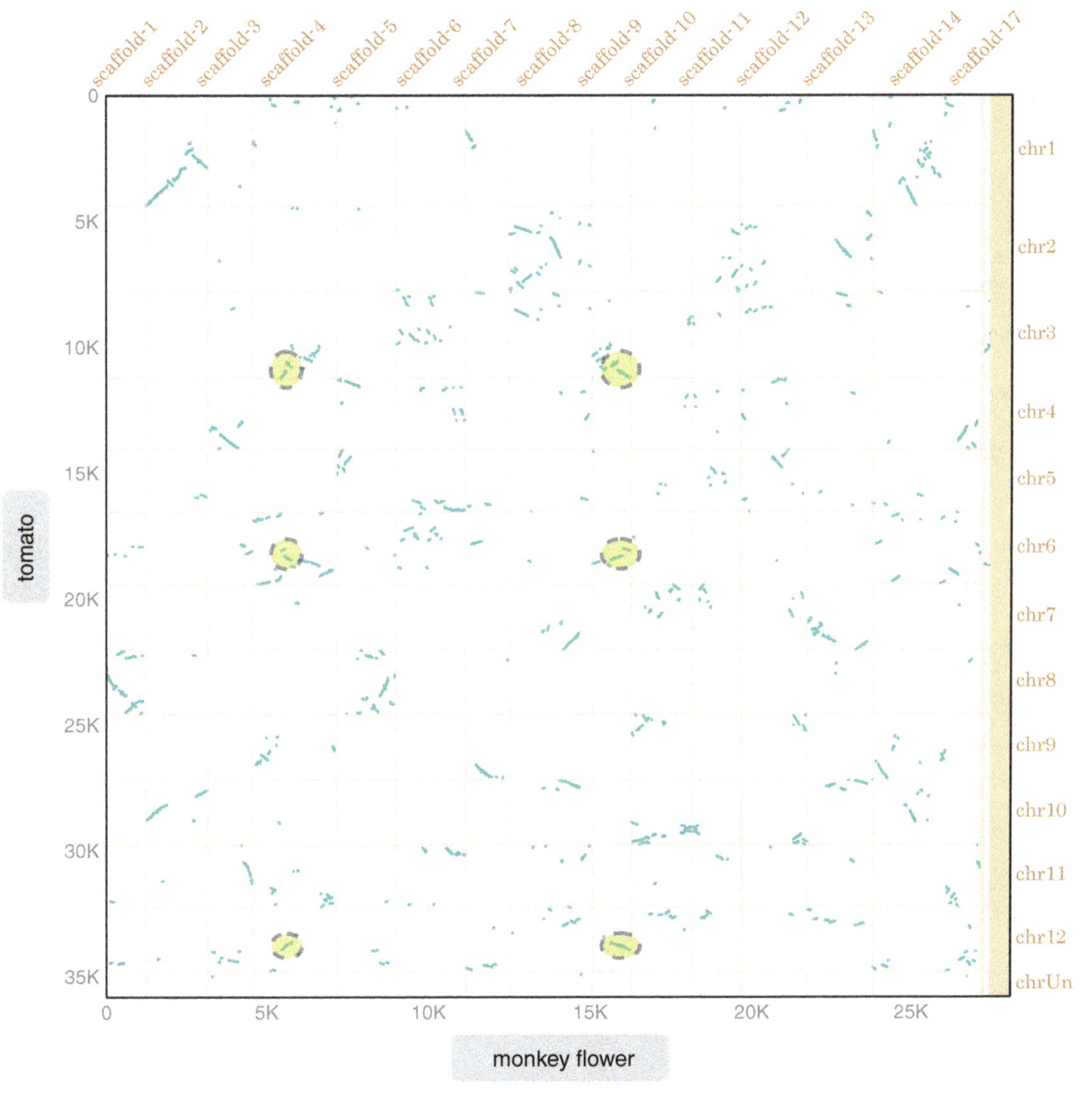

Fig. 11.3 Alignment of tomato and monkey flower genomes. Segment labeled as "chrUn" in tomato genome (y-axis) contains un-anchored scaffolds in the genome assembly. *Each dot* represents a pair of syntenic genes. Continuous stretches of syntenic matching are broken down by gene loss and rearrangement. *Yellow circles* with dashed borders highlight an exemplary set of syntenic regions with multiple-to-multiple (in this case 3 tomato–2 monkey flower) correspondences, reflecting lineage-specific triplication T (3×) in tomato and duplication (2×) in monkey flower

The coffee plant *Coffea arabica* belongs to the asterid order Gentianales, which is thought to have separated with the Solanales after their common ancestor diverged from the Lamiales (Moore et al. 2010; Soltis et al. 2011). As of this writing, there is no published genome sequence in Gentianales, but there are six coffee BACs in NCBI (GU123894–GU123899) coming from a contiguous region of ~900 Kb. Sequence alignment and colinearity analysis revealed that this region is syntenic to three tomato regions triplicated in T: Chr3:0.13–0.35 Mb, Chr6:33.0–33.4 Mb, Chr9:63.7–64.7 Mb (Fig. 11.4). The region on tomato Chr9 has significantly more hits to the coffee region than those on chr6 or chr3 (198, 86, 108 respectively, Chi-square test

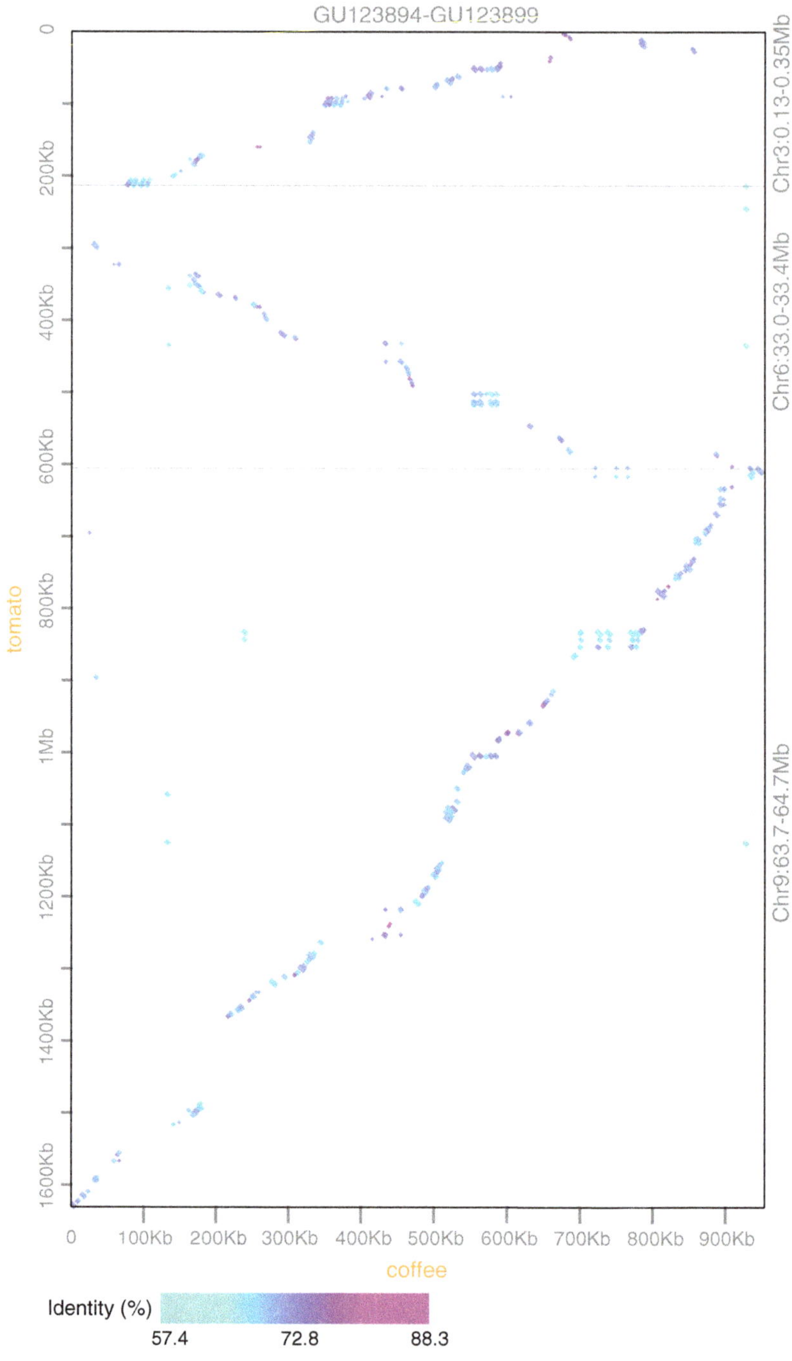

Fig. 11.4 LASTZ alignment of a ~900 kb coffee BAC sequenced region (Cenci et al. 2010) and its three syntenic regions on tomato chromosomes 3, 6, 9 (triplet from the paleo-hexaploidy T). The hits are represented by stretches of lines on the plot, with colors coded by percent identity, and line width proportional to the logarithm of hit length

$P = 1.79e{-}11$), favoring the "WGD shared" model, i.e., tomato–coffee divergence postdated triplication T. Analysis of two additional BACs (MA29G21 and MA17P03) from a pair of orthologous regions in a recent allo-tetraploid *Coffea arabica* strain also supported the model of triplication shared, with both of the BACs showing differentiated distance to the tomato triplets, and synteny between at least one pair of the homeologous regions lost or diminished beyond detection. Although biased fractionation of the T paleo-subgenomes could be an alternative

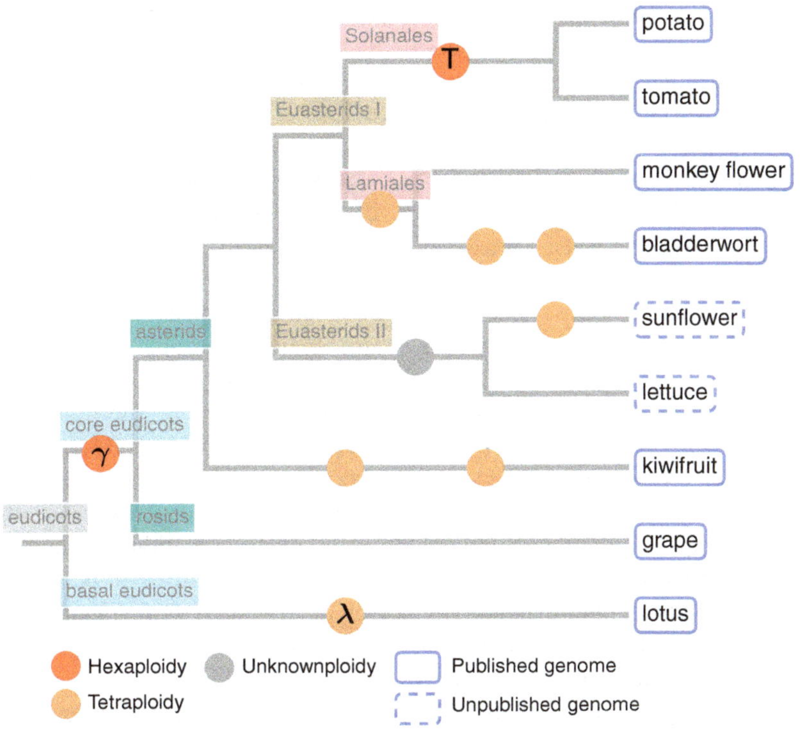

Fig. 11.5 Simplified cladogram of some representative asterid and outgroup lineages. The phylogenetic relationships are according to APG III (The Angiosperm Phylogeny Group 2009) and to our current best knowledge are unambiguous. Branch length has no meaning. Paleo-polyploidy events identified in those lineages are represented by *circles*, labeled with their names if given. The WGD event in the ancestor of sunflower and lettuce may be a triplication (Truco et al. 2013). The main references for the paleo-polyploidy events are: (Truco et al. 2013; Ming et al. 2013; Ibarra-Laclette et al. 2013; Huang et al. 2013; Tomato Genome Consortium 2012; Tang et al. 2008b; Barker et al. 2008; Jaillon et al. 2007; Hellsten et al. 2013)

explanation, such levels of difference in synteny retention as seen in the coffee–tomato comparisons are not usually seen among orthologous regions, but often seen between orthologous and out-paralogous regions, hence favoring the hypothesis that T was shared by ancestors of tomato and coffee. On the other hand, percentage identity of hits is not significantly different among the three alignments (Fig. 11.4, pairwise Wilcoxon rank sum test *P* values are: Chr3 hits and Chr6 hits: 0.277; Chr3 hits and Chr9 hits: 0.008; Chr6 hits and Chr9 hits: 0.212), supporting the alternative hypotheses that coffee did not share T, or that tomato and coffee diverged shortly after sharing T. A definitive inference will be possible when the genome sequences of coffee or other Gentianales become available.

In summary our current best inference is that the T event likely occurred near the Gentianales-Solanales split, a rough estimation of which is 108–71 MYA (Hedges et al. 2006). The exact distribution of asterid lineages that have experienced the paleo-hexaploidy T will become clear when more genomes are sequenced from this clade.

A More Ancient Hexaploidy γ Predated Divergence of Rosid and Asterid Plants

When comparing the first plant genome of *A. thaliana* with a soybean genetic map (Grant et al. 2000) and a 105 Kb tomato BAC region (Ku et al. 2000) it was suggested that the compact *A. thaliana* genome may nonetheless contain more than two paleo-subgenomes, possibly resulting from two or more paleo-polyploidies (Ku et al. 2000). Indeed, using a sensitive phylogenomic approach 34 paralogous regions covering a total of 89 % of the *A. thaliana* genome were circumscribed into three WGD events, named "γ," "β," and "α" (Bowers et al. 2003), the first of which turned out to be a hexaploidy (Jaillon et al. 2007; Tang et al. 2008b). Through several studies in recent years, the γ event has been found to be shared by most or all core eudicot lineages.

Synteny comparison between tomato and grape revealed that γ predated the asterid-rosid divergence. In an analysis of 72 tomato BACs and the sequenced grape genome, each individual tomato BAC has primary association to only one of the triplicate regions rather than showing equal matches to each of the three γ regions in grape, suggesting that γ likely predated tomato–grape divergence (Tang et al. 2008b). This inference was later supported by analysis of the tomato genome, in which individual regions correspond most closely to only one of the triplicated regions in grape, and no grape region is orthologous to more than one set of re-triplicated regions in tomato (Tomato Genome Consortium 2012).

On the other hand, the genome of the first sequenced basal eudicots, Sacred lotus (*Nelumbo nucifera*) of the order Proteales, did not share γ (but rather had a lineage-specific paleo-tetraploidy event "λ") (Ming et al. 2013), placing γ somewhere on the basal eudicot branches after the Proteales lineage branched off. Two recent studies have further confined the timing of the γ paleo-hexaploidy to a narrow window shortly predating the divergence of the earliest core eudicot lineages. Phylogenetic analysis of 769 gene families from a large collection of angiosperm species dated γ after the divergence of the Ranunculales (a basal eudicot) and core eudicots (Jiao et al. 2012). Phylogenetic analysis of subfamilies of MADS-box genes and transcriptomes from several basal eudicot species further placed γ after the divergence of two basal eudicot orders (Buxales and Trochodendrales) and the rest of eudicots, but before the branching of the Gunnerales (basal core eudicots) (Vekemans et al. 2012).

The Nature and Consequences of the γ and T Paleo-Hexaploidy Events

Subgenomes joined in a polyploidization event are typically "diploidized," i.e., gradually restoring diploid inheritance through processes of fractionation (loss of duplicated genes) (Thomas et al. 2006; Force et al. 1999; Lynch and Conery 2000) and structural rearrangement (Wolfe 2001; Tang et al. 2008a). Substantial difference in the levels of fractionation among subgenomes is sometimes indicative of possible ancient allo-polyploidy. Study of fractionation patterns in the three grape subgenomes produced in the γ paleo-hexaploidy showed that two subgenomes are more fractionated with respect to each other than to the third subgenome, suggesting that γ possibly involved hybridization between two somewhat divergent species, one of which had been previously autotetraploidized (Lyons et al. 2008). However, hybridization of differentiated progenitors is not a necessary condition for differentiated fractionation patterns between subgenomes, which could also be the results of post-polyploidy evolution. Phylogenetic trees constructed from triplets of γ paralogs and outgroup genes lack one dominant topology, suggesting that γ may also have been an autohexaploidy formed from a single progenitor, or an allo-hexaploidy formed from fusions of three moderately diverged genomes (Tang et al. 2008b). More knowledge of the ancestral karyotypes will be needed to distinguish between those evolutionary scenarios.

Much reminiscent of the case of γ, on one hand T triplets in tomato produce a mixed population of phylogenetic trees with all the possible topologies, indicating lack of sequence divergence in the T progenitor genomes. On the other hand there is fractionation difference between the three subgenomes: T1 and T2 are less fractionated with respect to each other than to the third subgenome T3 (data not shown). These results suggested that T was possibly an autohexaploidy or an allo-hexaploidy of two closely related species and one more distant species. Allo-polyploidy is often thought to be more frequent in nature due to advantages in the establishment of the polyploid strains resulting from factors such as heterosis, homeostasis, and fewer meiotic irregularities. However, the frequency of natural auto-polyploidy and its effects on species diversity may be higher than traditionally thought (Ramsey and Schemske 1998). As with γ, because of the antiquity of the T event, a definitive conclusion cannot be drawn due to degradation of molecular signatures and loss of the progenitor genomes. However, current data are in support of T having a higher possibility to have been an auto-polyploidy than the other two paleo-hexaploidies, the γ event (discussed above) and the *Brassica* paleo-hexaploidy which appears to have been an allo-hexaploidy (Tang et al. 2012). This would also be consistent with the fact that Solanaceae species do form autopolyploids in agricultural and natural settings. If T were indeed a paleo-autohexaploidy, it would be the only one known so far. Genome sequences from closely related sister taxa will aid in the test of this hypothesis (Fig. 11.5).

Comparison between the tomato and potato genomes showed that about 91 % of post-T gene loss is orthologous, indicating that these genes had been lost before tomato–potato divergence. Paleo-polyploidy events are usually followed by a phase of rapid genome evolution, including structural, sequence, and regulatory changes (Adams and Wendel 2005; Lynch and Conery 2000; Song et al. 1995). Therefore it is possible that many of the shared changes in tomato and potato occurred in their common ancestor shortly after T. On the other hand, evolution of genetic content in the triplicated paleo-genome of the Solanaceae ancestor continued long after the paleo-polyploidy event. The xyloglucan endotransglucosylase/hydrolase (XTH) family gene *XTH10* that was triplicated in the T event showed differential loss between the tomato and potato genomes which diverged ~65 MY after T (Tomato Genome Consortium 2012). Although tomato and potato genomes have maintained very similar karyotypes in ~7.3 MY of separate evolution, and 70–80 % of their genes have remained orthologous (Fig. 11.6 left panels), there has been continuous rearrangement of the ancestral genome content in the two lineages. The present-day tomato and potato chromosomes differ by nine major and several smaller inversions, and numerous local micro-synteny differences. About 4.8 % (tomato) and 4.6 % (potato) of the orthologous loci triplicated in T have been differentially lost between tomato and potato after their divergence. Ancestral subgenomes produced in the pan-core eudicot γ triplication had undergone extensive rearrangement before tomato–potato divergence, but have continued to be restructured independently in their recent independent lineage histories (Fig. 11.6 right panels). Therefore paleo-polyploidy poses both immediate and long-term effects on the evolution and diversity of genome structure.

In addition to the widespread effects of paleo-polyploidy, there are also important lineage-specific effects of the individual events. For example, the two ancient genome triplications in tomato have produced new gene family members that mediate important functions in its fruit ripening control, such as some transcription factors and enzymes necessary for red light photoreceptors influencing fruit quality (*PHYB1/PHYB2*) (expended in T), ethylene- and light-regulated genes mediating lycopene biosynthesis (*PSY1/PSY2*) (expended in T), and ethylene biosynthesis (*RIN, CNR, ACS*) (expended in T) and perception (*ETR3/NR, ETR4*) (expanded in γ) (Tomato Genome Consortium 2012). More case studies like this are a clear future research interest in revealing how the expanded genetic repertoire from paleo-polyploidy events contribute to biological diversity and the evolution of unique characteristics of individual lineages.

All paleo-hexaploidy events identified so far are in eudicot lineages, including one in the core eudicot stem lineage (γ), one near the origin of the asterid Solanaceae family (T), one in the rosid *Brassica* lineages (Wang et al. 2011), possibly one in the *Gossypium* lineages (Paterson et al. 2012) and one in the ancestral Compositae

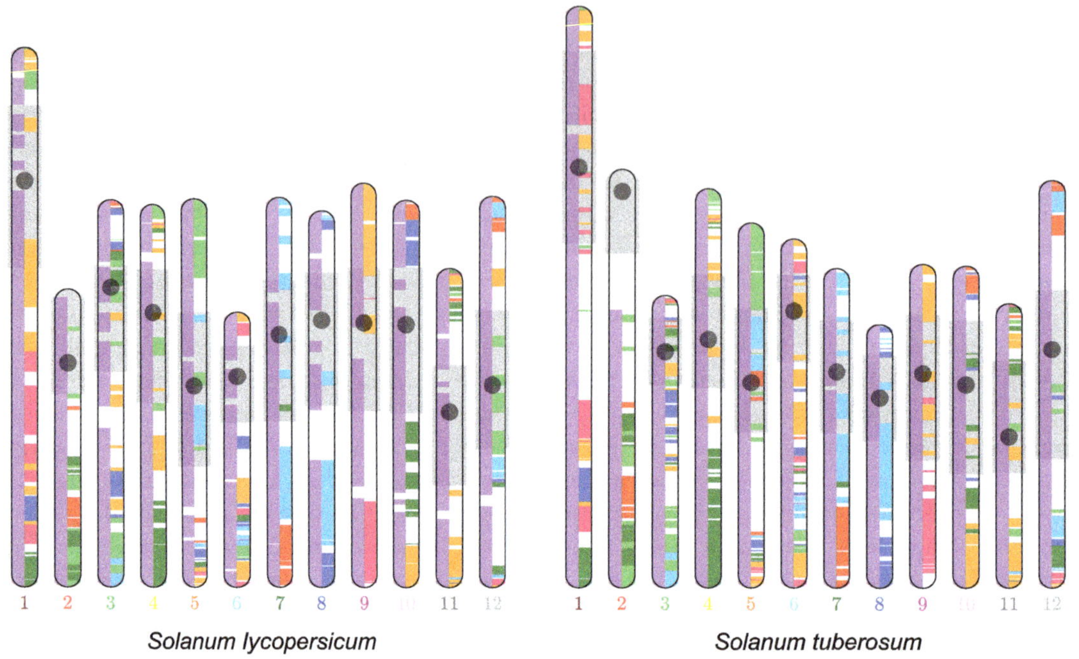

Fig. 11.6 Schematic representation of orthologous and paralogous regions in tomato (*S. lycopersicum*) and potato (*S. tuberosum*) genomes. On the *left side* of the chromosome bars the *purple regions* are orthologous between tomato and potato. On the *right side*, *7 colors* are used to paint genomic regions corresponding to 7 chromosomes in the inferred pan-core eudicot ancestral genomes (pre-γ) using grape genome data (Jaillon et al. 2007). Each of the γ-triplicated (3×) ancestral regions later underwent the T triplication (3×), resulting in their dispersed and multiplied (up to 9×) pattern in today's tomato and potato genomes. The *gray shades* and *dark gray circles* mark estimated heterochromatin regions and centromeres, respectively, from cytological experiments. Corresponding linkage groups (chromosomes) between tomato and potato are labeled with *same color*

lineages (Truco et al. 2013). Although some wild monocot plants such as the grass "Timothy" (*Phleum pratense*) (NordenskiÖLd 1953), and crops such as the bread wheat (*Triticum aestivum*) are neo-hexaploids, paleo-hexaploidy has not been found in any monocot genome studied so far. This raises curious questions about possible reasons and consequences associated with these events in the evolutionary history of some or all eudicot lineages, or alternatively, possible factors for suppressing such events in the evolution of other lineages.

Summary and Perspective

Sequencing of the tomato genome was very valuable in many ways, as detailed elsewhere in this volume. With regard to angiosperm evolution, the tomato genome sequence revealed the third paleo-hexaploidy identified in plants, and the first one in asterids, adding an important sample to the small collection of paleo-hexaploids. It confirmed that the γ event shared by all sequenced rosids was also shared by asterids, unmasking a new clade for studying the effects and consequences of γ. The T paleo-hexaploidy is possibly associated with the Solanaceae–Rubiaceae divergence, and divergence of early Solanaceae lineages, by triplicating the whole ancestral genome content, creating great potentials for subsequent diversification of homologous genomic associations and development of lineage-specific traits such as fruit ripening in tomato. Comparison of the tomato and potato genomes, both currently included in the genus *Solanum*, revealed continuous restructuring of paleo-triplicated ancestral loci long after the paleo-polyploidy events. The *Solanum* lineage is the first identified angiosperm lineage experiencing two paleo-hexaploidies but no paleo-tetraploidy. The consecutive paleo-hexaploidies γ and T are also valuable for comparative studies of the mechanisms and effects of paleo-hexaploidy and paleo-tetraploidy. Many questions about paleo-polyploidy have been answered, which nevertheless opened the door to more interesting questions.

References

Adams KL, Wendel JF (2005) Polyploidy and genome evolution in plants. Curr Opin Plant Biol 8(2):135–141. doi:10.1016/j.pbi.2005.01.001

Arabidopsis Genome Initiative (2000) Analysis of the genome sequence of the flowering plant Arabidopsis thaliana. Nature 408(6814):796–815. doi:10.1038/35048692

Aury JM, Jaillon O, Duret L, Noel B, Jubin C, Porcel BM, Segurens B, Daubin V, Anthouard V, Aiach N, Arnaiz O, Billaut A, Beisson J, Blanc I, Bouhouche K, Camara F, Duharcourt S, Guigo R, Gogendeau D, Katinka M, Keller AM, Kissmehl R, Klotz C, Koll F, Le Mouel A, Lepere G, Malinsky S, Nowacki M, Nowak JK, Plattner H, Poulain J, Ruiz F, Serrano V, Zagulski M, Dessen P, Betermier M, Weissenbach J, Scarpelli C, Schachter V, Sperling L, Meyer E, Cohen J, Wincker P (2006) Global trends of whole-genome duplications revealed by the ciliate Paramecium tetraurelia. Nature 444(7116):171–178. doi:10.1038/nature05230

Barker MS, Kane NC, Matvienko M, Kozik A, Michelmore RW, Knapp SJ, Rieseberg LH (2008) Multiple paleopolyploidizations during the evolution of the Compositae reveal parallel patterns of duplicate gene retention after millions of years. Mol Biol Evol 25(11):2445–2455. doi:10.1093/molbev/msn187

Bell CD, Soltis DE, Soltis PS (2010) The age and diversification of the angiosperms re-revisited. Am J Bot 97(8):1296–1303. doi:10.3732/ajb.0900346

Blanc G, Wolfe KH (2004) Widespread paleopolyploidy in model plant species inferred from age distributions of duplicate genes. Plant Cell 16(7):1667–1678. doi:10.1105/tpc.021345

Bowers JE, Chapman BA, Rong J, Paterson AH (2003) Unravelling angiosperm genome evolution by phylogenetic analysis of chromosomal duplication events. Nature 422(6930):433–438. doi:10.1038/nature01521

Bremer K, Friis EM, Bremer B (2004) Molecular phylogenetic dating of asterid flowering plants shows early Cretaceous diversification. Syst Biol 53(3):496–505

Cenci A, Combes M-C, Lashermes P (2010) Comparative sequence analyses indicate that Coffea (Asterids) and Vitis (Rosids) derive from the same paleo-hexaploid ancestral genome. Mol Genet Genomics 283(5):493–501

Cui L, Wall PK, Leebens-Mack JH, Lindsay BG, Soltis DE, Doyle JJ, Soltis PS, Carlson JE, Arumuganathan K, Barakat A, Albert VA, Ma H, dePamphilis CW (2006) Widespread genome duplications throughout the history of flowering plants. Genome Res 16(6):738–749. doi:10.1101/gr.4825606

Dehal P, Boore JL (2005) Two rounds of whole genome duplication in the ancestral vertebrate. PLoS Biol 3(10):e314. doi:10.1371/journal.pbio.0030314

Force A, Lynch M, Pickett FB, Amores A, Yan YL, Postlethwait J (1999) Preservation of duplicate genes by complementary, degenerative mutations. Genetics 151(4):1531–1545

Gebhardt C, Walkemeier B, Henselewski H, Barakat A, Delseny M, Stüber K (2003) Comparative mapping between potato (*Solanum tuberosum*) and Arabidopsis thaliana reveals structurally conserved domains and ancient duplications in the potato genome. Plant J 34(4):529–541

Grant D, Cregan P, Shoemaker RC (2000) Genome organization in dicots: Genome duplication in Arabidopsis and synteny between soybean and Arabidopsis. Proc Natl Acad Sci USA 97(8):4168–4173

Hedges SB, Dudley J, Kumar S (2006) TimeTree: a public knowledge-base of divergence times among organisms. Bioinformatics 22(23):2971–2972. doi:10.1093/bioinformatics/btl505

Hedges SB, Kumar S (2009) The timetree of life. OUP, Oxford

Hellsten U, Wright KM, Jenkins J, Shu S, Yuan Y, Wessler SR, Schmutz J, Willis JH, Rokhsar DS (2013) Fine-scale variation in meiotic recombination in Mimulus inferred from population shotgun sequencing. Proc Natl Acad Sci USA. doi:10.1073/pnas.1319032110

Huang S, Ding J, Deng D, Tang W, Sun H, Liu D, Zhang L, Niu X, Zhang X, Meng M, Yu J, Liu J, Han Y, Shi W, Zhang D, Cao S, Wei Z, Cui Y, Xia Y, Zeng H, Bao K, Lin L, Min Y, Zhang H, Miao M, Tang X, Zhu Y, Sui Y, Li G, Sun H, Yue J, Sun J, Liu F, Zhou L, Lei L, Zheng X, Liu M, Huang L, Song J, Xu C, Li J, Ye K, Zhong S, Lu BR, He G, Xiao F, Wang HL, Zheng H, Fei Z, Liu Y (2013) Draft genome of the kiwifruit Actinidia chinensis. Nat Commun 4:2640. doi:10.1038/ncomms3640

Ibarra-Laclette E, Lyons E, Hernandez-Guzman G, Perez-Torres CA, Carretero-Paulet L, Chang TH, Lan T, Welch AJ, Juarez MJ, Simpson J, Fernandez-Cortes A, Arteaga-Vazquez M, Gongora-Castillo E, Acevedo-Hernandez G, Schuster SC, Himmelbauer H, Minoche AE, Xu S, Lynch M, Oropeza-Aburto A, Cervantes-Perez SA, de Jesus O-EM, Cervantes-Luevano JI, Michael TP, Mockler T, Bryant D, Herrera-Estrella A, Albert VA, Herrera-Estrella L (2013) Architecture and evolution of a minute plant genome. Nature. doi:10.1038/nature12132

Jaillon O, Aury JM, Noel B, Policriti A, Clepet C, Casagrande A, Choisne N, Aubourg S, Vitulo N, Jubin C, Vezzi A, Legeai F, Hugueney P, Dasilva C, Horner D, Mica E, Jublot D, Poulain J, Bruyere C, Billault A, Segurens B, Gouyvenoux M, Ugarte E, Cattonaro F, Anthouard V, Vico V, Del Fabbro C, Alaux M, Di Gaspero G, Dumas V, Felice N, Paillard S, Juman I, Moroldo M, Scalabrin S, Canaguier A, Le Clainche I, Malacrida G, Durand E, Pesole G, Laucou V, Chatelet P, Merdinoglu D, Delledonne M, Pezzotti M, Lecharny A, Scarpelli C, Artiguenave F, Pe ME, Valle G, Morgante M, Caboche M, Adam-Blondon AF, Weissenbach J, Quetier F, Wincker P (2007) The grapevine genome sequence suggests ancestral hexaploidization in major angiosperm phyla. Nature 449(7161):463–467. doi:10.1038/nature06148

Jiao Y, Leebens-Mack J, Ayyampalayam S, Bowers JE, McKain MR, McNeal J, Rolf M, Ruzicka DR, Wafula E, Wickett NJ, Wu X, Zhang Y, Wang J, Zhang Y, Carpenter EJ, Deyholos MK, Kutchan TM, Chanderbali AS, Soltis PS, Stevenson DW, McCombie R, Pires JC, Wong GK, Soltis DE, Depamphilis CW (2012) A genome triplication associated with early diversification of the core eudicots. Genome Biol 13(1):R3. doi:10.1186/gb-2012-13-1-r3

Jiao Y, Wickett NJ, Ayyampalayam S, Chanderbali AS, Landherr L, Ralph PE, Tomsho LP, Hu Y, Liang H, Soltis PS, Soltis DE, Clifton SW, Schlarbaum SE, Schuster SC, Ma H, Leebens-Mack J, dePamphilis CW (2011) Ancestral polyploidy in seed plants and angiosperms. Nature 473(7345):97–100. doi:10.1038/nature09916

Kellis M, Birren BW, Lander ES (2004) Proof and evolutionary analysis of ancient genome duplication in the yeast *Saccharomyces cerevisiae*. Nature 428(6983):617–624

Kowalski SP, Lan TH, Feldmann KA, Paterson AH (1994) Comparative mapping of *Arabidopsis thaliana* and *Brassica oleracea* chromosomes reveals islands of conserved organization. Genetics 138(2):499–510

Ku HM, Vision T, Liu J, Tanksley SD (2000) Comparing sequenced segments of the tomato and Arabidopsis genomes: large-scale duplication followed by selective gene loss creates a network of synteny. Proc Natl Acad Sci USA 97(16):9121–9126. doi:10.1073/pnas.160271297

Lynch M, Conery JS (2000) The evolutionary fate and consequences of duplicate genes. Science 290(5494):1151–1155. doi:10.1126/science.290.5494.1151

Lyons E, Pedersen B, Kane J, Freeling M (2008) The value of nonmodel genomes and an example using SynMap within CoGe to dissect the hexaploidy that predates the rosids. Trop Plant Biol 1(3):181–190. doi:10.1007/s12042-008-9017-y

Masterson J (1994) Stomatal size in fossil plants: evidence for polyploidy in majority of angiosperms. Science 264(5157):421–424. doi:10.1126/science.264.5157.421

Ming R, Hou S, Feng Y, Yu Q, Dionne-Laporte A, Saw JH, Senin P, Wang W, Ly BV, Lewis KL, Salzberg SL, Feng L, Jones MR, Skelton RL, Murray JE, Chen C, Qian W, Shen J, Du P, Eustice M, Tong E, Tang H, Lyons E, Paull RE, Michael TP, Wall K, Rice DW, Albert H, Wang ML, Zhu YJ, Schatz M, Nagarajan N, Acob RA, Guan P, Blas A, Wai CM, Ackerman CM, Ren Y, Liu C, Wang J, Na JK, Shakirov EV, Haas B, Thimmapuram J, Nelson D, Wang X, Bowers JE, Gschwend AR, Delcher AL, Singh R, Suzuki JY, Tripathi S, Neupane K, Wei H, Irikura B, Paidi M, Jiang N, Zhang W, Presting G, Windsor A, Navajas-Perez R, Torres MJ, Feltus FA, Porter B, Li Y, Burroughs AM, Luo MC, Liu L, Christopher DA, Mount SM, Moore PH, Sugimura T, Jiang J, Schuler MA, Friedman V, Mitchell-Olds T, Shippen DE, dePamphilis CW, Palmer JD, Freeling M, Paterson AH, Gonsalves D, Wang L, Alam M (2008) The draft genome of the transgenic tropical fruit tree papaya (*Carica papaya* Linnaeus). Nature 452(7190):991–996. doi:10.1038/nature06856

Ming R, Vanburen R, Liu Y, Yang M, Han Y, Li LT, Zhang Q, Kim MJ, Schatz MC, Campbell M, Li J, Bowers JE, Tang H, Lyons E, Ferguson AA, Narzisi G, Nelson DR, Blaby-Haas CE, Gschwend AR, Jiao Y, Der JP, Zeng F, Han J, Min XJ, Hudson KA, Singh R, Grennan AK, Karpowicz SJ, Watling JR, Ito K, Robinson SA, Hudson ME, Yu Q, Mockler TC, Carroll A, Zheng Y, Sunkar R, Jia R, Chen N, Arro J, Wai CM, Wafula E, Spence A, Han Y, Xu L, Zhang J, Peery R, Haus MJ, Xiong W, Walsh JA, Wu J, Wang ML, Zhu YJ, Paull RE, Britt AB, Du C, Downie SR, Schuler MA, Michael TP, Long SP, Ort DR, William Schopf J, Gang DR, Jiang N, Yandell M, Depamphilis CW, Merchant SS, Paterson AH, Buchanan BB, Li S, Shen-Miller J (2013) Genome of the long-living sacred lotus (*Nelumbo nucifera* Gaertn.). Genome Biol 14(5):R41. doi:10.1186/gb-2013-14-5-r41

Moore MJ, Soltis PS, Bell CD, Burleigh JG, Soltis DE (2010) Phylogenetic analysis of 83 plastid genes further resolves the early diversification of eudicots. Proc Natl Acad Sci USA 107(10):4623–4628. doi:10.1073/pnas.0907801107

NordenskiÖLd H (1953) A genetical study in the mode of segregation in hexaploid phleum pratense. Hereditas 39(3–4):469–488. doi:10.1111/j.1601-5223.1953.tb03431.x

Ohno S (1970) Evolution by gene duplication. Springer, Berlin

Paterson AH (2014) Advances in botanical research, vol 69. Genomes of herbaceous land plants, 1st edn. Elsevier, Amsterdam

Paterson AH, Bowers JE, Burow MD, Draye X, Elsik CG, Jiang CX, Katsar CS, Lan TH, Lin YR, Ming R, Wright RJ (2000) Comparative genomics of plant chromosomes. Plant Cell 12(9):1523–1540

Paterson AH, Freeling M, Tang H, Wang X (2010) Insights from the comparison of plant genome sequences. Annu Rev Plant Biol 61:349–372. doi:10.1146/annurev-arplant-042809-112235

Paterson AH, Lan TH, Reischmann KP, Chang C, Lin YR, Liu SC, Burow MD, Kowalski SP, Katsar CS, DelMonte TA, Feldmann KA, Schertz KF, Wendel JF (1996) Toward a unified genetic map of higher plants, transcending the monocot–dicot divergence. Nat Genet 14(4):380–382. doi:10.1038/ng1296-380

Paterson AH, Wendel JF, Gundlach H, Guo H, Jenkins J, Jin D, Llewellyn D, Showmaker KC, Shu S, Udall J, Yoo MJ, Byers R, Chen W, Doron-Faigenboim A, Duke MV, Gong L, Grimwood J, Grover C, Grupp K, Hu G, Lee TH, Li J, Lin L, Liu T, Marler BS, Page JT, Roberts AW, Romanel E, Sanders WS, Szadkowski E, Tan X, Tang H, Xu C, Wang J, Wang Z, Zhang D, Zhang L, Ashrafi H, Bedon F, Bowers JE, Brubaker CL, Chee PW, Das S, Gingle AR, Haigler CH, Harker D, Hoffmann LV, Hovav R, Jones DC, Lemke C, Mansoor S, ur Rahman M, Rainville LN, Rambani A, Reddy UK, Rong JK, Saranga Y, Scheffler BE, Scheffler JA, Stelly DM, Triplett BA, Van Deynze A, Vaslin MF, Waghmare VN, Walford SA, Wright RJ, Zaki EA, Zhang T, Dennis ES, Mayer KF, Peterson DG, Rokhsar DS, Wang X, Schmutz J (2012) Repeated polyploidization of Gossypium genomes and the evolution of spinnable cotton fibres. Nature 492 (7429):423–427. doi:10.1038/nature11798

Potato Genome Sequencing C, Xu X, Pan S, Cheng S, Zhang B, Mu D, Ni P, Zhang G, Yang S, Li R, Wang J, Orjeda G, Guzman F, Torres M, Lozano R, Ponce O, Martinez D, De la Cruz G, Chakrabarti SK, Patil VU, Skryabin KG, Kuznetsov BB, Ravin NV, Kolganova TV, Beletsky AV, Mardanov AV, Di Genova A, Bolser DM, Martin DM, Li G, Yang Y, Kuang H, Hu Q, Xiong X, Bishop GJ, Sagredo B, Mejia N, Zagorski W, Gromadka R, Gawor J, Szczesny P, Huang S, Zhang Z, Liang C, He J, Li Y, He Y, Xu J, Zhang Y, Xie B, Du Y, Qu D, Bonierbale M, Ghislain M, Herrera Mdel R, Giuliano G, Pietrella M, Perrotta G, Facella P, O'Brien K, Feingold SE, Barreiro LE, Massa GA, Diambra L, Whitty BR, Vaillancourt B, Lin H, Massa AN, Geoffroy M, Lundback S, DellaPenna D, Buell CR, Sharma SK, Marshall DF, Waugh R, Bryan GJ, Destefanis M, Nagy I, Milbourne D, Thomson SJ, Fiers M, Jacobs JM, Nielsen KL, Sonderkaer M, Iovene M, Torres GA, Jiang J, Veilleux RE, Bachem CW, de Boer J, Borm T, Kloosterman B, van Eck H, Datema E, Hekkert B, Goverse A, van Ham RC, Visser RG (2011) Genome sequence and analysis of the tuber crop potato. Nature 475 (7355):189–195. doi:10.1038/nature10158

Proost S, Fostier J, De WD, Dhoedt B, Demeester P, Van de Peer Y, Vandepoele K (2012) i-ADHoRe 3.0–fast and sensitive detection of genomic homology in extremely large data sets. Nucleic Acids Res 40(2): e11. doi:10.1093/nar/gkr955

Ramsey J, Schemske DW (1998) Pathways, mechanisms, and rates of polyploid formation in flowering plants. Annu Rev Ecol Syst 29(1):467–501. doi:10.1146/annurev.ecolsys.29.1.467

Sankoff D, Zheng C (2012) Fractionation, rearrangement and subgenome dominance. Bioinformatics 28(18): i402–i408. doi:10.1093/bioinformatics/bts392

Schlueter J, Dixon P, Granger C, Grant D, Clark L, Doyle J, Shoemaker R (2004) Mining EST databases to resolve evolutionary events in major crop species. Genome 47(5):868–876. doi:10.1139/g04-047

Schnable JC, Springer NM, Freeling M (2011) Differentiation of the maize subgenomes by genome dominance and both ancient and ongoing gene loss. Proc Natl Acad Sci USA 108(10):4069–4074. doi:10.1073/pnas.1101368108

Simillion C, Janssens K, Sterck L, Van de Peer Y (2008) i-ADHoRe 2.0: an improved tool to detect degenerated genomic homology using genomic profiles. Bioinformatics 24(1):127–128. doi:10.1093/bioinformatics/btm449

Simillion C, Vandepoele K, Van Montagu MC, Zabeau M, Van de Peer Y (2002) The hidden duplication past of Arabidopsis thaliana. Proc Natl Acad Sci USA 99(21):13627–13632

Smith JJ, Kuraku S, Holt C, Sauka-Spengler T, Jiang N, Campbell MS, Yandell MD, Manousaki T, Meyer A, Bloom OE, Morgan JR, Buxbaum JD, Sachidanandam R, Sims C, Garruss AS, Cook M, Krumlauf R, Wiedemann LM, Sower SA, Decatur WA, Hall JA, Amemiya CT, Saha NR, Buckley KM, Rast JP, Das S, Hirano M, McCurley N, Guo P, Rohner N, Tabin CJ, Piccinelli P, Elgar G, Ruffier M, Aken BL, Searle SM, Muffato M, Pignatelli M, Herrero J, Jones M, Brown CT, Chung-Davidson YW, Nanlohy KG, Libants SV, Yeh CY, McCauley DW, Langeland JA, Pancer Z, Fritzsch B, de Jong PJ, Zhu B, Fulton LL, Theising B, Flicek P, Bronner ME, Warren WC, Clifton SW, Wilson RK, Li W (2013) Sequencing of the sea lamprey (*Petromyzon marinus*) genome provides insights into vertebrate evolution. Nat Genet. doi:10.1038/ng.2568

Soltis DE, Albert VA, Leebens-Mack J, Bell CD, Paterson AH, Zheng C, Sankoff D, Depamphilis CW, Wall PK, Soltis PS (2009) Polyploidy and angiosperm diversification. Am J Bot 96(1):336–348. doi:10.3732/ajb.0800079

Soltis DE, Smith SA, Cellinese N, Wurdack KJ, Tank DC, Brockington SF, Refulio-Rodriguez NF, Walker JB, Moore MJ, Carlsward BS, Bell CD, Latvis M, Crawley S, Black C, Diouf D, Xi Z, Rushworth CA, Gitzendanner MA, Sytsma KJ, Qiu YL, Hilu KW, Davis CC, Sanderson MJ, Beaman RS, Olmstead RG, Judd WS, Donoghue MJ, Soltis PS (2011) Angiosperm phylogeny: 17 genes, 640 taxa. Am J Bot 98(4):704–730. doi:10.3732/ajb.1000404

Song K, Lu P, Tang K, Osborn TC (1995) Rapid genome change in synthetic polyploids of Brassica and its implications for polyploid evolution. Proc Natl Acad Sci USA 92(17):7719–7723

Stebbins GL (1966) Chromosomal variation and evolution. Science 152(3728):1463–1469. doi:10.1126/science.152.3728.1463

Stevens PF (2012) Angiosperm phylogeny website. Version 12. Missouri Botanical Garden. http://www.mobot.org/MOBOT/research/APweb/

Tang H, Bowers JE, Wang X, Ming R, Alam M, Paterson AH (2008a) Synteny and collinearity in plant genomes. Science 320(5875):486–488. doi:10.1126/science.1153917

Tang H, Wang X, Bowers JE, Ming R, Alam M, Paterson AH (2008b) Unraveling ancient hexaploidy through multiply-aligned angiosperm gene maps. Genome Res 18(12):1944–1954. doi:10.1101/gr.080978.108

Tang H, Woodhouse MR, Cheng F, Schnable JC, Pedersen BS, Conant G, Wang X, Freeling M, Pires JC (2012) Altered patterns of fractionation and exon deletions in Brassica rapa support a two-step model of paleohexaploidy. Genetics 190(4):1563–1574. doi:10.1534/genetics.111.137349

The Angiosperm Phylogeny Group (2009) An update of the Angiosperm Phylogeny Group classification for the orders and families of flowering plants: APG III. Bot J Linn Soc 161(2):105–121. doi:10.1111/j.1095-8339.2009.00996.x

Thomas BC, Pedersen B, Freeling M (2006) Following tetraploidy in an Arabidopsis ancestor, genes were removed preferentially from one homeolog leaving clusters enriched in dose-sensitive genes. Genome Res 16(7):934–946

Tomato Genome Consortium (2012) The tomato genome sequence provides insights into fleshy fruit evolution. Nature 485(7400):635–641. doi:10.1038/nature11119

Truco MJ, Ashrafi H, Kozik A, van Leeuwen H, Bowers J, Reyes Chin Wo S, Stoffel K, Xu H, Hill T, Van Deynze A, Michelmore RW (2013) An ultra high-density, transcript-based, genetic map of lettuce. G3 (Bethesda). doi:10.1534/g3.112.004929

Vekemans D, Proost S, Vanneste K, Coenen H, Viaene T, Ruelens P, Maere S, Van de Peer Y, Geuten K (2012) Gamma paleohexaploidy in the stem lineage of core eudicots: significance for MADS-box gene and species diversification. Mol Biol Evol 29(12):3793–3806. doi:10.1093/molbev/mss183

Verde I, Abbott AG, Scalabrin S, Jung S, Shu S, Marroni F, Zhebentyayeva T, Dettori MT, Grimwood J, Cattonaro F, Zuccolo A, Rossini L, Jenkins J, Vendramin E, Meisel LA, Decroocq V, Sosinski B, Prochnik S, Mitros T, Policriti A, Cipriani G, Dondini L, Ficklin S, Goodstein DM, Xuan P, Fabbro CD, Aramini V, Copetti D, Gonzalez S, Horner DS, Falchi R, Lucas S, Mica E, Maldonado J, Lazzari B, Bielenberg D, Pirona R, Miculan M, Barakat A, Testolin R, Stella A, Tartarini S, Tonutti P, Arus P, Orellana A, Wells C, Main D, Vizzotto G, Silva H, Salamini F, Schmutz J, Morgante M, Rokhsar DS (2013) The high-quality draft genome of peach (*Prunus persica*) identifies unique patterns of genetic diversity, domestication and genome evolution. Nat Genet 45(5):487–494. doi:10.1038/ng.2586

Vision TJ, Brown DG, Tanksley SD (2000) The origins of genomic duplications in Arabidopsis. Science 290 (5499):2114–2117

Wang H, Moore MJ, Soltis PS, Bell CD, Brockington SF, Alexandre R, Davis CC, Latvis M, Manchester SR, Soltis DE (2009) Rosid radiation and the rapid rise of angiosperm-dominated forests. Proc Natl Acad Sci USA 106(10):3853–3858. doi:10.1073/pnas.0813376106

Wang X, Wang H, Wang J, Sun R, Wu J, Liu S, Bai Y, Mun JH, Bancroft I, Cheng F, Huang S, Li X, Hua W, Freeling M, Pires JC, Paterson AH, Chalhoub B, Wang B, Hayward A, Sharpe AG, Park BS, Weisshaar B, Liu B, Li B, Tong C, Song C, Duran C, Peng C, Geng C, Koh C, Lin C, Edwards D, Mu D, Shen D, Soumpourou E, Li F, Fraser F, Conant G, Lassalle G, King GJ, Bonnema G, Tang H, Belcram H, Zhou H, Hirakawa H, Abe H, Guo H, Jin H, Parkin IA, Batley J, Kim JS, Just J, Li J, Xu J, Deng J, Kim JA, Yu J, Meng J, Min J, Poulain J, Hatakeyama K, Wu K, Wang L, Fang L, Trick M, Links MG, Zhao M, Jin M, Ramchiary N, Drou N, Berkman PJ, Cai Q, Huang Q, Li R, Tabata S, Cheng S, Zhang S, Sato S, Sun S, Kwon SJ, Choi SR, Lee TH, Fan W, Zhao X, Tan X, Xu X, Wang Y, Qiu Y, Yin Y, Li Y, Du Y, Liao Y, Lim Y, Narusaka Y, Wang Z, Li Z, Xiong Z, Zhang Z (2011) The genome of the mesopolyploid crop species *Brassica rapa*. Nat Genet 43(10):1035–1039. doi:10.1038/ng.919

Wang Y, Tang H, Debarry JD, Tan X, Li J, Wang X, Lee TH, Jin H, Marler B, Guo H, Kissinger JC, Paterson AH (2012) MCScanX: a toolkit for detection and evolutionary analysis of gene synteny and collinearity. Nucleic Acids Res 40(7):e49. doi:10.1093/nar/gkr1293

Wolfe KH (2001) Yesterday's polyploids and the mystery of diploidization. Nat Rev Genet 2(5):333–341. doi:10.1038/35072009

Wolfe KH, Shields DC (1997) Molecular evidence for an ancient duplication of the entire yeast genome. Nature 387(6634):708–713

Synteny Among Solanaceae Genomes

12

Amy Frary, Sami Doganlar and Anne Frary

Abstract

The Solanaceae was among the first plant families to be analyzed via comparative mapping and thus was a pioneer in the realm of synteny studies. Analyses of chromosome content and organization have employed a range of techniques, including linkage mapping of genes and molecular markers, physical mapping via fluorescence in situ hybridization, and sequencing of relatively small genomic segments as well as the complete sequencing of the tomato genome. Early comparisons in the family involved tomato and its close relative potato and have extended outward to include eggplant, pepper, tobacco, and petunia. Not surprisingly, the degree of synteny among these species is a function of the time since their divergence, with inversion, translocation, and transposition being the chief mechanisms of chromosome rearrangement. The results of this work provide important insight into the modes and tempo of plant genome evolution while serving a practical purpose as well: knowledge of genome synteny and colinearity makes it easier to leverage resources from one species to another in this agronomically important family.

Keywords

Tomato · Eggplant · Pepper · Synteny · Solanaceae

Introduction

The term 'synteny' was originally used in genetics to describe the presence of two or more genes on the same chromosome, however, its meaning has evolved with changes in the discipline (McCouch 2001). Today the terms 'synteny,' 'conserved synteny,' and 'shared synteny' are all used to indicate co-localization of genes or

A. Frary
Department of Biological Sciences, Mount Holyoke College, South Hadley, MA 01075, USA

S. Doganlar · A. Frary (✉)
Department of Molecular Biology & Genetics, Izmir Institute of Technology, Urla, Izmir 35430, Turkey
e-mail: annefrary@iyte.edu.tr

M. Causse et al. (eds.), *The Tomato Genome*, Compendium of Plant Genomes,
DOI 10.1007/978-3-662-53389-5_12

markers on chromosomes of two or more species derived from a common ancestor (Abrouk et al. 2010). The terms 'colinear' (var. collinear) and 'conserved syntenic segments' (CSSs; Nadeau and Taylor 1984) are more specific and indicate the shared order of loci in syntenic regions (Abrouk et al. 2010). 'Macrosynteny' describes synteny for a large number of loci over a whole chromosome, while 'microsynteny' describes the detailed relationships between smaller CSSs.

Examination of shared synteny in plant genomes followed soon after the appearance of the first molecular linkage map in tomato (Bernatzky and Tanksley 1986). Only 3 years later, maps comparing the tomato, potato, and pepper genomes were published (Bonierbale et al. 1988; Tanksley et al. 1988). Since this pioneering work in Solanaceae, comparative genome mapping of molecular markers and genes has revealed much about macrosynteny in plant genomes. Another technique used in synteny studies is fluorescence in situ hybridization (FISH) which involves localization of specific probes on pachytene chromosomes. FISH analyses can reveal chromosomal rearrangements such as inversions and translocations. DNA sequencing projects including the complete sequencing of the tomato genome (Tomato Genome Consortium 2012) have allowed comparison of genomes on a finer, microsyntenic, scale.

The study of shared synteny can shed light on the evolution of individual chromosomes and whole genomes. As synteny is the result of descent from a common ancestor, disruption in CSSs can be used to deduce the mechanisms of chromosome rearrangement that accompanied species divergence. Examination of synteny also helps to identify orthologous regions in different species' genomes. This can be useful for determining gene function or for isolating genes in non-model plant species. Shared synteny is also important in the study of paleogenomics, the use of extant species to reconstruct ancestral genomes (Abrouk et al. 2010). More practical applications of shared synteny include the ability to map genes or markers in silico and to leverage resources developed for model species in lesser-studied genomes.

Tomato (*Solanum lycopersicum*) is the model species of the Solanaceae. As a result, most studies of synteny in this family have entailed comparisons with tomato. The species discussed in the following review of synteny research are thereby organized according to their relationship to tomato, beginning with comparisons between tomato and its wild relatives and moving to more distant species within the Solanaceae.

Cultivated Tomato

Examination of synteny within *S. lycopersicum* is extremely limited. Asamizu et al. (2012) compared bacterial artificial chromosome (BAC) end sequences from the cultivar 'Micro-Tom' with the sequence of 'Heinz 1706.' 'Heinz 1706' is the inbred tomato cultivar whose genome was sequenced by the Tomato Genome Consortium (2012). 'Micro-Tom' is a dwarf cultivar which is used as a model because of its small size, relatively short lifecycle, and ease of genetic transformation. Examination of microsynteny between the two cultivars indicated two possible rearrangements. Chromosome 2 contains an inversion of 20–220 kb, its size depending on the orientation of the inversion. Chromosome 3 contains an intrachromosomal translocation and inversion. The presence of a putative reverse transcriptase within the region allowed the authors to hypothesize that the rearrangement was due to retrotransposon activity.

Wild Tomato

The closest relatives of domesticated tomato include nine wild tomato species that can be crossed with *S. lycopersicum*. These species are a rich source of genetic diversity (Tanksley and McCouch 1997; Bai and Lindhout 2007) and have been widely exploited for improvement of tomato including the introgression of over 40 disease resistance alleles from wild germplasm to cultivated tomato (Hajjar and Hodgkin 2007). Moreover, by providing DNA polymorphism which is limited within *S. lycopersicum*, interspecific populations derived from the wild species have allowed identification and mapping of many qualitative and quantitative traits (Lippman et al. 2007). Fine mapping of disease resistance and morphological genes in interspecific populations of tomato have also revealed genomic rearrangements that distinguish cultivated tomato from its closest wild relatives. Reduced recombination within introgressed segments is often a preliminary indicator of altered synteny. For example, in fine mapping the *Cf-4*/*Cf-9* leaf mold resistance gene cluster on chromosome 1 of tomato, Bonnema et al. (1997) noted that a *S. pennellii*-derived population had a highly suppressed recombination rate as compared to a *S. peruvianum*-derived one. The authors surmised that small inversions in the region might be responsible for this discrepancy.

Lack of recombination in a *S. pennelli*-derived population also hindered high resolution mapping of the *sun* locus on the short arm of chromosome 7 (van der Knaap et al. 2004). This led to the identification of a paracentric inversion in *S. pennellii* relative to cultivated tomato. The same inversion was not detected in *S. pimpinellifolium* (van der Knaap et al. 2004), *S. peruvianum* (van Heusden et al. 1999) or potato (Gebhardt et al. 1991) but is present in eggplant (Doganlar et al. 2002). These results were confirmed by FISH analysis of chromosome 7S which suggested that the *S. pennelli*/eggplant arrangement is ancestral (Szinay et al. 2012). Thus, the inversions occurred independently in the *S. pennellii* and eggplant lineages suggesting that this region of the genome may be subject to frequent rearrangements during evolution. Interestingly, the region containing *sun* is 30 kb shorter in *S. pimpinellifolium* than in *S. lycopersicum* (van der Knaap et al. 2004). Further investigation indicated that the size discrepancy is due to a 24.7 kb duplication at the *sun* locus in cultivated tomato which confers an elongated phenotype to fruit (Xiao et al. 2008). This duplication was attributed to the activity of a long terminal repeat retrotransposon, *Rider*. Thus, a lack of microsynteny between two genomic regions helped to elucidate the identity and mechanism of the *sun* locus in tomato.

Another inversion distinguishing cultivated and wild tomato was detected on chromosome 6 in the region of a root knot nematode resistance gene (*Mi-1*) (Seah et al. 2004). The region contains two clusters of homologous genes which are arranged similarly in both *S. lycopersicum* and *S. peruvianum*, the original source of *Mi-1*. Physical mapping revealed that the clusters are inverted relative to each other in the two species. Examination of microsynteny in the region indicated that simple inversion alone could not explain the arrangement and sequence identity of homologues (Seah et al. 2007). Instead the authors proposed the occurrence of several rearrangements (inversion and/or intra- or inter-chromosomal recombination) as well as gene conversion but did not specify the events or their timing during evolution. Interestingly this chromosome 6 inversion in *S. peruvianum* was not detected by Szinay et al. (2012) using BAC-FISH. However, they did identify an inversion at the top of 6S in *S. pennelli*. This research also showed that a portion of *S. chilense* chromosome 12S is inverted relative to tomato and wild tomato species.

Physical mapping and sequence analysis were also used to compare large portions of the *S. lycopersicum* and *S. pennellii* genomes (Kamenetzky et al. 2010). With QTLs for metabolic traits as the starting point for their comparisons, the authors examined five regions of the genome and produced a detailed physical map of 1 % of the wild species' genome. *S. pennellii* and cultivated tomato were found to be mostly colinear in these regions. In addition, over 1 million bp of DNA were sequenced and functionally annotated. Examination of the microsynteny in this region revealed that gene order, orientation and exon/intron structure were conserved between the two species with small differences in transposable element insertion and the size of intergenic regions. A divergence time of 2.7 million years ago (MYA) was estimated based on the rate of amino acid substitution for *S. lycopersicum* and *S. pennellii*.

Tomato-Like Nightshades

Nightshade is often used as a general term to refer to any member of the Solanaceae. However a more specific definition, 'tomato-like nightshades,' includes only those species closely related to tomato: *S. ochranthum*, *S. juglandifolium*, *S. sitiens*, and *S. lycopersicoides* (Rick 1979). These species are of interest because both morphological and molecular phylogenetic studies place them between tomato and potato (Peralta and Spooner 2001; Albrecht and Chetelat 2009). In addition, the tomato-like nightshades are expected to contain more diversity for useful traits such as biotic and abiotic stress tolerance than their domesticated relatives (Albrecht and Chetelat 2009). Some species of tomato-like nightshades can be hybridized, albeit with some difficulty, to tomato, therefore, these species represent a potential genepool of novel traits for tomato improvement.

Molecular genetic mapping in *S. lycopersicum* × *S. lycopersicoides* BC1 and BC2 populations revealed a high degree of synteny between the two species' genomes (Chetelat et al. 2000; Chetelat and Meglic 2000). A total of 139 RFLP, isozyme and morphological markers previously mapped in tomato indicated complete colinearity with the tomato genome except on chromosome 10L. These results suggested an inversion of this arm in *S. lycopersicoides* relative to tomato, a paracentric inversion that is also observed in potato. The same rearrangement of 10L was detected in a pseudo F2 population derived from a cross between two related nightshades, *S. lycopersicoides* and *S. sitiens*, and mapped with 101 RFLP markers (Pertuze et al. 2002). Because this arrangement is common to these tomato-like nightshades, potato, pepper, and eggplant, the inversion must have occurred during the divergence of tomato from these other species. However, it is important to note that BAC-FISH analysis in the same species did not confirm the 10L inversion, instead inversions were detected on chromosomes 6S and 7S of the *S. lycopersicoides* genome relative to tomato. (Szinay et al. 2012). Thus, more detailed analysis of these regions is merited.

The chromosome 10L inversion was also not detected in an F2 population derived from the cross *S. ochranthum* × *S. juglandifolium* (Albrecht and Chetelat 2009). Mapping using 132 tomato COS, COSII, RFLP, and SSR markers revealed overall synteny between these nightshades and tomato with a shared arrangement of chromosome 10L. This finding was confirmed by FISH analysis of 10L in *S. ochranthum* (Szinay et al. 2012). Overall, these results agree with the molecular phylogeny which indicates that the section Juglandifolia nightshades (*S. ochranthum* and *S. juglandifolium*) are more closely related to tomato than the section Lycopersicoides nightshades (*S. sitiens* and *S. lycopersicoides*) (Peralta and Spooner 2001). Szinay et al. (2012) also described an inversion on 6S of *S. ochranthum* relative to tomato which is larger than that in *S. lycopersicoides* and is also shared by potato. The *S. ochranthum* and *S. juglandifolium* genomes were found to differ by a reciprocal translocation of chromosomes 8 and 12. Although other inversions were detected, they might be artifacts as they were only supported by single marker deviations from colinearity.

In other work, the first linkage map for a non-tomato-like nightshade was constructed (D'Agostino et al. 2013). *S. dulcamara*, also known as bittersweet or climbing nightshade, is native to Europe and may be a source of useful abiotic and biotic stress resistances for related crop species. D'Agostino et al. (2013) compared this species' genome with those of tomato, potato, and eggplant. Five *S. dulcamara* chromosomes (1, 3, 6, 8, and 9) were completely colinear with the respective tomato chromosomes indicating that the tomato/bittersweet chromosomes represent the ancestral arrangement. Chromosomes 2, 5, 7, and 10 of *S. dulcamara* contain inversions relative to their tomato counterparts with some of these inversions also observed in potato, eggplant, and/or pepper. Translocations were seen on chromosomes 4, 11, and 12 as has also been observed in solanaceous crop species but with different combinations of chromosome arms. This re-use of chromosome breakpoints suggests that certain chromosomes are unstable and have been rearranged more than once over evolutionary time.

Potato

In economic and agricultural terms, potato (*Solanum tuberosum*) is the most important member of the Solanaceae. The genetics of potato is more complex than tomato, owing to its autotetraploid nature. The existence of diploid wild relatives as well as potato's close kinship with tomato provided an essential foundation for molecular genetic analyses in the crop. The construction of molecular genetic linkage maps for potato using genomic and cDNA clones derived from tomato permitted some of the first explorations of synteny in dicot plant genomes (Bonierbale et al. 1988; Gebhardt et al. 1991; Tanksley et al. 1992). The initial molecular map of potato was developed by examining the segregation of 134 RFLP and isozyme markers in offspring from an interspecific cross of diploid *Solanum* parents: *S. phureja* × (*S. tuberosum* × *S. chacoense*) (Bonierbale et al. 1988). The use of common markers revealed homologous relationships between the 12 linkage groups of potato and tomato and demonstrated that marker content and order are highly conserved between the two species. Four paracentric inversions were identified as disrupting the karyotypic similarity between the species. Three of these chromosomal rearrangements were confirmed and an additional two inversions were discovered as a result of the subsequent parallel construction of high-resolution maps of the tomato and potato genomes (Tanksley et al. 1992). Based on a cross between *S. tuberosum* and *S. berthaultii*, this potato map provided evidence that the entire short arms of chromosomes 5, 9, 11, and 12 and the long arm of chromosome 10 are inverted relative to tomato. Synteny and colinearity of markers between the two species was otherwise strongly conserved leading the authors to surmise that chromosome breakage followed by inversion was the principle mechanism of genome rearrangement during divergence of the two lineages.

Wu and Tanksley (2010) performed a comprehensive review of the data from COSII marker and other comparative mapping studies to ascertain the nature of structural changes distinguishing the genomes of tomato, potato, eggplant, pepper, and *Nicotiana* (Fig. 12.1). This work confirmed the positions of six inversions in potato relative to tomato and deduced that two of these inversion events had occurred along the potato line whereas the rest were specific to tomato (Fig. 12.2). They estimated that the last common ancestor (LCA) of potato–tomato lived 7.3 MYA and that the karyotype of the ancestral genome resembled the following extant chromosomes (where T = tomato and Pt = potato): T1/Pt1, Pt2, T3/Pt3, T4/Pt4, Pt5, T6/Pt6, T7/Pt7, T8/Pt8, Pt9, Pt10, T11, T12 (Wu et al. 2010).

Another approach to elucidating the synteny of the potato and tomato genomes has involved comparative mapping of disease resistance and pathogen recognition genes (Grube et al. 2000; Huang et al. 2004, 2005). In a genome-wide survey of resistance genes (R genes) in tomato, potato, and pepper, Grube et al. (2000) discovered that clustering of R genes at homologous positions is a common phenomenon: four such clusters were found on potato–tomato chromosomes 6, 9, 10, and 12. The authors point out that it would, however, be difficult to exploit this synteny for the purposes of isolating orthologous R genes as corresponding R gene clusters typically contain genes with different pathogen specificity. Nevertheless, Huang et al. (2004, 2005) found strong conservation in marker content and order between the *I2* region of tomato (which confers *Fusarium* wilt resistance) and the *R3* region of potato (confers late blight resistance) that proved useful in isolating *R3a*. Based on their protein sequences, *I2* and *R3a* belong to the same gene family and may have evolved from an R gene locus present in their LCA. Interestingly, the tomato *I2* region is half the size of the potato *R3* region. Thus in addition to the change in pathogen specificity, a greater number of R genes have evolved at this particular locus in potato than in tomato (Huang et al. 2005).

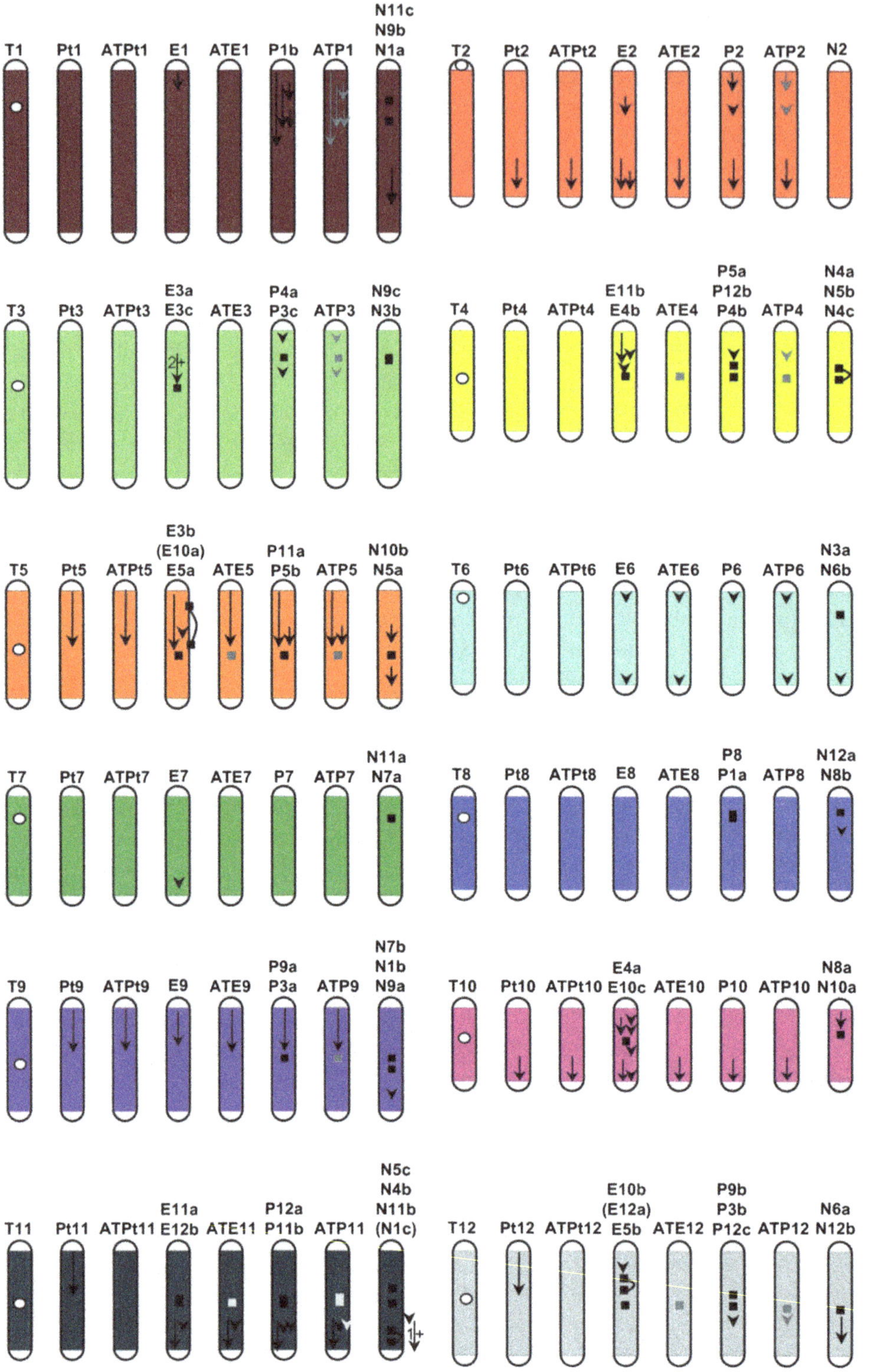
T1 Pt1 ATPt1 E1 ATE1 P1b ATP1 N11c N9b N1a
T2 Pt2 ATPt2 E2 ATE2 P2 ATP2 N2
T3 Pt3 ATPt3 E3a E3c ATE3 P4a P3c ATP3 N9c N3b
T4 Pt4 ATPt4 E11b E4b ATE4 P5a P12b P4b ATP4 N4a N5b N4c
T5 Pt5 ATPt5 E3b (E10a) E5a ATE5 P11a P5b ATP5 N10b N5a
T6 Pt6 ATPt6 E6 ATE6 P6 ATP6 N3a N6b
T7 Pt7 ATPt7 E7 ATE7 P7 ATP7 N11a N7a
T8 Pt8 ATPt8 E8 ATE8 P8 P1a ATP8 N12a N8b
T9 Pt9 ATPt9 E9 ATE9 P9a P3a ATP9 N7b N1b N9a
T10 Pt10 ATPt10 E4a E10c ATE10 P10 ATP10 N8a N10a
T11 Pt11 ATPt11 E11a E12b ATE11 P12a P11b ATP11 N5c N4b N11b (N1c)
T12 Pt12 ATPt12 E10b (E12a) E5b ATE12 P9b P3b P12c ATP12 N6a N12b

◀ **Fig. 12.1** Comparative maps of tomato (T), potato (Pt), eggplant (E), pepper (P), and tobacco (N) chromosomes and their most recent ancestors (chromosomes with AT prefixed to name) as determined by COSII mapping. *White circles* indicate positions of tomato centromeres. *Black arrows* and bars indicate inversions and breakpoints relative to tomato. *Grey symbols* indicate uncertain chromosomal rearrangements (used with permission from Wu and Tanksley 2010)

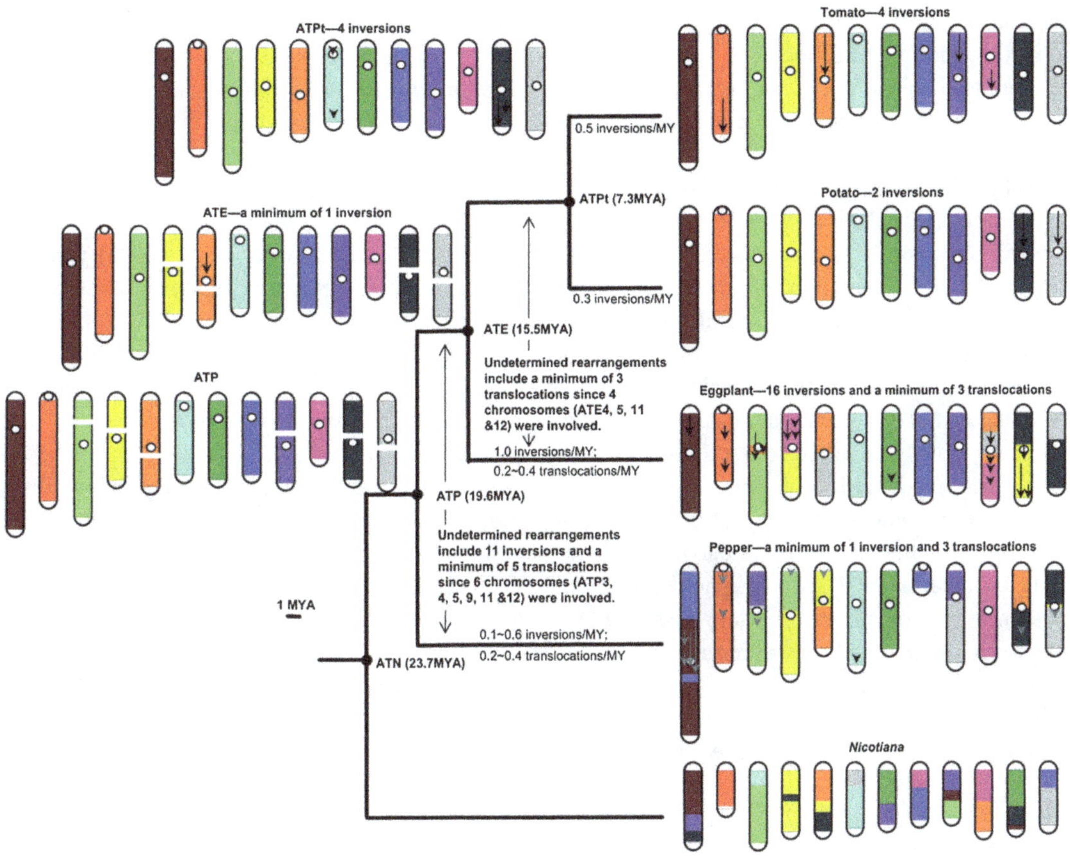

Fig. 12.2 Karyotypes of tomato, potato, eggplant, pepper, and tobacco and their most recent ancestors as determined by COSII mapping. Tomato chromosomes are *color-coded*. *Symbols* and abbreviations are as described for Fig. 12.1. Chromosome breaks indicate areas that require further study (used with permission from Wu and Tanksley 2010)

With the availability of the complete genome sequence of tomato (Tomato Genomics Consortium 2012), it has become easier to conduct genome-wide surveys and phylogenetic analyses of disease-resistance genes. An analysis of the bHLH transcription factor family uncovered 152 members distributed across the entire tomato genome, with evidence suggesting that one was upregulated in a resistant line following infection by tomato yellow leaf curl virus (Wang et al. 2015). Comparison with potato enabled the identification of over 160 orthologous gene pairs, each single copy tomato gene being represented in potato by up to four genes, reflecting the polyploid origins of *S. tuberosum*. Physical localization of these gene pairs revealed a high degree of synteny between the potato and tomato genomes. Similarly, Andolfo et al. (2013) performed a comprehensive analysis of pathogen recognition genes (specifically, nucleotide-binding site, receptor-like protein, and receptor-like kinase genes) in the tomato genome, in an effort to localize the genes and trace their evolutionary origins via gene duplication. More than 300 putative orthologs of the pathogen recognition genes were obtained from potato, using synteny (comparative genome positions) as a key criterion for orthology. Slight differences in the genomic positions of many of the orthologous pairs were noticed in addition to differences in gene number between tomato and potato, suggesting that these genes are evolving independently in the two lineages.

Although comparative genetic mapping has revealed a great deal about the syntenic relationship between genomes in the Solanaceae, suppressed recombination in heterochromatic genomic regions limits the ability of genetic maps to fully resolve genome structure. Recognizing this shortcoming, physical mapping approaches have become increasingly popular especially now that extensive BAC libraries have been made available by the work of the tomato and potato genome sequencing consortia (http://www.sgn.cornell.edu/; http://potatogenome.net). Thus localization of BACs on pachytene chromosomes of tomato and potato using FISH uncovered two structural differences between tomato and potato chromosome 6 (Iovene et al. 2008; Tang et al. 2008). Iovene et al. (2008) reported that, while the pachytene chromosomes of tomato and potato are morphologically similar, an interstitial heterochromatic knob is specific to potato 6L. BAC colinearity on 6L is, however, conserved. In addition, both studies (Iovene et al. 2008; Tang et al. 2008) confirmed the existence of a large inversion encompassing the euchromatic portion of 6S that had been suggested by a previous molecular genetic analysis of chromosome 6 (van Wordragen et al. 1994) but was not apparent on the high-density tomato–potato map (Tanksley et al. 1992).

Comparative sequencing offers an avenue for elucidating microsyntenic relationships between potato and tomato. Wang et al. (2008) included potato in their sequence analysis of a 105 kb CSS in five solanaceous species (tomato, potato, eggplant, pepper, petunia). Of the 17 genes contained within this region, two showed a reversed orientation in potato as compared with tomato (Fig. 12.3). Because the potato orientation was also seen in eggplant and petunia, it was judged to be the ancestral condition. These authors also calculated an approximate date of 6.2 MYA for the divergence of potato and tomato (Wang et al. 2008). Zhu et al. (2008) generated almost 90 Mb of potato genomic sequence from 77,000 BAC ends and 22 BACs. BLAST searches in Genbank and the SGN database (solgenomics.net) were then used to

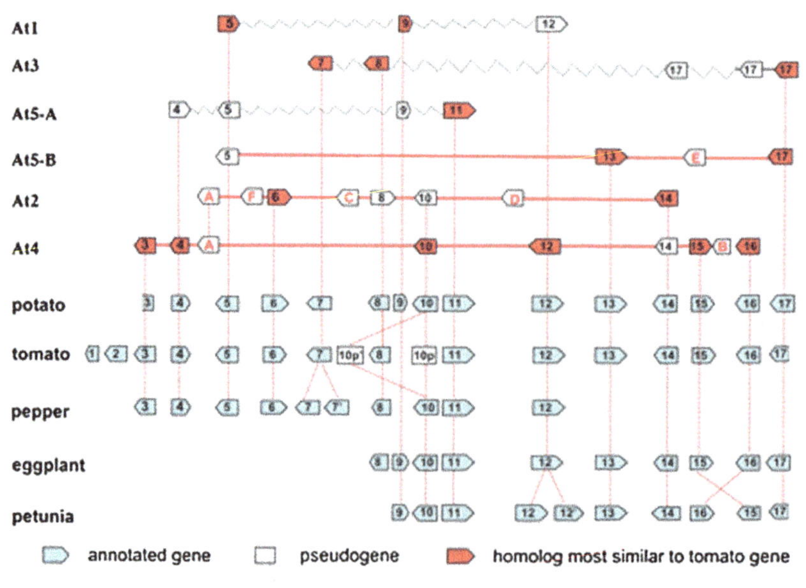

Fig. 12.3 Organization of a 105 kb conserved syntenic segment (CSS) in potato, tomato, pepper, eggplant, and petunia containing 17 annotated genes. Positions of the genes in the arabidopsis (At) genome are also shown. Putative orthologs are connected by *dashed red lines* (used with permission from Wang et al. 2008)

identify segments syntenic to tomato. In some instances, the conserved segments spanned more than 100 kb and sequence coverage ranged from 13 to 73 %. Although macrosynteny between potato and tomato was apparent, evidence of small-scale rearrangements such as insertions/deletions and micro-inversions were also seen. Nevertheless, protein sequence alignments as well as a comparison of the length of genes, exons, and introns indicated that genic synteny was maintained. A comprehensive analysis of repeated DNA sequences within the BACs suggests that transposition, alongside chromosome inversion, is a key contributor to genome restructuring between potato and tomato (Zhu et al. 2008).

In a multipronged approach employing cross-species BAC-FISH and comparative sequencing, Peters et al. (2012) analyzed 7 Mb of the euchromatic portion of the long arm of chromosome 2. Six major rearrangements including inversions ranging in size from 20 kb to 3 Mb as well as several translocations were identified. These structural changes appear to have occurred along the lineage leading to tomato as they are absent from pepper, eggplant, and potato. This work also revealed that the rearrangements affecting 6S, 10L, and 11L are more complex than previously suspected. The inversion on 6S involved several reversals, deletions, and translocations. A second inversion was pinpointed on 10L. In addition, three inversions, three deletions, and an inverted translocation occurred on 11L. Microsynteny in potato and tomato was also explored by examining the adjacency of orthologous genes on 2L. Within 664 ortholog groups, the vast majority (96 %) consisted of homologous gene pairs that mapped to corresponding colinear positions. However, gene adjacencies were not conserved between potato and tomato for 46 % of these ortholog pairs. In many cases, the insertion of putative retrotransposons appears to have disrupted microcolinearity. Sequences similar to transposable elements were also found near rearrangement junctions suggesting that repeat-mediated recombination is a plausible mechanism for genome reorganization. Accordingly, the authors hypothesized that a series of intra-strand and ectopic recombination events transformed 2L from the ancestral state found in potato to that found in tomato (Peters et al. 2012). Thus, what has emerged from this and other physical mapping studies is a far clearer picture of chromosome evolution in *Solanum* as well as the importance of examining synteny and colinearity on a finer scale.

Synteny studies in potato have extended beyond tomato to encompass a number of other *Solanum* species. In a comparison of tuber-bearing *Solanums*, a BC1 population was derived from a cross between two Mexican diploid species *S. pinnatisectum* (a source of late blight resistance) and *S. cardiophyllum* ssp. *cardiophyllum* (Kuhl et al. 2001). The resulting molecular map, albeit low resolution (99 markers derived from tomato) and incomplete (13 linkage groups), showed good overall synteny and colinearity with previously published potato linkage maps (Bonierbale et al. 1988; Tanksley et al. 1992; Perez et al. 1999). Interestingly, despite the morphological similarity among potato and its non-tuber-bearing relatives (section *Etuberosum*) (Contreras-M and Spooner 1999), a wide range of evolutionary mechanisms has operated to distinguish the A genome of cultivated potato from the E genome of section *Etuberosum* species (Perez et al. 1999). Using established tomato/potato markers on a F2 population derived from an interspecific cross between *S. palustre* and *S. etuberosum*, Perez et al. (1999) placed 80 loci in 19 linkage groups. While the excess of linkage groups indicates that the E genome was not completely mapped, this work did reveal general synteny in that markers usually mapped to homeologous chromosomes in both genomes. However, the linear order of markers was frequently disrupted by putative translocations, inversions, and occasional transpositions. Thus, it is not surprising that attempts to cross A and E genome *Solanum* species have not been successful: the extent of chromosome rearrangement could explain the lack of chromosome pairing and hybrid sterility that are typically observed (Ramanna and Hermsen 1979; Watanabe et al. 1995). The comparative mapping results of Perez et al. (1999) provide additional support for the phylogenetic placement of *S. tuberosum* and *S. lycopersicum* as sister groups on a lineage separate from section *Etuberosum* (Spooner et al. 1993). However, more recent research (Szinay et al. 2012; described below) places *S. etuberosum* closer to the tomato clade.

Research in potato has also examined synteny within the species. Tang et al. (2008) extended the analysis of the chromosome 6S inversion across six potato genotypes and found evidence of a single minor structural rearrangement of 6S in one potato line. This, combined with their failure to observe the 6L interstitial knob identified by Iovene et al. (2008) in any of their lines, led them to speculate that a certain degree of chromosomal rearrangement has occurred within *S. tuberosum* (Tang et al. 2008). Lou et al. (2010) broadened the cytogenetic comparison of chromosome 6 to include a total of seven *Solanum* species: cultivated potato (A genome), two wild potato species (*S. bulbocastanum* and *S. chromatophilum* representing the B and P genomes, respectively), the E genome species *S. etuberosum*, as well as tomato, eggplant, and its relative *S. caripense*. Synteny in BAC position and orientation was found across all species with the exception of the aforementioned paracentric short arm inversion in tomato and a large pericentric inversion in *S. etuberosum*. The paracentric 6S inversion was deemed to have occurred after the divergence of tomato from the other *Solanum* species. The pericentric inversion is noteworthy as being the first such inversion identified in the genus. Interestingly, Perez et al. (1999) failed to detect this inversion in their cross-species comparison of the A and E genomes, once again highlighting the shortcomings of linkage analysis as the sole approach to synteny studies.

In addition to revealing hidden structural changes in genomes, the BAC-FISH approach has been useful as a means of understanding evolutionary relationships within *Solanum*. Szinay et al. (2012) used BAC-FISH signal order to perform a phylogenetic analysis of 18 *Solanum* species/accessions. BACs specific to seven chromosome arms known to harbor inversions among the selected species (5S, 6S, 7S, 9S, 10L, 11S, 12S) were isolated and mapped via FISH. Two syntenic species groups (composed of species with identical hybridization patterns) emerged: group A comprising potato and its relatives within section *Petota* and group B comprising several members of section *Lycopersicon*, including tomato. The genome of *S. etuberosum* differed from that of syntenic species group A due to inversions on three of the studied chromosome arms: 7S, 9S, and 10L (this latter inversion is apparently shared with group B). As a result, the phylogenetic tree based on these results places *S. etuberosum* closer to tomato and its relatives than to potato (Szinay et al. 2012), a topology that differs slightly from that deduced by Perez et al. (1999) based on their comparison of the A and E genomes. Finally, the authors hypothesize that cultivated and wild potato species must have diverged in the recent past as no structural differences among their genomes were detected (Szinay et al. 2012). In a broader comparison within the clade, over 300 COSII markers were assessed in eight potato accessions (including the wild species *S. berthaultii*, *S. chomatophilum,* and *S. paucissectum* as well as two diploid landraces of *S. tuberosum*) (Lindqvist-Kreuze et al. 2013). Only a small number of the COSII markers did not map in their predicted locations based on the established synteny between potato and tomato.

In contrast, by aligning diversity arrays technology (DArT) marker sequences derived from the wild tuber-bearing species *S. commersonii* and *S. bulbocastanum* with genome sequences from cultivated potato and tomato, Traini et al. (2013) discovered a greater amount of variability between the genomes. The failure of a proportion of the markers to align with the potato (8 %) or the tomato (21 %) genome sequences was taken as evidence of this heterogeneity. The existence of gaps between the markers provided additional support of small-scale structural divergence between the genomes of wild and cultivated potato. The use of DArT markers to construct medium-density genetic linkage maps for *S. bulbocastanum* has shed additional light on the degree of divergence between the A genome of

S. tuberosum and the B genome of *S. bulbocastanum* (Iorizzo et al. 2014). The wild potato genome shows the same nine chromosomal rearrangements previously described as distinguishing tomato and cultivated potato. Moreover, two additional, albeit small (5–10 cM), inversions on 2S and 8S appear to be specific to *S. bulbocastanum*. The results of these two studies are suggestive of the sorts of microscale, lineage-specific rearrangements that have accompanied the diversification of potato as a clade and which should become more and more visible as new strategies for mining whole genome sequence data are developed.

Eggplant

Eggplant (*S. melongena*), unlike many other solanaceous species, was domesticated in the Old World and its current importance as a crop is primarily limited to the Mediterranean Basin and Asia. In the past decade, genetic studies in eggplant have been facilitated by genome similarity with tomato. Synteny between eggplant and tomato was first investigated by Doganlar et al. (2002) who mapped 233 single copy tomato RFLP markers in an interspecific (*S. linnaeanum* × *S. melongena*) eggplant population. This work indicated that the eggplant and tomato genomes share large colinear regions and differ by 28 rearrangements encompassing 23 inversions, 4 reciprocal, and 1 non-reciprocal translocation. A total of 36 CSSs were identified in the genomes with an average size of 34 cM. Two chromosomes, 1 and 8, were found to be completely syntenic between eggplant and tomato. The use of tomato RFLP markers that had also been mapped in potato allowed comparisons between eggplant and potato and revealed similar high levels of synteny with 24 rearrangements differentiating these two genomes. Examination of the synteny between eggplant and these two species indicated that eggplant and tomato are five to six times more diverged than tomato and potato in terms of numbers of rearrangements. The work also provided insight into mechanisms of chromosome evolution in the Solanaceae. Results indicated a moderate rate of chromosome evolution (0.19 rearrangements per chromosome per million years) and that paracentric inversions of CSSs were the primary mechanism of rearrangement. Translocations were of secondary importance in divergence of eggplant and tomato/potato. Translocations and inversions generally occurred at or near the centromeres as indicated by the presence of telomeric sequences at the centromeres of affected chromosomes (Presting et al. 1996).

Further work with the same population by Wu et al. (2009a), added 110 COSII markers to the eggplant map. Their results were very similar to those of Doganlar et al. (2002) indicating that the eggplant and tomato genomes share 37 CSSs and differ by 24 inversions and five translocations with some differences detected in the locations of the inversions. Wu et al. also identified five single markers with altered positions suggesting possible transposable element activity during the divergence of eggplant and tomato from their LCA. The authors took advantage of the high degree of synteny of the species to infer the locations of 522 additional COSII markers thus producing a virtual eggplant map containing 869 markers. Comparison of this map with potato and pepper maps indicated that several rearrangements are shared by eggplant and pepper only and not by tomato and potato. These results indicated that the eggplant-pepper arrangements seen at the bottoms of chromosomes E2, E10, and E12 and at the tops of chromosomes E6 and E9 are ancestral.

In more recent work, Doganlar et al. (2014) mapped an additional 192 RFLP, 6 COSII, and 400 AFLP markers on the interspecific eggplant population. This work confirmed the established syntentic relationships between eggplant and tomato with 33 CSSs identified. However, the higher resolution map indicated more translocations (19) and fewer inversions (14) than previously hypothesized. Thus, translocation appears to be a more common mechanism of chromosome evolution in the Solanaceae than previously thought. Eleven marker transpositions were also detected confirming the role of transposable elements in genome evolution of the family as suggested by Wu et al. (2009a).

Wu and Tanksley (2010) compared COSII maps for eggplant and four other solanaceous species to deduce ancestral chromosome arrangements (Fig. 12.1) and their timing. Based on this work, they hypothesized that 16 inversions and 3 translocations occurred along the eggplant lineage (Fig. 12.2). They calculated the divergence time of tomato and eggplant as 15.5 MYA. This value allowed them to estimate the rate of chromosomal evolution in the eggplant lineage. Thus, they determined that this lineage experienced 1 inversion and 0.2–0.4 translocations every MY. This rate of inversion is higher than those calculated for the potato and pepper lineages, however, the authors cautioned that the number of inversions in the eggplant lineage may have been overestimated due to difficulties in assigning some inversions to specific lineages. The rate of translocation was the same as that estimated for the pepper lineage.

In the past, mapping in intraspecific populations of eggplant was constrained by limited polymorphism. However, this has changed with the advent of SNP and InDel markers. Fukuoka et al. (2012) integrated results from two intraspecific eggplant populations to obtain a map with 952 markers. Of these, 469 were SNP and InDel markers derived from *Solanum* ortholog gene sets (SOL markers). These are orthologous unigene markers identified in the eggplant, tomato and potato genomes. As 70 % of these markers had also been mapped in the tomato genome, the authors were able to observe synteny between the two genomes. Although detailed comparisons were not made, the results indicated several rearrangements with overall agreement with the findings of Wu et al. (2009a).

Gramazio et al. (2014) have recently developed an interspecific genetic linkage map for a BC1 population generated by backcrossing a *S. melongena* × *S. incanum* F_1 to the *S. melongena* parent. Of the molecular markers mapped (a combination of COSII, SSR, AFLP, CAPS and SNPs), 123 had been positioned on the maps of Nunome et al. (2009), Wu et al. (2009a), Barchi et al. (2012) and Fukuoka et al. (2012). These anchor loci revealed good correspondence in marker order between the new map and the four previous eggplant maps (Fig. 12.4). In addition, comparing the map positions of 130 markers shared with tomato (Fulton et al. 2002) uncovered regions of chromosome shuffling that were consistent with those reported by Wu et al. (2009a).

The well-established synteny between the eggplant and tomato genomes has been used to infer gene position and to identify candidate genes controlling quantitative traits. Thus, Gramazio et al. (2014) mapped a number of genes involved in chlorogenic acid synthesis and phenol oxidation using orthologous sequences from tomato. In all cases, the genes mapped to the eggplant linkage groups in regions corresponding to their syntenic positions on the tomato map. Similarly, synteny with tomato was used to identify candidate genes in genomic regions harboring QTL for anthocyanin levels and fruit color (Cericola et al. 2014) and a range of agronomic traits (Portis et al. 2014) in eggplant.

Synteny in eggplant and other Solanaceae has also been examined with FISH mapping. Using this technique, Lou et al. (2010) localized 17 BAC clones on chromosome 6 of tomato, eggplant, potato, and wild potatoes. Based on this analysis, the authors concluded that the arrangement of chromosome 6 in eggplant represents the ancestral condition. Moreover, the paracentric inversion of 6S, which was detected by Doganlar et al. (2002) in their comparison of tomato and eggplant, was not identified in any of the other species and suggesting that it only occurred in the tomato lineage. More extensive BAC-FISH analysis examined seven previously identified inversions in the *Solanum* genome (Szinay et al. 2012). This work indicated that the *S. melongena* genome represents the ancestral state for chromosomes 6S, 7S, 9S, and 11S. Therefore, the inversions in these regions occurred in the tomato–potato lineage. In contrast the inversion described on chromosome 10L is derived and occurred only in the eggplant lineage.

To date, limited insight has been gained from sequencing analyses of eggplant. Wang et al. (2008) included eggplant in their comparison of a

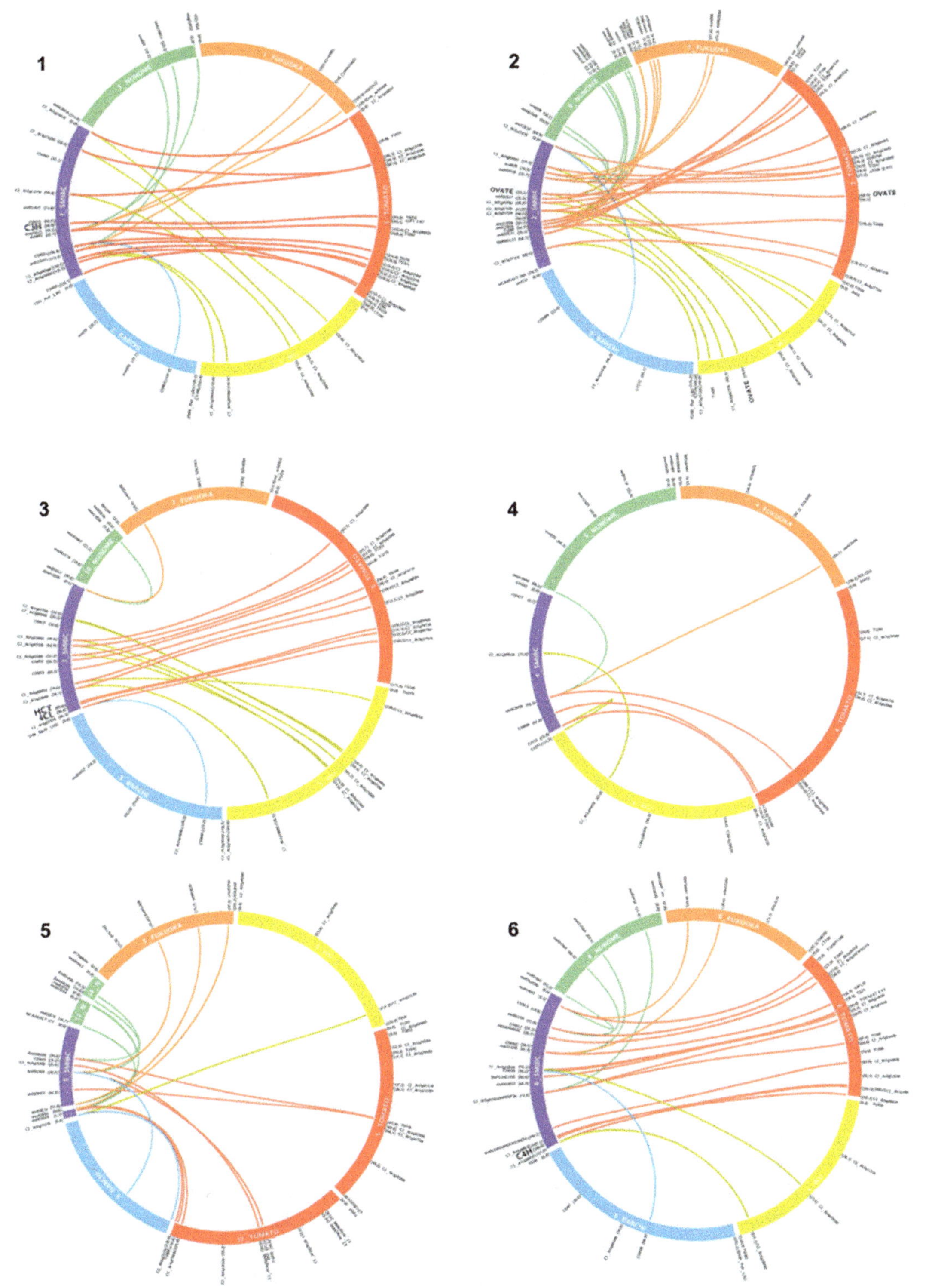

Fig. 12.4 Macro-syntenic relationships between five eggplant maps and tomato for linkage groups E1–E6. Each eggplant linkage map is *color-coded*: Gramazio et al. (2014) in *purple*, Barchi et al. (2012) in *blue*, Fukuoka et al. (2012) in *orange*, Nunome et al. (2009) in *green* and Wu et al. (2009a) in *yellow*. The tomato map (Fulton et al. 2002) is in *red*. Marker names and positions appear on the outside of the *circles* (used with permission from Gramazio et al. 2014)

105 kb CSS in solanaceous species. They found that gene orientation and position in the segment was very similar with only two differences between eggplant and tomato (Fig. 12.3). In the case of one gene, tomato had a reverse orientation which was not shared by any of the other species studied (potato, pepper, petunia, eggplant). In the other example, eggplant, potato, and petunia all contained a gene which was absent in tomato and pepper. Based on their examination of rates of evolution in the CSS, the authors hypothesized that 13.7 MY separate eggplant and tomato from their LCA which is similar to the value estimated by Wu and Tanksley (2010).

The recent release of a draft genome of eggplant that covers an estimated 74 % of the genome promises to reveal much more about chromosomal evolution in the Solanaceae (Hirakawa et al. 2014). Mapping nearly 10,000 eggplant sequence super-scaffolds over >98 % of the tomato genome revealed 56 conserved synteny blocks and 44 synteny break points between the genomes of the two species. Newly identified rearrangements included inversions on chromosomes 1 and 8. Given its higher resolution, it is not surprising that whole genome sequence analysis has enabled detection of a greater number of syntenic blocks and chromosome rearrangements than genetic linkage analysis (Doganlar et al. 2002, 2014). Refinement of the draft genome therefore promises to reveal much more about how the eggplant and tomato genomes have diverged.

Pepper

The genus *Capsicum* contains five domesticated species, *C. annuum*, *C. chinense*, *C. frutescens*, *C. baccatum*, and *C. pubescens* (Heiser and Pickersgill 1969), collectively referred to as peppers. Of these, *C. annuum* is the most commonly cultivated and has the most characterized genome. In fact, some of the first comparative molecular mapping done in plants was performed between *C. annuum* and tomato (Tanksley et al. 1988). In this early work, an interspecific map (85 tomato RFLPs and isozyme loci) constructed from a cross between *C. annuum* and *C. chinense* indicated that at least 32 chromosome breaks distinguished the tomato and pepper genomes. Using a similar *C. annuum* × *C. chinense* population, Prince et al. (1993) mapped nearly 200 RFLP markers which showed that 32 % of marker order was conserved between the two species. The map also indicated that far fewer breaks, at least 15, could explain the differences between the pepper and tomato genomes. Livingstone et al. (1999) extended this map to include 352 markers that could be used for comparisons of synteny between the pepper and tomato genomes. With these markers, they identified 13 linkage groups and 18 homeologous segments which corresponded to 95 and 98 % of the pepper and tomato genomes, respectively. Four chromosome pairs were entirely syntenic between tomato and pepper: T2/P2, T6/P6, T7/P7, and T10/P10. The remaining linkage groups were substantially rearranged with at least 30 breaks required to explain the differences between the two genomes. Livingstone et al. (1999) proposed the occurrence of 5 translocations, 10 paracentric inversions, 2 pericentric inversions, and 4 other changes since divergence of tomato and pepper from their LCA.

Synteny between pepper and tomato was further examined by Wu et al. (2009b) who primarily used COSII markers (263 COSII, 36 RFLP markers). For the first time, the number of linkage groups (12) corresponded to pepper's base chromosome number. These linkage groups included 35 CSSs which covered 67 % of the pepper map and had an average length of 32 cM. Based on this work, at least six translocations, 19 inversions, and many single gene transpositions were required to explain the genomic differences between tomato and pepper. As in other comparisons, most of the rearrangements seemed to involve breakpoints at or near centromeres. Moreover, individual markers that moved among non-homologous chromosomes tended to be located near the centromere. Because pericentromeric regions of the tomato genome have been found to be rich in retrotransposons (Wang et al. 2006), this can be taken as evidence of transposon activity. Comparison of the pepper and tomato maps with those of potato and eggplant allowed estimation of the timing of some rearrangements (Wu et al. 2009b). Thus it was found that inversions in the lower part of P11 occurred in the tomato/potato lineage while four other inversions most likely occurred in the tomato lineage after divergence of potato and tomato.

Further comparisons of the pepper COSII map with those of tomato, potato, eggplant, and tobacco allowed determination of the timing of other chromosomal rearrangements (Wu and Tanksley 2010). According to this work, only one inversion and three translocations are specific to the pepper lineage (Figs. 12.1 and 12.2). The divergence time between tomato and pepper was estimated as 19.6 MYA, allowing the rate of chromosomal rearrangements in the pepper lineage to be estimated as 0.1–0.6 inversions and 0.2–0.3 translocations per MY.

Use of interspecific populations for comparative mapping has also allowed rearrangements in the genomes of different *Capsicum* species to be identified. Namely, two translocations (one reciprocal, one nonreciprocal) and a duplication/deletion were found to distinguish *C. annuum* and *C. chinense* (Livingstone et al. 1999). In addition, the work of Wu et al. (2009b) proposed a model to explain the karyotypic differences between *C. annuum* and *C. chinense*/*C. frutescens*. Cultivated *C. annuum* has 2 acrocentric and 10 metacentric chromosomes while wild *C. annuum*, *C. chinense,* and *C. frutescens* have only 1 acrocentric and 11 metacentric chromosomes (Lanteri and Pickersgill 1993). According to Wu et al. (2009b) this difference can be explained by illegitimate recombination between ribosomal RNA (R45S) gene clusters on chromosomes 1 and 8 in the wild *C. annuum* genome resulting in a reciprocal translocation that altered chromosome arm length as seen in chromosomes I and XII in cultivated *C. annuum*.

Synteny of gene, rather than marker, location was examined by Grube et al. (2000) in their study of the genomic organization of disease resistance genes in tomato, pepper, and potato. They found that homologues of the tomato *N*, *Pto*, *Prf*, *Sw-5,* and *I2C* genes had syntenic positions in the pepper and tomato genomes. This work also showed that resistance genes clustered at homeologous positions in tomato, pepper, and potato on chromosomes T3, T4, T9, and T11. Resistance gene clusters on T1 and T7 were syntenic between tomato and pepper while a cluster on T8 had synteny between pepper and potato. (Tomato–potato synteny is discussed in the section on potato.)

Mazourek et al. (2009) focused on the orthologous disease resistance genes *Bs2* and *Rx/Gpa2* in pepper and potato, respectively. They demonstrated that the orthology between *Bs2* and the potato genes was disrupted by recombination, duplication, and deletion events, at least some of which involved retrotransposons. *Bs2* was found to map to chromosome P9 in a region syntenic to the top of potato chromosome 12 (XII) which contains *Rx* and *Gpa2*. This region is colinear in tomato (T12), pepper,and potato; however, this part of chromosome 12 in potato is inverted. Moreover, although *Rx* and *Gpa2* are tightly clustered in the potato genome, the resistance genes in the syntenic regions of the pepper and tomato genomes are more numerous and more dispersed. In fact, an examination of the entire pepper and tomato genomes indicates a close correspondence between the locations of R genes and chromosome breakpoints (Fig. 12.5). These results reinforce previous work which showed that chromosome breakpoints were associated with resistance gene duplication and dispersal in arabidopsis and *Medicago truncatula* (Baumgarten et al. 2003; Ameline-Torregrosa et al. 2008).

BAC-FISH analysis was also used to examine macrosynteny in pepper, tomato, and potato (Peters et al. 2012). FISH analysis on chromosomes 2L, 6S, 10L, and 11L revealed that the pepper arrangement differs by inversion from tomato but is colinear to potato on 2L, 6S, and 10L. In contrast, tomato and potato share an arrangement of 11L which is interrupted by an inverted translocation in pepper (Yang et al. 2009; Peters et al. 2012). These results disagree with those of Livingstone et al. (1999) who found complete colinearity between pepper and tomato

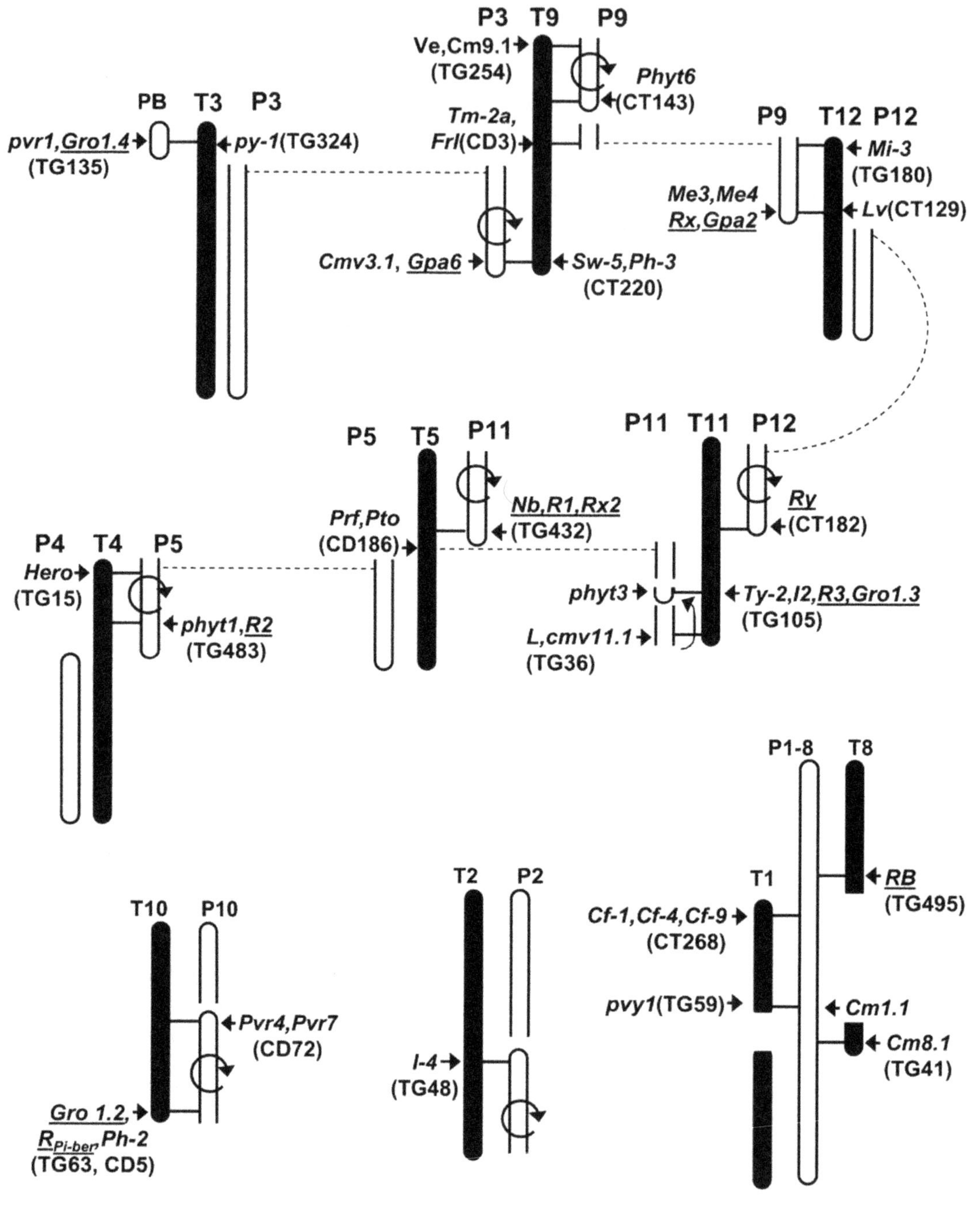

Fig. 12.5 Locations of selected resistance (R) genes in the tomato (T) and pepper (P) genomes. Potato genes are *underlined*. *Circular arrows* indicate putative inversions while *dotted lines* indicate translocations between chromosomes (used with permission from Mazourek et al. 2009)

chromosomes 2, 6, and 10. This discrepancy highlights the limitations of mapping for detailed analyses of synteny. While molecular mapping and FISH analyses allow examination of gross chromosomal rearrangements, other techniques are required to study microsynteny. One such technique is sequencing. As previously mentioned, Wang et al. (2008) sequenced a 105 kb CSS in tomato, pepper, potato, eggplant, and petunia. They detected only two differences between pepper and tomato in gene arrangement in the region (Fig. 12.3). One gene had a reverse orientation in tomato as compared to all of the other species indicating that an inversion occurred along the tomato lineage. In addition, pepper had a duplication of one of the genes which was hypothesized to have occurred by tandem duplication. Based on their analyses, Wang et al. (2008) calculated that pepper and tomato diverged from their LCA approximately 19.1 MYA, which is nearly identical to the divergence time (19.6 MYA) calculated by Wu and Tanksley (2010).

Although gene order and repertoire are conserved in the Solanaceae, genome size is variable. The pepper genome contains fourfold more DNA than the tomato genome (Kim et al. 2014). To determine the cause of this difference, Park et al. (2011) compared nearly 36 Mb of euchromatic pepper DNA sequence with its orthologous region in tomato. They found that the number and identities of predicted genes in the genomic sequences were similar. Gene length differences were mainly due to longer introns in pepper (1815 bp on average as compared to 1459 bp in tomato). In addition, pepper contained many more transposons between genes. These were mostly *Ty3/Gypsy*-like LTR retrotransposons. FISH with one of these transposons, Tat, showed that this element is primarily located in heterochromatic regions of the tomato genome while it is dispersed in both euchromatin and heterochromatin in pepper. Thus, as found in other plant and animal species, transposable elements play a major role in genome size determination in pepper with a lesser, but still significant role played by intron size (reviewed by Gregory 2005).

A more comprehensive comparison of the pepper and tomato genomes was made possible by sequencing the pepper genome (Kim et al. 2014; Qin et al. 2014). Sequencing of hot pepper cultivar CM334 (Kim et al. 2014) and nonpungent cultivar Zunla-1 (Qin et al. 2014) provided nearly full coverage of pepper's 3.48 Gb genome. In both studies, predicted gene number was similar to that for tomato, approximately 35,000 protein-coding sequences with nearly 18,000 orthologous gene sets shared by the pepper and tomato genomes. Both genomes were found to have many large blocks syntenic with tomato. However, the pepper genome is fourfold larger due to the accumulation of repetitive sequences which make up 81 % of the genome (Qin et al. 2014). As in previous work (Park et al. 2011), these repetitive sequences were found to be mostly *Gypsy*-like LTR retrotransposons which are not seen to such an extent in tomato (Kim et al. 2014; Qin et al. 2014). Based on these results and an estimate of the timing of transposon activity, the authors hypothesized that the accumulation of transposable elements in the pepper genome was quite recent (0.3 Mya) (Qin et al. 2014) and that the concomitant alteration and increase in heterochromatin were involved in pepper speciation (Kim et al. 2014). Qin et al. (2014) also report that translocations were the main drivers of chromosomal rearrangement in the Solanaceae with 612 and 430 translocations differentiating pepper from the tomato and potato genomes, respectively. Extensive inversions also occurred with 468 and 367 inversion events distinguishing pepper from tomato and potato, respectively.

Kim et al. (2014) used the whole genome sequence and microsynteny to identify capsaicinoid pathway orthologs in the pepper, tomato, and potato genomes. The orthologs were found to be expressed during placenta development in pepper but in tomato or potato. Microsynteny was also used to analyze the region surrounding the capsaicin synthase gene in hot pepper and the corresponding area in tomato. The region was found to contain seven acyltransferase genes in pepper but only four in tomato. Phylogenetic analyses of these genes indicated that the capsaicin synthase gene emerged after speciation. More dramatic gene family expansion was observed for the *Bs2*-containing subclass of NBS-LRR (nucleotide-binding site-leucine-rich repeat) disease resistance genes. Hot pepper contains 82 such genes in its genome while tomato and potato have only three and one, respectively. The expansion of this gene family has resulted in a loss of colinearity in the affected genomic regions of the three species. Thus, both retrotransposon amplification and gene family expansion were found to be significant factors in the divergence and speciation of hot pepper.

Nicotiana

Very few comparative genetic mapping studies have been published in *Nicotiana*, a genus of 66 species that includes the agricultural commodity and model organism for genetic engineering, cultivated tobacco (*N. tabacum*). While tobacco is an allotetraploid with a base chromosome number of 12, karyotypic variability is found within section *Alatae*. Chromosome numbers of nine and ten are found in several species, including *N. alata*, *N. bonariensis*, *N. forgetiana* and *N. langsdorffii* (x = 9) as well as *N. longiflora,* and *N. plumbaginifolia* (x = 10). While this variability in chromosome number makes *Nicotiana* an attractive system for chromosome evolution and synteny studies, the late development (within the last decade) of genetic maps in the genus means that such work remains to be done.

The first analysis of genome synteny in *Nicotiana* was made possible by the construction of a RFLP/RAPD linkage map for an interspecific (*N. plumbaginifolia* × *N. longiflora*) population (Lin et al. 2001). Only nine linkage groups were obtained, thus genome coverage was not complete. Nevertheless, comparison of linkage group assignments of 20 RFLP markers derived from *N. sylvestris* (Suen et al. 1997) revealed a lack of synteny between the mapped portions of the *plumbaginifolia* and *sylvestris* genomes. Given the difference in chromosome number between these species (x = 10 vs. x = 12 in *sylvestris*), this indication of chromosome disruption was not unexpected. Unfortunately, the positions of the markers relative to one another were unknown in *sylvestris*. Thus, colinearity between the two genomes could not be explored. However, based on evidence that several duplicate and triplicate loci in *sylvestris* were single copy in *plumbaginifolia*, the authors hypothesized that gene loss from multigene families may have contributed significantly to chromosome evolution and reduction in the *Nicotiana* genome.

While its large genome size (4500 Mbp; Arumuganathan and Earle 1991) and polyploid nature make *N. tabacum* less amenable to genomic research than other *Nicotiana* species, a concerted effort has been made to overcome these difficulties in recent years. A microsatellite map comprising 293 loci was published for a cross between the varieties 'Hicks Broadleaf' and 'Red Russian' (Bindler et al. 2007). The authors deemed the initial map incomplete due to the presence of large gaps and unlinked markers and subsequently published a high resolution map containing 2317 microsatellite markers (Bindler et al. 2011). The tenfold increase in marker number was accomplished by screening EST sequences generated by the Tobacco Genome Initiative (Gadani et al. 2003). As the authors suggest, this strategy of targeting single copy sequences for SSR marker development should facilitate the localization of homologous regions in other solanaceous genomes. However, explorations of tobacco-tomato synteny based on these markers have not yet been reported.

An allotetraploid that behaves as a diploid, tobacco is thought to have arisen as an interspecific hybrid between *N. sylvestris* and *N. tomentosiformis* (Kenton et al. 1993; Lim et al. 2004). Thanks to the availability of the aforementioned tobacco map (Bindler et al. 2011) as well as COSII/SSR maps (Wu et al. 2010) chromosome evolution within the two genomes of tobacco (the S- and T-genomes) is starting to be revealed. The mapping study conducted by Wu et al. (2010) compared the genomes of *N. tomentosiformis* and *N. acuminata* to each other and to that of *N. tabacum*. Extremely low polymorphism prevented mapping of *N. sylvestris*. *N. acuminata* was chosen as a substitute because it is evolutionarily closer to *N. sylvestris* than is *N. tomentosiformis*. Comparative analysis of COSII marker positions indicated that a minimum of seven chromosomal inversions and one reciprocal translocation distinguish the *tomentosiformis* (Tmf) and *acuminata* (Acn) genomes.

Chromosomes 6, 7, 9, and 11 show conservation of gene content and order in the diploid *Nicotiana* genomes. The timing of these structural changes relative to the polyploidization event leading to tobacco was determined by using a set of SSR markers from the Bindler et al. (2007) map. Mapping the SSR markers in the Tmf and Acn genomes also allowed the 24 tobacco linkage groups to be assigned to their respective ancestral genomes (T-genome for those markers that mapped to Tmf and S-genome for those in Acn). Since the divergence of the Tmf and Acn genomes, four inversions occurred in the lineage leading to *tomensiformis*, with the majority (3 of 4) pre-dating the tetraploidization event leading to tobacco. Of the two inversions specific to the Acn genome, one occurred before its split from *sylvestris* and one after. Based on these changes, it was estimated that the rate of chromosome evolution in *N. tomentosiformis* has varied from 0.5 to 2.1 rearrangements/MY before tetraploidization to 0.6 rearrangements/MY after that event. Similarly, a slower rate of evolution is evident in *N. acuminata* since its divergence from *N. sylvestris*: 0.4–1.3 rearrangements/MY before to 0.2 rearrangements/MY after divergence.

Comparison of SSR marker positions between the diploid *Nicotiana* species and tobacco revealed a minimum of 12 rearrangements: 9 inversions and single occurrences of chromosome breakage, fusion, and reciprocal translocation (Wu et al. 2010). The inversions were split almost equally between the T- and S- genomes (4 and 5 events, respectively). Since the speciation event, which the authors estimated as occurring less than one MY ago, the sub-genomes of tobacco are evolving at a faster rate than their diploid relatives: six changes in the T-genome as compared to just one in the Tmf genome. Estimated chromosomal evolution rates provide another measure of accelerated evolution following interspecific hybridization and polyploidization: 3.5 rearrangements/MY in the T-genome (as compared to 0.6 in *N. tomentosiformis*) and as many as 1.2 rearrangements/MY in the S-genome (vs. 0.2 in the Acn genome). Bindler et al. (2011) also found evidence for chromosomal rearrangement in tobacco, so much so that they were unable to identify homeologous chromosomes: more than 90 % of their SSR markers were specific to just one of the ancestral genomes. Nevertheless, the fact that markers specific to *N. sylvestris* and *N. tomentosiformis* mapped to the same linkage group in tobacco was interpreted as evidence of translocation between homeologous chromosomes. Thus the findings of Wu et al. (2010) and Bindler et al. (2011) provide valuable insight into how plant genomes reorganize after polyploidization.

Wu et al. (2010) also used COSII marker position to investigate the synteny of the *Nicotiana* and tomato genomes. Colinearity of markers was observed in 25 CSSs. With an average size of 15 cM, these CSSs spanned 34 % of the Tmf map. Outside of these regions of conservation, at least 11 reciprocal translocations and three (and perhaps as many as 10) inversions have occurred since the divergence of tomato and the LCA of the *Nicotiana* species. The relative frequency of each type of structural change was a bit of a surprise; previous analyses in the Solanaceae had prepared the authors to expect a greater number of inversions relative to translocations. Given the length of evolutionary time separating tomato and *Nicotiana* (some 27.7 MYA as calculated in this study), a number of inversions may have been obscured by subsequent changes in chromosome structure. In addition, they also noted that comparing tomato to the extant species *N. tomentosiformis* shows a slightly different picture: 14 inversions and 11 translocations, supporting the hypothesis that inversion is the predominant mode of rearrangement. As with the analysis of the ancestral genomes of tobacco, numerous instances of single marker transpositions were found between tomato and the *Nicotiana* species (36 in Tmf and 13 in Acn), suggesting that this is another important mechanism contributing to loss of synteny as genomes diverge.

Rapid progress has been made in sequencing *Nicotiana* genomes as evidenced by the recent publication of draft genome sequences of the diploid species *N. sylvestris* and

N. tomentosiformis (Sierro et al. 2013) as well as the allotetraploid species *N. benthamiana* (Bombarely et al. 2012) and *N. tabacum* (Sierro et al. 2014). These sequences should provide considerable insight into synteny within the genus and the family, however such analyses remain in their infancy. Thus, comparison of the diploid species genomes with those of other solanceous species was limited to localizing COSII markers from the *N. acuminata* and *N. tomentosiformis* genetic maps on the genome assemblies of *N. sylvestris* and *N. tomentosiformis*, respectively (Sierro et al. 2013). Only one-third of COSII markers could be mapped, highlighting the fragmented and incomplete nature of these assemblies. The lack of a genetic map for *N. benthamiana* has restricted comparative genomic studies with this species despite its popularity as a model system for plant–pathogen interations. However analysis of the draft *N. benthamiana* genome sequence revealed microsynteny with tomato in the *Pto-Prf* gene cluster, suggesting that this region evolved before the divergence of *Nicotiana* and *Solanum* (Bombarely et al. 2012).

The sequencing of three varieties of commercial tobacco (*N. tabacum*) has provided the greatest insight into genome evolution within the genus (Sierro et al. 2014). Comparisons with the genetic maps generated by Wu et al. (2009b) and Bindler et al. (2011) show the extent of chromosomal rearrangements that have occurred within the diploid and allotetraploid species. Moreover, use of the sequence data provided by the *sylvestris* and *tomentosiformis* draft genomes (Sierro et al. 2013) helped verify these species as the putative ancestors of *N. tabacum* and revealed that only 4–8 % of the ancestral genomes was lost subsequent to the hybridization event that created *N. tabacum* (Sierro et al. 2014). Mapping protein sequences of tomato and potato onto the tobacco genome sequence revealed considerable genome reorganization, however syntenic regions do exist and gene content is strongly conserved across the ~ 30 MY of evolution separating these solanaceous species (Fig. 12.6).

Petunia

Analyses of synteny between the horticultural plant *Petunia* and tomato have been hampered by the fact that genome mapping efforts in petunia have been limited. Until recently, the most comprehensive genetic linkage map for petunia contained just 36 RFLP markers spread across the plant's seven chromosomes (Strommer et al. 2000). For this reason, early reports of synteny between petunia and tomato arose from research focused on specific genomic regions.

It is not surprising that the first report of synteny between the petunia and tomato genomes originated from an analysis of the self-incompatibility (SI) locus. Members of the Solanaceae, including petunia, have served as model organisms for the analysis of gametophytic SI for several decades. While mapping the self-incompatibility locus (*S* locus), ten Hoopen et al. (1998) established a syntenic relationship between petunia chromosome III and chromosome 1 of tomato and potato. The position of the *S* locus was initially determined through T-DNA tagging and further substantiated by the cosegregation of an *S*-linked potato RFLP marker (CP100) with a peroxidase isozyme locus (*PrxA*) that had been previously mapped to chromosome III. Citing similar linkage between the *S*-locus and a peroxidase isozyme in *Nicotiana alata* (Labroche et al. 1983), the authors suggested that synteny of the self-incompatibility locus may be conserved in the Solanaceae (ten Hoopen et al. 1998).

In contrast, a region of the *Petunia* genome in which five floral traits (color, UV absorption, scent, and pistil and stamen length) are tightly linked appears not to be conserved in the family (Hermann et al. 2013). BLAST searches revealed that homologs of ten *Petunia* markers spanning this pollination syndrome gene cluster were widely distributed in the tomato and potato genomes, mapping to six chromosomes in tomato and five in potato. The authors speculate that this difference between *Petunia* and *Solanum* species may reflect different evolutionary pressures on these genes in the two genera. *Solanum*

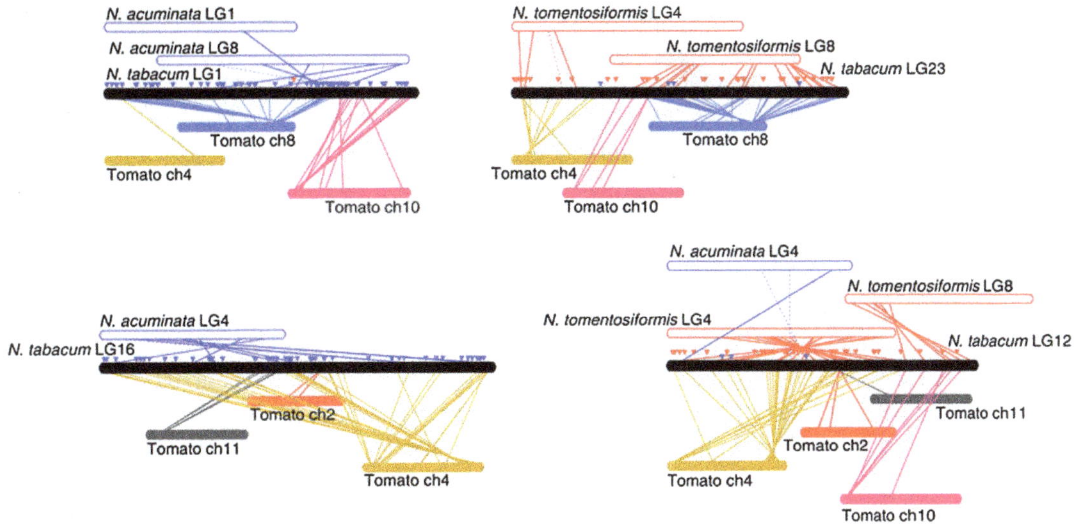

Fig. 12.6 Synteny between *Nicotiana* species and tomato for selected tobacco linkage groups. The comparison of *N. tabacum* and tomato is based on mapping tomato proteins. Links between tobacco and *N. acuminata* and *N. tomentosiformis* are derived from shared SSR (*solid lines*) and COSII (*dotted lines*) markers (used with permission from Sierro et al. 2014)

species tend to be bee-pollinated, whereas *Petunia* species exhibit frequent pollinator changes which could be facilitated by tight linkage of the genes underlying pollinator attraction.

On a broader scale, the conservation of a 17-gene region in the Solanaceae was revealed in sequence analysis of a 105 kb CSS in five species including petunia (Wang et al. 2008). While gene order and orientation were largely maintained, a small number of petunia-specific evolutionary changes were identified including a relatively recent tandem duplication of gene 12 and a 20 kb inversion involving genes 15 and 16 (Fig. 12.3). Comparison of gene structure revealed that ORFs and exon/intron positions in four out of seven genes were conserved across lineages. The degree of conservation of gene content was somewhat surprising given that homologous regions of the Arabidopsis genome have evolved at a considerably faster rate (Ku et al. 2000). However, the authors attributed this difference to the fact that no whole genome duplication events have occurred within the solanaceous lineage over the ~ 30 million years since the divergence of petunia and tomato.

The recent construction of linkage maps for wild *Petunia* species (Bossolini et al. 2011) has made it possible to compare synteny within *Petunia* as well as between petunia and tomato on a genome-wide basis. Such comparative mapping studies provide valuable insight into patterns of chromosomal evolution throughout the Solanaceae. A question of particular interest which the work of Bossolini et al. (2011) begins to answer is how chromosome number in the family was reduced from an ancestral value of $x = 12$ to $x = 7$ in the lineage leading to petunia. Bossolini et al. (2011) mapped a total of 207 CAPs and AFLP markers in two interspecific populations: *P. exserta* × *P. parodii* and *P. axillaris* × *P. inflata*. Thirty-seven shared markers revealed complete preservation of marker order between the two petunia maps. In order to compare the petunia and tomato genomes, BLASTN searches were used to position the petunia marker sequences on the physical map of tomato. A large amount of rearrangement was uncovered, with the degree of macrosynteny varying depending on the chromosome. Petunia chromosomes 5 and 7 were found to be syntenic with tomato chromosomes T12 and T8, respectively. Petunia chromosomes 1 and 6 were composite in nature. Petunia chromosome 1 has segments specific to T5 and T6 while chromosome 6 carries markers shared with T1 and

T9. While portions of chromosomes 3 and 4 are syntenic with T3 and T4, the synteny is limited to a segment of the long arms of the tomato chromosomes. An even more complex pattern is seen in the make up of chromosome 2 as it comprises markers found on T2, T7, T8, and T10. Thus, as compared to the localized synteny revealed by ten Hoopen et al. (1998) and Wang et al. (2008), the genomes of petunia and tomato show evidence of extensive structural differentiation when viewed on a larger scale. This loss of synteny makes it difficult to use the abundant genomic resources of tomato to assist petunia genetics but is indicative of the complex patterns of evolutionary change that occur during genome evolution.

Conclusions

The Solanaceae has been the subject of pioneering work in studies of genome synteny. By investigating the differences in genome size, content, and organization, this work has provided insight into the ways in which the structural rearrangement of chromosomes can lead to reproductive isolation and, ultimately, speciation. Moreover, it has helped to extend the utility of the extensive genomic resources of tomato to other members of this economically important family of plants.

Both DNA content and chromosome number vary in the Solanaceae. It has been hypothesized that the primary mechanisms responsible for genome size variation in plants and animals are transposable element replication, polyploidy, intron size, gene, and chromosome loss (Gregory 2005). Species in the Solanaceae provide examples of each of these mechanisms. There is ample evidence of transposable element, especially retrotransposon, activity in the genomes of tomato (Asamizu et al. 2012), wild tomato (Xiao et al. 2008), potato (Peters et al. 2012), eggplant (Wu et al. 2009a) and pepper (Mazourek et al. 2009; Park et al. 2011). For example, the pepper genome is three times larger than that of tomato and large scale (35.6 Mb) sequence comparisons indicated that many LTR retrotransposons (primarily Tat and Athila) have been inserted in the pepper genome while gene order and content are conserved across the two species (Park et al. 2011). Transposable element activity is also evident in the displacement of single markers/genes in genomes that are otherwise syntenic as observed in potato (Perez et al. 1999), eggplant (Wu et al. 2009a; Doganlar et al. 2014), pepper (Mazourek et al. 2009; Wu et al. 2009b), and tobacco (Wu et al. 2010). Examination of repetitive DNA at rearrangement junctions in potato indicates that transposition has played an important role in larger breaks from synteny including translocation and inversion (Zhu et al. 2008; Peters et al. 2012). In addition, transposable element activity was found to be responsible for a duplication resulting in the *sun* locus phenotype in tomato (Xiao et al. 2008). Polyploidy is observed in both the potato and tobacco genomes. Studies of synteny in these species provide insight into how genomes are rearranged after polyploidization (Bindler et al. 2011; Wu et al. 2010). Intron size has not yet been examined extensively in the Solanaceae, however, preliminary work in pepper indicates that increased gene size in pepper is mainly due to the presence of longer introns as compared to tomato (Park et al. 2011). Gene and chromosome loss are observed in some species of tobacco which have fewer members within multigene families and only ten chromosomes (Suen et al. 1997; Lin et al. 2001). Chromosome loss is also apparent in the petunia lineage as this species has only seven chromosomes (Bossolini et al. 2011).

While genome size is affected by a number of factors, only a few mechanisms are responsible for restructuring chromosomes: inversions, translocations, and transpositions (discussed in the previous paragraph). In the Solanaceae, paracentric inversions are usually reported as the most common type of rearrangement in potato, pepper, nightshade, and eggplant (Tanksley et al. 1992; Livingstone et al. 1999; Chetelat et al. 2000; Doganlar et al. 2002). A definite pericentric inversion was reported for *S. etuberosum* (Lou et al. 2010) and others have been hypothesized for pepper (Livingstone et al. 1999; Wu et al. 2009a, b). The preponderance of paracentric as compared to pericentric inversions supports the hypothesis that these inversions have fewer harmful effects on fertility than pericentric inversions (Burnham 1962), and therefore, may be less detrimental to organism fitness. Similarly, translocations may be more likely than inversions to interfere with chromosome pairing and are usually less frequent than inversions in the Solanaceae (Wu and Tanksley 2010). Interestingly, during the evolution of the Solanaceae, a few recurrent chromosomal breakpoints seem to have been primarily responsible for genome restructuring (Wu and Tanksley 2010). These breakpoints are commonly found in pericentromeric regions (Wu and Tanksley 2010), areas of the tomato genome that are known to be rich in retrotransposons (Wang et al. 2006) and other repetitive sequences (Presting et al. 1996). Thus, these observations in Solanaceae are in accordance with the general finding that rearrangements involving repetitive DNA are important in plant speciation (Raskina et al. 2008).

Examination of synteny among tomato, potato, eggplant, pepper, and tobacco using COSII markers has allowed rates of rearrangement and divergence times along these lineages to be calculated (Fig. 12.2). Such estimates allow comparison of rates of evolution within the family, as well as with other plant families. Thus it has been hypothesized that the eggplant lineage has undergone the more frequent rearrangements than the tomato, potato or pepper lineages (Wu and Tanksley 2010). The same work also indicates that genomes in the Solanaceae are evolving at similar rates as genomes in the Poaceae (the true grasses, including all the major cereals), Malvaceae (the mallows, including cotton and cacao) and Brassicaceae (the crucifers including rapeseed and cabbage) (Wu and Tanksley 2010).

In addition to impacting our understanding of the modes and tempo of plant genome evolution, synteny studies in solanaceous species have had several practical applications. Discovery of shared synteny in the family has allowed the use of RFLP and COSII markers originally developed for tomato (Bernatzky and Tanksley 1986; Wu et al. 2006) in related species (Tanksley et al. 1992; Livingstone et al. 1999; Doganlar et al. 2002; Albrecht and Chetelat 2009; Wu et al. 2009a, b, 2010; Doganlar et al. 2014). This has been a tremendous advantage for studies of the less important solanaceous species such as eggplant and nightshade. Shared synteny also enables in silico mapping of markers (Wu et al. 2009a, b) and facilitates gene cloning. For example, late blight resistance genes in both potato (Huang et al. 2005) and wild potato (Pel et al. 2009) were isolated with the help of comparative genomic analyses made possible by conserved genome organization in these species. In addition, knowledge of syntenic relationships between donor and recipient genomes can be extremely useful when attempting to minimize linkage drag while introgressing traits (Peters et al. 2012). Thus, synteny studies in the Solanaceae have provided fruitful results in both basic and applied plant genetics and will continue to do so as genome sequence data become more readily available.

Acknowledgments We are grateful to The Scientific and Technological Research Council of Turkey (Project No. 104T224) for support of our work in eggplant.

References

Abrouk M, Murat F, Pont C, Messing J, Jackson S, Faraut T, Tannier E, Plomion C, Cooke R, Feuillet C, Salse J (2010) Paleogenomics of plants: synteny-based modelling of extinct ancestors. Trends Plant Sci 15:479–487

Albrecht E, Chetelat RT (2009) Comparative genetic linkage map of *Solanum* sect. Juglandifolia: evidence of chromosomal rearrangements and overall synteny with the tomatoes and related nightshades. Theor Appl Genet 118:831–847

Ameline-Torregrosa C, Wang B-B, O'Bleness MS, Deshpande S, Zhu H, Roe B, Young ND, Cannon SB (2008) Identification and characterization of nucleotide-binding site-leucine-rich repeat genes in the model plant *Medicago truncatula*. Plant Physiol 146:5–21

Andolfo G, Sanseverino W, Rombauts S, Van de Peer Y, Bradeen JM, Carputo D, Frusciante L, Ercolano MR (2013) Overview of tomato (*Solanum lycopersicum*) candidate pathogen recognition genes reveals important *Solanum* R locus dynamics. New Phytol 197:223–237

Arumuganathan K, Earle ED (1991) Nuclear DNA content of some important plant species. Plant Mol Biol Rep 9:208–218

Asamizu E, Shirasawa K, Hirakawa H, Sato S, Tabata S, Yano K, Ariizumi T, Shibata D, Ezura H (2012) Mapping of Micro-Tom BAC-end sequences to the reference tomato genome reveals possible genome rearrangements and polymorphisms. Int J Plant Genomics. doi:10.1155/2012/437026

Bai Y, Lindhout P (2007) Domestication and breeding of tomatoes: what have we gained and what can we gain in the future? Ann Bot 100:1085–1094

Barchi L, Lanteri S, Portis E, Val G, Volante A, Pulcini L, Ciriaci T, Acciarri N, Barbierato V, Toppino L (2012) A RAD tag derived marker based eggplant linkage map and the location of QTLs determining anthocyanin pigmentation. PLoS ONE 7:e43740

Baumgarten A, Cannon S, Spangler R, May G (2003) Genome-level evolution of resistance genes in *Arabidopsis thaliana*. Genetics 165:309–319

Bernatzky R, Tanksley SD (1986) Toward a saturated linkage map in tomato based on isozymes and random cDNA sequences. Genetics 112:887–898

Bindler G, van der Hoeven R, Gunduz I, Plieske J, Ganal M, Rossi L, Gadani F, Donini P (2007) A microsatellite marker based linkage map of tobacco. Theor Appl Genet 114:341–349

Bindler G, Plieske J, Bakaher N, Gunduz I, Ivanov N, van der Hoeven R, Ganal M, Donini P (2011) A high density genetic map of tobacco (*Nicotiana tabacum* L.) obtained from large scale microsatellite marker development. Theor Appl Genet 123:219–230

Bombarely A, Rosli HG, Vrebaliv J, Moffett P, Mueller LA, Martin GB (2012) A draft genome sequence of *Nicotiana benthamiana* to enhance molecular plant-microbe biology research. Mol Plant Microbe Interact 25:1523–1530

Bonierbale MW, Plaisted RL, Tanksley SD (1988) RFLP maps based on a common set of clones reveal modes of chromosomal evolution in potato and tomato. Genetics 120:1095–1103

Bonnema G, Schipper D, van Heusden S, Zabel P, Lindhout P (1997) Tomato chromosome 1: high-resolution genetic and physical mapping of the short arm in an interspecific *Lycopersicon esculentum* × *L. peruvianum* cross. Mol Gen Genet 253:455–462

Bossolini E, Klahre U, Brandenburg A, Reinhardt D, Kuhlemeier C (2011) High resolution linkage maps of the model organism *Petunia* reveal substantial synteny decay with the related genome of tomato. Genome 54:327–340

Burnham CR (1962) Discussions in cytogenetics. Burgess, Minneapolis

Cericola F, Portis E, Lanteri S, Toppino L, Barchi L, Acciarri N, Pulcini L, Sala T, Rotino GL (2014) Linkage disequilibrium and genome-wide association analysis for anthocyanin pigmentation and friut color in eggplant. BMC Genom 15:896–911

Chetelat RT, Meglic V (2000) Molecular mapping of chromosomes segments introgressed from *Solanum lycopersicoides* into cultivated tomato (*Lycopersicon esculentum*). Theor Appl Genet 100:232–241

Chetelat RT, Meglic V, Cisneros P (2000) A genetic map of tomato based on BC1 *Lycopersicon esculentum* × *Solanum lycopersicoides* reveals overall synteny but supressed recombination between these homeologous genomes. Genetics 154:857–867

Contreras-M A, Spooner DM (1999) Revision of *Solanum* section *Etuberosum* (subgenus *Potatoe*). In: Nee M, Symon DE, Lester RN, Jessop JP (eds) Solanaceae IV. Royal Botanic Gardens, Kew, pp 227–245

D'Agostino N, Golas T, van der Geest H, Bombarely A, Dawood T, Zethof J, Driedonks N, Wijnker E, Bargsten J, Nap J-P, Mariani C, Rieu I (2013) Genomic analysis of the native European *Solanum* species, *S. dulcamara*. BMC Genom 15:356–370

Doganlar S, Frary A, Daunay M-C, Lester RN, Tanksley SD (2002) A comparative genetic linkage map of eggplant (*Solanum melongena*) and its implications for genome evolution in the Solanaceae. Genetics 161:1697–1711

Doganlar S, Frary A, Daunay M-C, Huvenaars K, Mank R, Frary A (2014) High resolution map of eggplant (*Solanum melongena*) reveals extensive chromosome rearrangement in domesticated members of the Solanaceae. Euphytica 198:231–241

Fukuoka H, Miyatake K, Nunome T, Negoro S, Shirasawa K, Isobe S, Asamizu E, Yamaguchi H, Ohyama A (2012) Development of gene-based markers and construction of an integrated linkage map in eggplant by using *Solanum* orthologous (SOL) gene sets. Theor Appl Genet 125:47–56

Fulton TM, Van der Hoeven R, Eannetta NT, Tanksley SD (2002) Identification, analysis, and utilization of conserved ortholog set markers for comparative genomics in higher plants. Plant Cell 14:1457–1467

Gadani F, Hayes A, Opperman CH, Lommel SA, Sosinski BR, Burke M, Hi L, Brierly R, Salstead A, Heer J, Fuelner G, Lakey N (2003) Large scale genome sequencing and analysis of *Nicotiana tabacum*: the tobacco genome initiative. In: Proceedings, 5èmes Journées Scientifiques du Tabac de Bergerac—5th Bergerac Tobacco Scientific Meeting, Bergerac, pp 117–130

Gebhardt C, Ritter E, Barone A, Debener T, Walkemeier B, Schachtschabel U, Kaufman H, Thompson RD, Bonierbale MW, Ganal MW, Tanskley SD, Salamini F (1991) RFLP maps of potato and their alignment with the homoeologous tomato genome. Theor Appl Genet 83:49–57

Gramazio P, Prohens J, Plazas M, Andjar I, Herraiz FJ, Castillo E, Knapp S, Meyer RS, Vilanova S (2014) Location of chlorogenic acid biosynthesis pathway and polyphenol oxidase genes in a new interspecific anchored linkage map of eggplant. BMC Plant Biol 14:350–365

Gregory TR (2005) The C-value enigma in plants and animals: a review of parallels and an appeal for partnership. Ann Rev Bot 95:133–146

Grube RC, Radwanski ER, Jahn M (2000) Comparative genetics of disease resistance within the Solanaceae. Genetics 155:873–887

Hajjar R, Hodgkin T (2007) The use of wild relatives in crop improvement: a survey of developments over the last 20 years. Euphytica 156:1–13

Heiser CB, Pickersgill B (1969) Names for the cultivated *Capsicum* species (Solanaceae). Taxon 18:277–283

Hermann K, Klahre U, Moser M, Sheehan H, Mandel T, Kuhlemeier C (2013) Tight genetic linkage of prezygotic barrier loci creates a multifunctional speciation island in *Petunia*. Curr Biol 23:873–877

Hirakawa H, Shirasawa K, Miyatake K, Nunome T, Negoro S, Ohyama A, Yamaguchi H, Sato S, Isobe S, Tabata S, Fukuoka H (2014) Draft genome sequence of eggplant (*Solanum melongena* L.): the representative *Solanum* species indigenous to the Old World. DNA Res doi:10.1093/dnares/dsu027

Huang S, Vleeshouwers VGAA, Werij JS, Hutten RC, van Eck HJ, Visser RGF, Jacobsen E (2004) The *R3* resistance to *Phytophthora infestans* in potato is conferred by two closely linked R genes with distinct specificities. Mol Plant Microbe Interact 17:428–435

Huang S, van der Vossen EAG, Kuang H, Vleeshouwers VGAA, Zhang N, Borm TJA, van Eck HJ, Baker B, Jacobsen E, Visser RGF (2005) Comparative genomics enabled the isolation of the *R3a* late blight resistance gene in potato. Plant J 42:251–261

Iorizzo M, Gao L, Mann H, Traini A, Chiusano ML, Kilian A, Aversano R, Carputo D, Bradeen JM (2014) A DArT marker-based linkage map for wild potato *Solanum bulbocastanum* facilitates structural comparisons between *Solanum* A and B genomes. BMC Genet 15:123–132

Iovene M, Wielgus SM, Simon PW, Buell CR, Jiang J (2008) Chromatin structure and physical mapping of chromosome 6 of potato and comparative analyses with tomato. Genetics 180:1307–1317

Kamenetzky L, Asis R, Bassi S. de Godoy F, Bermudez L, Fernie AR, Van Sluys MA, Vrebalov J, Giovannoni JJ, Rossi M, Carrari F (2010) Genomic analysis of wild tomato introgressions determining metabolism- and yield-associated traits. Plant Physiol 152:1772–1786

Kenton A, Parokonny AS, Gleba YY, Bennett MD (1993) Characterization of the *Nicotiana tabacum* L. genome by molecular cytogenetics. Mol Gen Genet 240:159–169

Kim S, Park M, Yeom S-I, Kim Y-M, Lee JM et al (2014) Genome sequence of the hot pepper provides insights into the evolution of pungency in *Capsicum* species. Nat Genet 46:270–279

Ku H-K, Vision T, Liu J, Tanksley SD (2000) Comparing sequenced segments of the tomato and Arabidopsis genomes: large-scale duplication followed by selective gene loss creates a network of synteny. Proc Natl Acad Sci USA 97:9121–9126

Kuhl JC, Hanneman RE, Havey MJ (2001) Characterization and mapping of *Rpi1*, a light-blight resistance locus from diploid (1EBN) Mexican *Solanum pennatisectum*. Mol Genet Genomics 265:977–985

Labroche P, Poirier-Hamon S, Pernes J (1983) Inheritance of leaf peroxidase isozymes in *Nicotiana alata* and linkage with the *S*-incompatibility locus. Theor Appl Genet 65:163–170

Lanteri S, Pickersgill B (1993) Chromosomal structural changes in *Capsicum annuum* L. and *C. chinense* Jacq. Euphytica 67:155–160

Lim KY, Matyasek R, Kovarik A, Leitch AR (2004) Genome evolution in allotetraploid *Nicotiana*. Biol J Linn Soc 82:599–606

Lin TY, Kao YY, Lin S, Lin RF, Chen CM, Huang CH, Wang CK, Lin YZ, Chen CC (2001) A genetic linkage map of *Nicotiana plumbaginifolia/Nicotiana longiflora* based on RFLP and RAPD markers. Theor Appl Genet 103:905–911

Lindqvist-Kreuze H, Cho K, Portal L, Rodriguez F, Simon R, Mueller LA, Spooner DM, Bonierbale M (2013) Linking the potato genome to the conserved ortholog set (COS) markers. BMC Genom 14:51–63

Lippman ZB, Semel Y, Zamir D (2007) An integrated view of quantitative trait variation using tomato interspecific introgression lines. Curr Opin Genet Dev 17:545–552

Livingstone KD, Lackney VK, Blauth JR, van Wijk R, Jahn MK (1999) Genome mapping in *Capsicum* and the evolution of genome structure in the Solanaceae. Genetics 152:1183–1202

Lou Q, Iovene M, Spooner DM, Buell CR, Jiang J (2010) Evolution of chromosome 6 of *Solanum* species revealed by comparative fluorescence in situ hybridization mapping. Chromosoma 119:435–442

Mazourek M, Cirulli ET, Collier SM, Landry LG, Kang B-C, Quirin EA, Bradeen JM, Moffett P, Jahn MM (2009) The fractionated orthology of *Bs2* and *Rx/Gpa2* supports shared synteny of disease resistance in the Solanaceae. Genetics 182:1351–1364

McCouch (2001) Genomics and synteny. Plant Physiol 125:152–155

Nadeau JH, Taylor BA (1984) Lengths of chromosomal segments conserved since divergence of man and mouse. Proc Natl Acad Sci USA 81:814–818

Nunome T, Negoro S, Kono I, Kanamori H, Miyatake K, Yamaguchi H, Ohyama A, Fukuoka H (2009) Development of SSR markers derived from SSR-enriched genomic library of eggplant (*Solanum melongena* L.). Theor Appl Genet 119:1143–1153

Park M, Jo SH, Kwon J-K, Park J, Ahn JH, Kim S, Lee Y-H, Yang T-J, Hur C-G, Kang B-C, Kim B-D, Choi D (2011) Comparative analysis of pepper and tomato reveals euchromatin expansion of pepper genome caused by differential accumulation of *Ty3/Gypsy*-like elements. BMC Genom 12:85

Pel MA, Foster SJ, Park T-H, Rietman H, van Arkel G, Jones JDG, Van Eck HJ, Jacobsen E, Visser RGF, Van der Vossen EAG (2009) Mapping and cloning of late blight rsistance genes from *Solanum venturii* using an interspecific candidate gene approach. Mol Plant Microb Interact 22:601–615

Peralta I, Spooner DM (2001) Granule-bound starch synthase (GBSSI) gene phylogeny of wild tomatoes (*Solanum L.* section *Lycopersicon* (Mill.) Wettst. subsection *Lycopersicon*). Am J Bot 88:1888–1902

Perez F, Menendez A, Dehal P, Quiros CF (1999) Genomic structural differentiation in *Solanum*: comparative mapping of the A and E genomes. Theor Appl Genet 98:1183–1193

Pertuze RA, Ji Y, Chetelat RT (2002) Comparative linkage map of the *Solanum lycopersicoides* and *S. sitiens* genomes and their differentiation from tomato. Genome 45:1003–1012

Peters SA, Bargsten JW, Szinay D, van de Belt J, Visser RGF, Bai Y, de Jong H (2012) Structural homology in the Solanaceae: analysis of genomic regions in support of synteny studies in tomato, potato and pepper. Plant J 71:602–614

Portis E, Barchi L, Toppino L, Lanteri S, Acciarri N, Felicioni N, Fusari F, Barbierato V, Cericola F, Vale G, Rotino GL (2014) QTL mapping in eggplant reveals clusters of yield-related loci and orthology with the tomato genome. PLoS ONE 9:e89499

Presting G, Frary A, Pillen K, Tanksley SD (1996) Telomere-homologous sequences occur near the centromeres of many tomato chromosomes. Mol Gen Genet 251:526–531

Prince JP, Pochard E, Tanksley SD (1993) Construction of a molecular linkage map of pepper and comparison of synteny with tomato. Genome 36:404–417

Qin C, Yu C, Shen Y, Fang X et al (2014) Whole-genome sequencing of cultivated and wild peppers provides insights into *Capsicum* domestication and specialization. Proc Natl Acad Sci 111:5135–5140

Ramanna M, Hermsen J (1979) Unique meiotic behavior in F1 plants from a cross between non-tuberous and tuberous *Solanum* species in section *Petota*. Euphytica 28:9–15

Raskina O, Barber JC, Nevo E, Belyayev A (2008) Repetitive DNA and chromosomal rearrangement: speciation-related events in plant genomes. Cytogenet Genome Res 120:351–357

Rick CM (1979) Biosystematic studies in *Lycopersicon* and closely related species of *Solanum*. In: Hawkes JG, Lester RN, Skelding AD (eds) The biology and taxonomy of the Solanaceae. Academic Press, New York, pp 667–678

Seah S, Yaghoobi J, Rossi M, Gleason CA, Williamson VM (2004) The nematode-resistance gene, *Mi-1*, is associated with an inverted chromosomal segment in susceptible compared to resistance tomato. Theor Appl Genet 108:1635–1642

Seah S, Telleen AC, Williamson VM (2007) Introgressed and endogenous *Mi-1* gene clusters in tomato differ by complex rearrangements in flanking sequences and show sequence exchange and diversifying selection among homologues. Theor Appl Genet 114:1289–1302

Sierro N, Battey JND, Ouadi S, Bovet L, Goepfert S, Bakaher N, Peitsch MC, Ivanov NV (2013) Reference genomes and transcriptomes of *Nicotiana sylvestris* and *Nicotiana tomentosiformis*. Genome Biol 14:R60

Sierro N, Battey JND, Ouadi S, Bakaher N, Bovet L, Willig A, Goepfert S, Peitsch MC, Ivanov NV (2014) The tobacco genome sequence and its comparison with those of tomato and potato. Nat Commun 5:3833

Spooner D, Anderson G, Jansen R (1993) Chloroplast DNA evidence for the interrelationships of tomatoes, potatoes and pepinos (Solanaceae). Am J Bot 80:676–688

Strommer J, Gerats AGM, Sanago M, Molnar SJ (2000) A gene-based RFLP map of *Petunia*. Theor Appl Genet 100:899–905

Suen DF, Wang CK, Lin RF, Kao YY, Lee FM, Chen CC (1997) Assignment of DNA markers to *Nicotiana sylvestris* chromosomes using monosomic alien addition lines. Theor Appl Genet 94:331–337

Szinay D, Wijnker E, van den Berg R, Visser RGF, de Jong H, Bai Y (2012) Chromosome evolution in *Solanum* traced by cross-species BAC-FISH. New Phytol 195:688–698

Tang X, Szinay D, Lang C, Ramanna MS, ven der Vossen EAG, Datema E, Lankhorst RK, de Boer J, Peters SA, Bachem C, Stiekema W, Visser RGF, de Jong J, Bai Y (2008) Cross-species bacterial artificial chromosome-fluorescence in situ hybridization painting of the tomato and potato chromosome 6 reveals undescribed chromosomal rearrangements. Genetics 180:1319–1328

Tanksley SD, McCouch SR (1997) Seed banks and molecular maps: unlocking genetic potential from the wild. Science 277:1063–1066

Tanksley SD, Bernatzky R, Lapitan NL, Prince JP (1988) Conservation of gene repertoire but not gene order in pepper. Proc Natl Acad Sci USA 85:6419–6423

Tanksley SD, Ganal MW, Prince JP, de Vicente MC, Bonierbale MW, Broun P, Fulton TM, Giovannoni JJ, Grandillo S, Martin GB, Messeguer R, Miller JC, Miller L, Paterson AH, Pineda O, Roder MS, Wing RA, Wu R, Young ND (1992) High density molecular linkage maps of the tomato and potato genomes. Genetics 132:1141–1160

ten Hoopen R, Harbord RM, Maes T, Nanninga N, Robbins TP (1998) The self-incompatibility (*S*) locus in *Petunia hybrida* is located on chromosome III in a region syntenic for the Solanaceae. Plant J 16:729–734

Tomato Genome Consortium (2012) The tomato genome sequence provides insights into fleshy fruit evolution. Nature 485:635–641

Traini A, Iorizzo M, Mann H, Bradeen JM, Carputo D, Frusciante L, Chiusano ML (2013) Genome microscale heterogeneity among wild potatoes revealed by diversity arrays technology marker sequences. Int J of Genomics 2013:257218

van der Knaap E, Sanyal A, Jackson SA, Tanksley SD (2004) High-resolution fine mapping and fluorescence *in situ* hybridization analysis of *sun*, a locus controlling tomato fruit shape, reveals a region of the tomato genome prone to DNA rearrangements. Genetics 168:2127–2140

van Heusden AW, Koornneef M, Voorrips RE, Bruggemann W, Pet G, Vrielink-van Ginkel R, Chen X, Lindhout P (1999) Three QTLs from *Lycopersicon peruvianum* confer a high level of resistance to *Clavibacter michiganensis* ssp. *michiganensis*. Theor Appl Genet 99:1068–1074

van Wordragen MF, Weide R, Liharska T, Vandersteen A, Koornneef M, Zabel P (1994) Genetic and molecular organization of the short arm and pericentromeric region of tomato chromosome 6. Euphytica 79:169–174

Wang J, Hu H, Zhao T, Yang Y, Chen T, Yang M, Yu W, Zhang B (2015) Genome-wide analysis of bHLH transcription factor and involvement in the infection by yellow leaf curl virus in tomato (*Solanum lycopersicum*). BMC Genom 16:39–53

Wang Y, Tang X, Cheng Z, Mueller L, Giovannoni J, Tanksley SD (2006) Euchromatin and pericentromeric heterochromatin: comparative composition in the tomato genome. Genetics 172:2529–2540

Wang Y, Diehl A, Wu F, Vrebalov J, Giovannoni J, Siepel A, Tanksley SD (2008) Sequencing and comparative analysis of a conserved syntenic segment in the Solanaceae. Genetics 180:391–408

Watanabe K, Orrillo M, Vega S, Valkonen J, Pehu E, Hurtado A, Tanksley S (1995) Overcoming crossing barriers between non-tuber-bearing and tuber-bearing Solanum species: towards potato genome enhancement with a broad spectrum of solanaceous genetic resources. Genome 38:27–35

Wu F, Eannetta NT, Xu Y, Tanksley SD (2009a) A detailed synteny map of the eggplant genome based on conserved ortholog set II (COSII) markers. Theor Appl Genet 118:927–935

Wu F, Eannetta NT, Xu Y, Durrett R, Mazourek M, Jahn MM, Tanksley SD (2009b) A COSII genetic map of the pepper genome provides a detailed picture of synteny with tomato and new insights into recent chromosome evolution in the genus *Capsicum*. Theor Appl Genet 118:1279–1293

Wu F, Eannetta NT, Xu Y, Plieske J, Ganal M, Pozzi C, Bakaher N, Tanksley SD (2010) COSII genetic maps of two diploid *Nicotiana* species provide a detailed picture of synteny with tomato and insights into chromosome evolution in tetraploid *N. tabacum*. Theor Appl Genet 120:809–827

Wu F, Mueller LA, Crouzillat D, Petiard V, Tanksley SD (2006) Combining bioinformatics and phylogenetics to identify large sets of single-copy orthologous genes (COSII) for comparative, evolutionary and systematic studies: a test case in the euasterid plant clade. Genetics 174:1407–1420

Wu F, Tanksley SD (2010) Chromosomal evolution in the plant family Solanaceae. BMC Genom 11:182–193

Xiao H, Jiang N, Schaffner E, Stockinger EJ, van der Knaap (2008) A retrotransposon mediated gene duplication underlies morphological variation of tomato fruit. Science 319:1527–1530

Yang H-B, Liu WY, Kang W-H, Jahn M, Kang B-C (2009) Development of SNP markers linked to the *L* locus in *Capsicum* spp. By comparative genetic analysis. Mol Breed 24:433–446

Zhu W, Ouyang S, Iovene M, O'Brien K, Vuong H, Jiang J, Buell CR (2008) Analysis of 90 Mb of the potato genome reveals conservation of gene structure and order with tomato but divergence in repetitive sequence composition. BMC Genom 9:286–300

Tomato Databases

13

Lukas Mueller and Noe Fernandez-Pozo

Abstract

Databases have become indispensible for conducting biological research. Here we present a brief overview of databases that focus on *Solanum lycopersicum*. The databases address different needs for researchers, and cover germplasm, mutant, gene expression, metabolism and genome sequence databases. Researchers should familiarize themselves with these resources to make the most out of tomato as a model system.

Keywords

Tomato · Databases · Germplasm · Sequence · Genome

Introduction

With the rapid accumulation of vast amounts of biological data, databases have become an indispensable pillar for research. The extent of data growth is particularly evident for sequence data, which now grows faster than computer storage capacity (Stein 2010). Other datatypes, such as genotypic and phenotypic data, are also growing rapidly, making it difficult to pursue biology without a bioinformatics and database infrastructure. Model species, such as tomato, for which many large datasets are produced, are intractable without appropriate databases. With novel methods and technologies, it is easy to predict that data growth will only increase in the future. For the tomato researcher it is therefore critical to have access to databases covering a wide range of topics that enable data to be queried and analyzed. Researchers need timely and easy access to up-to-date gene annotation information, gene expression data, and data associated with germplasm (i.e., phenotypic data, passport data, and ordering information). Databases are a key element in the toolkit of an organism that makes it an attractive model, and tomato is no exception. Fortunately, there are a large number of databases available for tomato, with the Sol Genomics Network (SGN, http://solgenomics.net/) website serving as a central hub linking out to, and incorporating much of, the sequence, phenotypic and genotypic data available through different sources, facilitating access for researchers through a "one stop shop" (Bombarely et al. 2011; Menda et al. 2008; Mueller et al. 2005; Tecle et al. 2010).

L. Mueller (✉) · N. Fernandez-Pozo
Boyce Thompson Institute, Cornell University, Tower Rd, Ithaca, NY 14853-2901, USA
e-mail: lam87@cornell.edu

M. Causse et al. (eds.), *The Tomato Genome*, Compendium of Plant Genomes,
DOI 10.1007/978-3-662-53389-5_13

This chapter gives a brief overview of tomato-centric databases and other generic resources highly significant for tomato data, which the reader is encouraged to further explore using the links provided.

Germplasm and Mutant Collections

An important resource for the community is a comprehensive germplasm collection with an associated database for searching and ordering mutants and wild accessions. Large-scale mutagenesis projects have produced a considerable amount of data on phenotypic variation in tomato. Several such projects exist for tomato, with a number of associated databases for web-based data access. Some representative examples are described here in more detail. Most of the germplasm resources have integrated seed ordering, but usually require a Material Transfer Agreement (MTA) to be signed before seed can be obtained.

An important aspect in germplasm and mutant collections is the description of phenotypes. The description of phenotypes is most useful if they are based on common vocabularies, such that phenotypes can be compared between different experiments, projects and information resources. In addition to the Plant Ontology (Ilic et al. 2006), which describes plant anatomy and plant developmental stages, other ontologies have been developed for tomato and the larger Solanaceae (Menda et al. 2004), to describe mutant phenotypes that are particularly also useful for breeders. This Solanaceae Phenotype (SP) ontology is available from SGN (Bombarely et al. 2011). The SP ontology has been widely used in a number of projects and expanded with terms describing very specific attributes, such as fruit shape (Brewer et al. 2006; Rodriguez et al. 2011). Most of the tomato mutant databases use the SP ontology, or a close derivative thereof, for the description of their phenotypes, making the data more easily comparable between databases.

TGRC (http://tgrc.ucdavis.edu/)

The tomato community has been fortunate to have a comprehensive resource for wild and mutant germplasm, the Tomato Genetic Resource Center (TGRC, http://tgrc.ucdavis.edu/), based on the collection of Charles Rick at UC Davis (Anonymous). This comprehensive germplasm collection provides a common standard for naming accessions and provides basic information, such as passport data, as well as seed ordering. TGRC has a website with searches for a number of criteria, such as accession number, phenotypic characteristics, ploidy level, and importantly, specific gene mutants. Seed can be ordered directly on the website. The TGRC data is also mirrored on SGN.

TOMATOMA (http://tomatoma.nbrp.jp/index.jsp)

Due to its small size, the MicroTom tomato variety can be grown comfortably in large numbers in growth chambers. MicroTom was developed as a model system in a number of projects, particularly in Japan and France, and has recently been sequenced (Kobayashi et al. 2014). Large-scale mutagenesis projects have been initiated in both countries. In Japan, tomato has been chosen as one of the Bioresource projects (http://tomato.nbrp.jp/indexEn.html), which includes a number of resources for mutants, sequences, and databases, with a focus on MicroTom. The TOMATOMA site describes more than 3300 mutants generated at the University of Tsukuba by EMS mutagenesis (3048 lines) and gamma radiation induced lines (289 lines) (Saito et al. 2011). Lines can be ordered on the site for a small fee and come with a Material Transfer Agreement. A TILLING platform for MicroTom has also been established (Okabe et al. 2011). The TOMATOMA lines are also available for searching on SGN, from where links to the TOMATOMA database are provided.

Tilling Site at Ucdavis (ctilling.ucdavis.edu/index.php/Tomato_Tilling)

In addition to the TGRC site, there is a tilling project database in Davis, available at http://tilling.ucdavis.edu/index.php/Tomato_Tilling. The database contains information about 4000 M2 lines, but awaits more funding to complete the work.

LycoTILL (http://www.agrobios.it/tilling/)

LycoTILL, based in Naples, Italy, is an EMS-mutation based tilling platform based on the Red Setter tomato variety. Red Setter is a processing tomato that has high yield and is amenable to mechanical harvesting. The mutant collection comprises 6677 M2 and 5872 M3 families that are searchable on the website and can be ordered with a Material Transfer Agreement (Minoia et al. 2010).

EU-SOL Database (https://www.eu-sol.wur.nl)

The EU-SOL database describes a core collection of about 7000 lines, which have been selected to represent a large fraction of the diversity found in tomato germplasm. The germplasm was sourced from a number of germplasm collections around the world, and phenotyped as well as genotyped. Seed can be ordered on-line, and a number of tools, such as the Marker to Sequence tool, are available on the site. This tool allows to locate markers and intervals from the tomato genome and retrieves the corresponding sequence. This database is maintained at the University of Wageningen in the Netherlands.

SolCAP (http://solcap.msu.edu/)

The SolCAP project phenotyped and genotyped panels of tomato and potato accessions. Genotyping was done using the SolCAP-developed SNP chip based on the Illumina Infinium platform; the chips are available to the community. Phenotyping focused on traits important to breeders. The data can be downloaded from the SolCAP site at http://solcap.msu.edu and have also been integrated into the SGN site.

SGN (http://solgenomics.net/)

As mentioned for the respective databases, SGN mirrors certain datasets from other databases in its germplasm database, complete with images, annotations, and links to the original database. The database currently comprises 25,000 accessions, the large majority of which (23,000) are *Solanum lycopersicum* accessions. Additionally, full-length protein kinase clones from TOKN 1.0 (Singh et al. 2014) can be requested at http://solgenomics.net/kinases/clones/form.

Genome and Sequence Databases

With the availability of genome reference sequences, databases have established themselves as an indispensable part of the research infrastructure. Reference genome sequences provide detailed information and a unified reference for genomic features, allowing the community to more easily share and compare data. The SGN site serves as a primary repository for the tomato reference sequence and provides versioned annotations from the International Tomato Annotation Group (ITAG). Based on this reference sequence, a number of tools and databases have been built for specific applications. Before the reference sequence was available, in the early 2000s, a number of EST-based databases were created which are still available and provide useful information (Aoki et al. 2010; Bombarely et al. 2011; Chiusano et al. 2008; Duvick et al. 2008; Quackenbush et al. 2000). However, newer technologies such as next-generation RNA-Seq technology made EST sequencing obsolete. RNA-Seq data, due to their massive size, are difficult to integrate into databases that are designed for EST data. Therefore, most of the sites have not been updated with data based on RNA-Seq.

SGN (http://solgenomics.net)

SGN is a comprehensive site for tomato genome information and a central "one stop shop" for tomato and the Solanaceae. As the repository of the tomato genome and its annotation, and providing sequence and genome data for many other Solanaceae, the SGN site offers a wide range of tools to query and analyze the data, covering standard tools such as genome viewers and BLAST (Altschul et al. 1990), but also a map viewer (Mueller et al. 2008), an interactive sequence alignment tool, tree viewers, mapping tools, and an expression database (Edwards et al. 2010) and a tomato expression atlas for tomato data from laser dissection is under development.

An important goal of SGN is to enable the links between genomes and phenomes (G2P). The sequence module is tightly linked to the phenotype module, which contains more than 25,000 accessions, over 90 % of which are currently accessions of *Solanum lycopersicum* (Bombarely et al. 2011; Mueller et al. 2005). Tools to link phenotype and genotype include the solQTL tool (Tecle et al. 2010), as well as a Genomic Selection tool currently under development. Linking phenotypes and genotypes is a key activity for breeders, and several new features are being added to the site to enable breeders to more easily take advantage of genomic tools, including the ability to run aspects of a breeding program directly from within the SGN database (Fig. 13.1).

Tomatogenome.net

Recently, the 150 tomato genomes project has made available a large number of genomic sequences from 150 tomato lines and wild relatives on the website tomatogenome.net. In total, 83 genotypes including 10 old varieties, 43 land races and 30 wild accessions were sequenced. Ten accessions of *S. lycopersicum* var. lycopersicum and *S. lycopersicum* var. cerasiforme were selected that represent the maximum range of expected genetic variation. The data are presented in a genome viewer showing the single nucleotide polymorphisms between the resequenced lines and the tomato reference genome (Causse et al. 2013). The data are also available from SGN.

The PGSB Tomato Genome Database (http://mips.helmholtz-muenchen.de/plant/tomato/index.jsp)

As a partner in the sequencing project, the MIPS database in Munich, Germany, has a site that presents the tomato genome sequence and annotations, hosted by the PGSB group (Plant Genome and System Biology) at the Helmholtz Zentrum Munchen at Germany.

ISOL@ (http://biosrv.cab.unina.it/isola/)

ISOL@ has a twofold focus: the genome and the transcriptome. It integrates the reference tomato (and potato) genome data with transcriptome data, currently mostly derived from EST sequences (Chiusano et al. 2008) as a tool to analyze expression. The EST data are included through the SOLESTdb site (D'Agostino et al. 2009). Since both tomato and potato datasets are in the database, a goal is to provide comparative genomics functionality on the site (Chiusano et al. 2008).

MiBASE TomatoDB (http://www.pgb.kazusa.or.jp/mibase/)

MiBASE TomatoDB is a site maintained at the Kazusa DNA research center in Japan that revolves mainly around the MicroTom variety, with information on unigene datasets, metabolic networks, and expression data (Aoki et al. 2010). The database can be searched by functional annotation of unigene sets, based on keywords and Gene Ontology terms (Ashburner et al. 2000), biochemical pathways, or genetic markers. BLAST (Altschul et al. 1990) and download tools are available, and a link for ordering clones generated at Kazusa is provided.

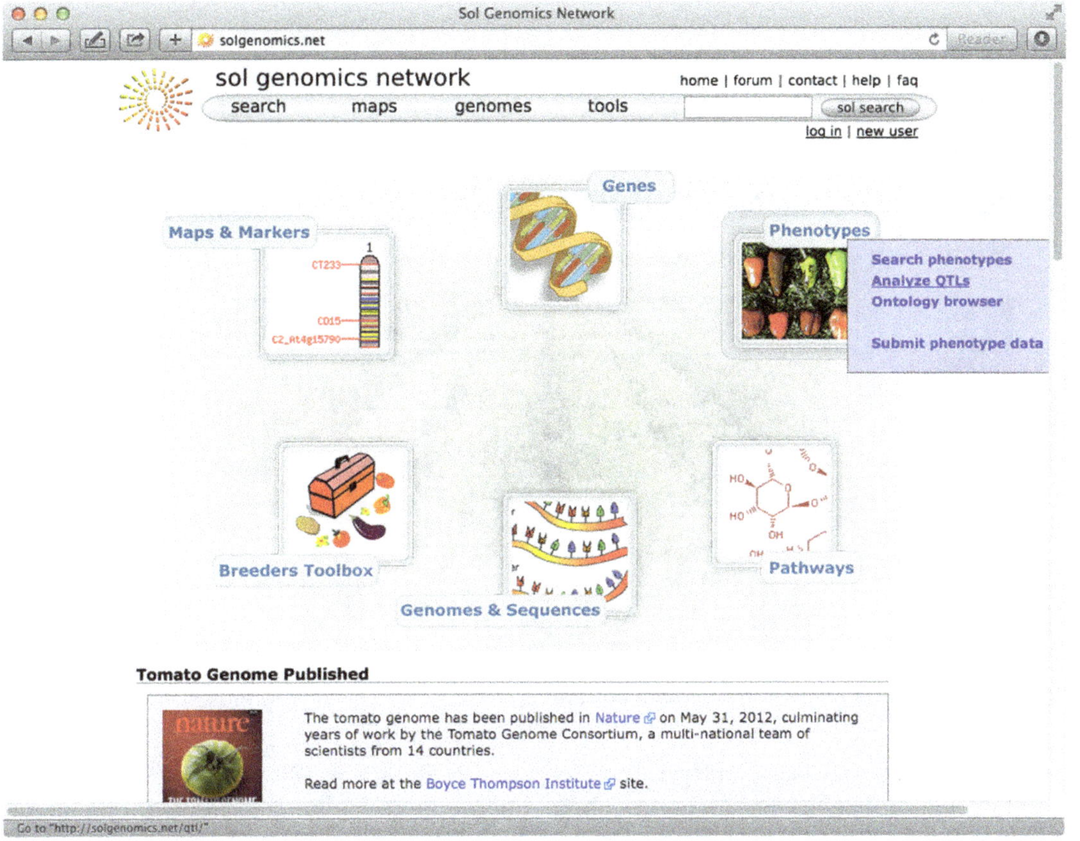

Fig. 13.1 The SGN database (http://solgenomics.net/) provides comprehensive information on Solanaceae genomes with an easy to use web-based interface

KafTom (http://www.pgb.kazusa.or.jp/kaftom/)

Created by the same team at Kazusa that also create MiBase, KafTom provides a database of full-length cDNA clones for tomato, based on the MicroTom variety (Aoki et al. 2010). Clone ordering is available for a small fee (Fig. 13.2).

Tomatomics (http://bioinf.mind.meiji.ac.jp/tomatomics/index.php)

This database integrates the data from miBASE and KafTom to create a new version of unigenes for the Micro-Tom variety. DNA markers, microarray data, gene expression networks and metabolic pathways are stored in the database together with the SNPs and InDels of Micro-Tom compared with Heinz. GBrowse is available to display Micro-Tom annotations (Fig. 13.3).

Genbank (http://www.ncbi.nlm.nih.gov/)

Genbank is the global repository for sequence and sequence-related information, and as such should contain all publicly available sequence data, including data for tomato. Genbank consists of several databases, including Nucleotide, EST, GSS (genome survey sequence), taxonomy, chemical compounds (PubChem), literature (PubMed), and the Short Read Archive (SRA) for next-generation sequencing (Benson et al. 2014). In the nucleotide section, which

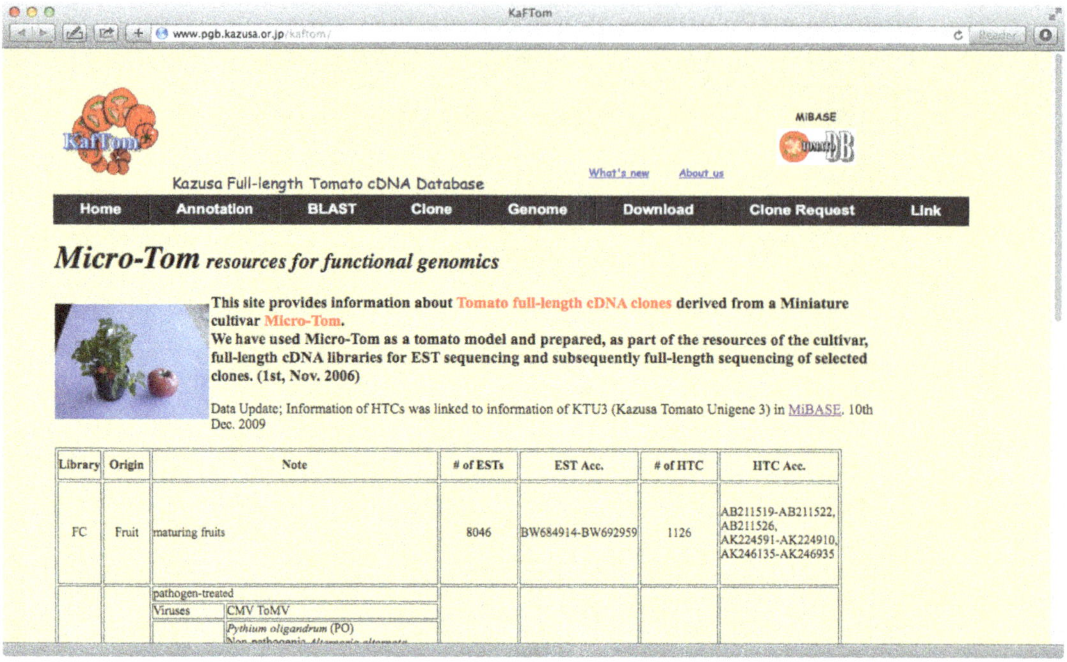

Fig. 13.2 The KafTom database at the Kazusa in Japan

comprises mRNA sequences and gene predictions from the genome reference sequence, there are currently 65,378 entries, while there are over 300,000 sequences in the EST database. The SRA database currently contains over 400 next-generation sequencing runs for tomato. It is important to note that the RefSeq database, which contains genome annotations, contains an annotation for tomato that has been done at Genbank and is different from the annotation presented in the tomato genome paper (Tomato Genome Consortium 2012) and available from SGN.

UniProt (http://www.uniprot.org)

UniProt is a generic protein repository for all kingdom species. It has two sections Swiss-Prot, including manual curated proteins, and TrEMBL, with automatic annotations (UniProt Consortium 2010). Protein entries can include many cross-references with other databases, publications, alternative sequences and annotations. The new UniProt web site allows the access to the tomato proteome by chromosome. Uniprot also include tools like BLAST and an aligner and many protein resources on its FTP site.

Phytozome (http://phytozome.jgi.doe.gov/pz/portal.html#!info?alias=org_slycopersicum)

Phytozome is a web portal for plant comparative genomics (Goodstein et al. 2012) which also contains tomato genomic data as annotated by the International Tomato Annotation Group (ITAG) and tools such as BLAST, BLAT and JBrowse. Phytozome also implements InterMine (Smith et al. 2012), a powerful tool to extract, filter and compare data from its database, useful to find genes or proteins by their annotations and annotations from a gene list. It is also possible to extract the sequence or part of the sequence from a list of genes (CDS, UTR, mRNA, flanking sequences, etc.) and retrieve the gene family components from a gene.

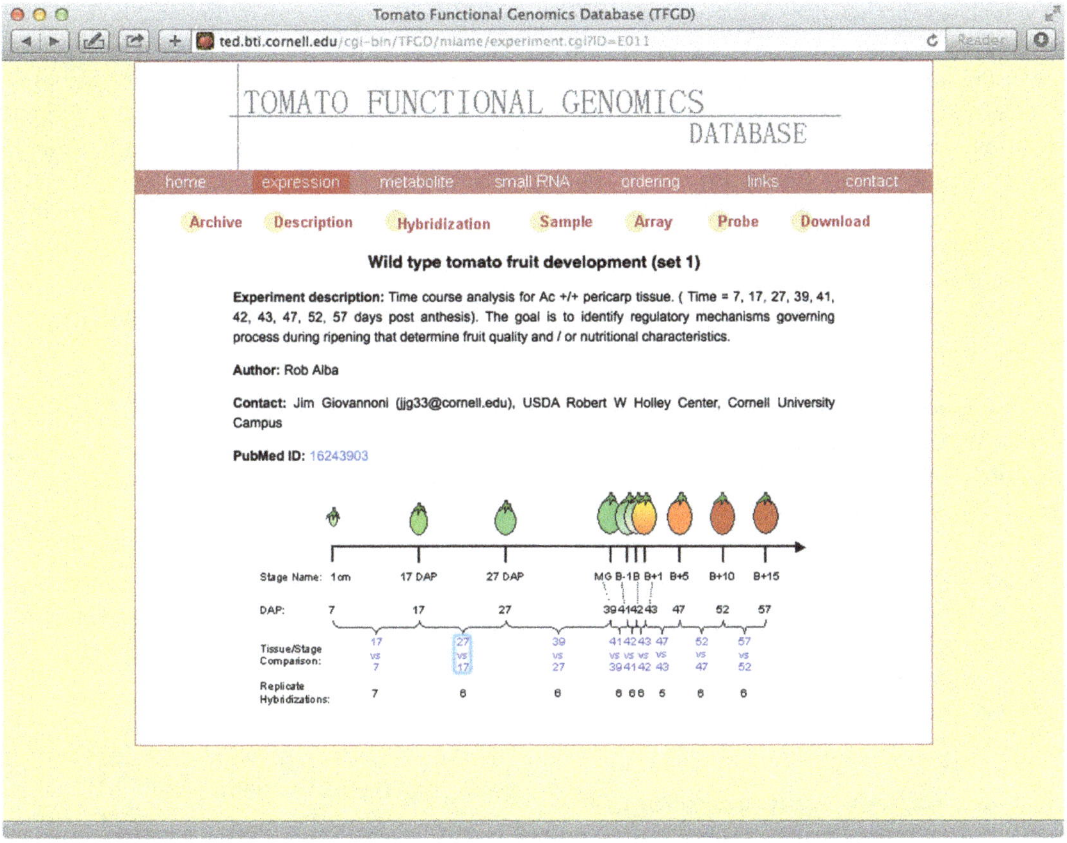

Fig. 13.3 Web interface at the Tomato Expression Database (TED; http://ted.bti.cornell.edu/, also accessible through http://solgenomics.net/ted/)

GreenPhyl V4 (http://www.greenphyl.org/cgi-bin/index.cgi)

Greenphyl is a database for gene families from green plants (Rouard et al. 2011). These gene families are automatic clustered and manually annotated. GreenPhyl tools allow the user to get useful data like phylogenetic trees or gene ontologies from gene families, check InterPro domains on genes or get homologs based on phylogeny and blast mutual hits.

Metabolic Databases

Knowledge of gene annotations can be powerful, but gene networks can give insights on the "system level" properties of a cell. Metabolic networks are a type of network that can be generated relatively easily, as metabolism is well studied, relatively well conserved between organisms, and a large number of tools are available. A number of databases with a metabolic interest have been created for tomato.

SolCyc (http://solcyc.solgenomics.net/)

The SolCyc databases are Pathway/Genome Databases (PGDBs) generated using the Pathway Tools software suite (Karp et al. 2002) for the Solanaceae. The species specific pathways are extracted from the MetaCyc reference database based on annotated genes using a Pathway Tools module called Pathologic. Databases for tomato, potato, tobacco, pepper, petunia, and

Nicotiana benthamiana have been created. The database for tomato (LycoCyc), potato (PotatoCyc), and *Nicotiana benthamiana* (BenthamianaCyc) and Pepper (CapCyc) were generated from their respective annotated genomes, while the other databases are based on datasets from annotated transcript assemblies. The SolCyc site can be searched for pathways, genes, enzymes, and compounds, and the results are displayed graphically. Pathways are displayed using zoom levels, with increasing zoom levels revealing more about the pathway. The Cellular Overview diagram shows the entire metabolism in that species' database, with the pathways represented as small glyphs. With the Omics Viewer, users can overlay data—for example, expression data and metabolomic data—on the Cellular Overview to identify pathways that have altered expression or metabolite levels. Pathway Tools have been used for many other plant species, as well as animals, fungi, and prokaryotes. The system is well supported and actively developed.

KEGG (http://www.genome.jp/kegg/)

The Kyoto Encyclopedia of Genes and Genomes (KEGG), similar in concept to the PGDBs of MetaCyc, is a resource that integrates metabolic pathways for many species, organizing the data in pathway maps including metabolites and enzymes (Aoki and Kanehisa 2005).

Tomato Expression Database (http://www.ted.bti.cornell.edu)

The Tomato Expression Database provides a comprehensive collection of expression data based on microarray data, mostly obtained with the TOM1 and the TOM2 array produced at the Boyce Thompson Institute, as well as a number of experiments based on the Affymetrix chip experiments and RNA-Seq. The web interface provides tools to query and view expression data by array id and can convert the widely used SGN unigene identifiers.

Tomato epigenome database is also part of the site (http://ted.bti.cornell.edu/epigenome/). It provides information on the methylation of tomato fruits in different stages as described in (Zhong et al. 2013).

Tomato EFP Browser (http://bar.utoronto.ca/efp_tomato/cgi-bin/efpweb.cgi)

The Tomato electronic Fluorescent Pictograph (eFP) Browser developed by the Provart group at the University of Toronto, displays gene expression values from RNA-Seq experiments in a graphical way over schematic pictures of the tomato plant, including data from several tomato tissues at different stages.

TomPLEX (http://www.plexdb.org/plex.php?database=tomato)

This database is part of PLEXdb (Plant Expression Database) and contains expression data from tomato microarrays data from several tissues, stages, and conditions, that are displayed in a graphical way (Winter et al. 2007).

Plant MetGenMap (http://bioinfo.bti.cornell.edu/cgi-bin/metgenmap/home.cgi)

Plant MetGenMAP is a web-based visualization and analysis software hosted at the Boyce Thompson Institute. It allows the identification of significant enrichment in genes or metabolites from an experiment. It uses LycoCyc pathways, making possible to visualize the profile data in a biochemical pathway context and can also identify enriched GO terms (Joung et al. 2009).

KOMICS (http://www.kazusa.or.jp/komics/en/)

The Kazusa Metabolomics Portal integrates databases and tools for metabolomics, including tools for annotation, data mining, and visualization (Sakurai et al. 2014).

Tomato MapMan

This popular tool is available in both a standalone tool and a web-based version that provide similar functionality. MapMan essentially allows to use or create custom diagrams for the overlay of expression or metabolomic data (Thimm et al. 2004). Tomato-related diagrams are available.

Phenome Networks (http://www.phenome-networks.com)

The Phenome Networks database pulls together phenotypic and genotypic data to create a tool that is appealing to the breeder. It is maintained by a company as a commercial product, while granting limited access to the interested user free of charge.

Genevestigator

Genevestigator is a database from a commercial provider that incorporates many species, including tomato, and specifically also clinical data. The site focuses on expression data and analysis, both based on microarray data and RNA-seq data (Hruz et al. 2008; Zimmermann et al. 2005). For academics, limited basic functionality is free, but more advanced features, such as analyzing many genes at the same time, requires a paid subscription.

Database of Transcription Factors

ITAK (http://bioinfo.bti.cornell.edu/cgi-bin/itak/db_browse.cgi)

Database hosted at the Boyce Thompson Institute for Plant Research. It includes transcription factors (TFs) and protein kinases (PKs) predicted by the iTAK program for tomato and other plant species. TFs are predicted following the rules described by (Perez-Rodriguez et al. 2010) and classified in TF families. PKs are identified and classify in gene families using Hidden Markov Models and the Pfam database (Finn et al. 2014), a database for protein domains.

PlantTFDB (http://planttfdb.cbi.edu.cn)

Database including transcription factors from tomato based on EST and unigenes. It contains 998 TFs, less than the half of the TFs predicted by iTAK based on the tomato genomic sequence (Perez-Rodriguez et al. 2010).

Value of the Databases

The databases represent a significant investment in terms of research dollars, but the wealth of data that these databases provide enable researchers to be far more efficient, and make model systems such as tomato far more compelling for future research. Today, databases are indispensable tools in advanced research. The research impact of databases is evident in terms of website use (page hits and number of users) as well as number of citations of papers describing databases. However, funding for databases is dwindling, and some databases have seen their funding disappear. While it is true that databases are expensive to maintain, in the absence of databases, fewer clearly versioned and well annotated datasets would be available, putting researchers in the awkward position to have to recreate datasets by themselves—which is time consuming, inefficient, and in the end even more costly.

As we have shown in this brief review, a rich variety of databases exist for tomato, As many of the databases use standardized structured vocabularies and versioned identifiers, researchers can easily exploit all the resources jumping from one resource to the next in the quest for new hypotheses.

References

Altschul SF, Gish W, Miller W, Myers EW, Lipman DJ (1990) Basic local alignment search tool. J Mol Biol 215(3):403–410

Aoki K, Yano K, Suzuki A, Kawamura S, Sakurai N, Suda K, Kurabayashi A, Suzuki T, Tsugane T, Watanabe M, Ooga K, Torii M, Narita T, Shin-I T, Kohara Y, Yamamoto N, Takahashi H, Watanabe Y, Egusa M, Kodama M, Ichinose Y, Kikuchi M, Fukushima S, Okabe A, Arie T, Sato Y, Yazawa K, Satoh S, Omura T, Ezura H, Shibata D (2010) Large-scale analysis of full-length cDNAs from the tomato (*Solanum lycopersicum*) cultivar micro-tom, a reference system for the solanaceae genomics. BMC Genom 11:210-2164-11-210

Aoki KF, Kanehisa M (2005) Using the KEGG database resource. Curr Protoc Bioinform, Chapter 1, Unit 1.12

Ashburner M, Ball CA, Blake JA, Botstein D, Butler H, Cherry JM, Davis AP, Dolinski K, Dwight SS, Eppig JT, Harris MA, Hill DP, Issel-Tarver L, Kasarskis A, Lewis S, Matese JC, Richardson JE, Ringwald M, Rubin GM, Sherlock G (2000) Gene ontology: tool for the unification of biology. The gene ontology consortium. Nat Genet 25(1):25–29

Benson DA, Clark K, Karsch-Mizrachi I, Lipman DJ, Ostell J, Sayers EW (2014) Genbank. Nucleic Acids Res 42(1):D32–D37

Bombarely A, Menda N, Tecle IY, Buels RM, Strickler S, Fischer-York T, Pujar A, Leto J, Gosselin J, Mueller LA (2011) The sol genomics network (solgenomics.net): growing tomatoes using perl. Nucleic Acids Res 39(Database issue):D1149–D1155

Brewer MT, Lang L, Fujimura K, Dujmovic N, Gray S, van der Knaap E (2006) Development of a controlled vocabulary and software application to analyze fruit shape variation in tomato and other plant species. Plant Physiol 141(1):15–25

Causse M, Desplat N, Pascual L, Le Paslier MC, Sauvage C, Bauchet G, Berard A, Bounon R, Tchoumakov M, Brunel D, Bouchet JP (2013) Whole genome resequencing in tomato reveals variation associated with introgression and breeding events. BMC Genom 14:791-2164-14-791

Chiusano ML, D'Agostino N, Traini A, Licciardello C, Raimondo E, Aversano M, Frusciante L, Monti L (2008) ISOL@: an italian SOLAnaceae genomics resource. BMC Bioinform 9:S7

D'Agostino N, Traini A, Frusciante L, Chiusano ML (2009) SolEST database: a "one-stop shop" approach to the study of solanaceae transcriptomes. BMC Plant Biol 9:142-2229-9-142

Duvick J, Fu A, Muppirala U, Sabharwal M, Wilkerson MD, Lawrence CJ, Lushbough C, Brendel V (2008) PlantGDB: A resource for comparative plant genomics. Nucleic Acids Res 36(Database issue): D959–D965

Edwards KD, Bombarely A, Story GW, Allen F, Mueller LA, Coates SA, Jones L (2010) TobEA: an atlas of tobacco gene expression from seed to senescence. BMC Genom 11:142

Finn RD, Bateman A, Clements J, Coggill P, Eberhardt RY, Eddy SR, Heger A, Hetherington K, Holm L, Mistry J, Sonnhammer EL, Tate J, Punta M (2014) Pfam: the protein families database. Nucleic Acids Res 42(Database issue):D222–D230

Goodstein DM, Shu S, Howson R, Neupane R, Hayes RD, Fazo J, Mitros T, Dirks W, Hellsten U, Putnam N, Rokhsar DS (2012) Phytozome: A comparative platform for green plant genomics. Nucleic Acids Res 40(Database issue):D1178–D1186

Hruz T, Laule O, Szabo G, Wessendorp F, Bleuler S, Oertle L, Widmayer P, Gruissem W, Zimmermann P (2008) Genevestigator v3: a reference expression database for the meta-analysis of transcriptomes. Adv Bioinform 2008:420747

Ilic K, Kellogg EA, Jaiswal P, Zapata F, Stevens PF, Vincent LP, Avraham S, Reiser L, Pujar A, Sachs MM, Whitman NT, McCouch SR, Schaeffer ML, Ware DH, Stein LD, Rhee SY (2006) Plant structure ontology. Unified vocabulary of anatomy and morphology of a flowering plant. Plant Physiol

Joung JG, Corbett AM, Fellman SM, Tieman DM, Klee HJ, Giovannoni JJ, Fei Z (2009) Plant MetGenMAP: An integrative analysis system for plant systems biology. Plant Physiol 151(4):1758–1768

Karp PD, Paley S, Romero P (2002) The pathway tools software. Bioinformatics 18(Suppl 1):S225–S232

Kobayashi M, Nagasaki H, Garcia V, Just D, Bres C, Mauxion JP, Le Paslier MC, Brunel D, Suda K, Minakuchi Y, Toyoda A, Fujiyama A, Toyoshima H, Suzuki T, Igarashi K, Rothan C, Kaminuma E, Nakamura Y, Yano K, Aoki K (2014) Genome-wide analysis of intraspecific DNA polymorphism in 'micro-tom', a model cultivar of tomato (solanum lycopersicum). Plant Cell Physiol

Menda N, Buels RM, Tecle I, Mueller LA (2008) A community-based annotation framework for linking solanaceae genomes with phenomes. Plant Physiol 147(4):1788–1799

Menda N, Semel Y, Peled D, Eshed Y, Zamir D (2004) In silico screening of a saturated mutation library of tomato. Plant J 38(5):861–872

Minoia S, Petrozza A, D'Onofrio O, Piron F, Mosca G, Sozio G, Cellini F, Bendahmane A, Carriero F (2010) A new mutant genetic resource for tomato crop improvement by TILLING technology. BMC Res. Notes 3:69-0500-3-69

Mueller LA, Mills AA, Skwarecki B, Buels RM, Menda N, Tanksley SD (2008) The SGN comparative map viewer. Bioinformatics 24(3):422–423

Mueller LA, Solow TH, Taylor N, Skwarecki B, Buels R, Binns J, Lin C, Wright MH, Ahrens R, Wang Y, Herbst EV, Keyder ER, Menda N, Zamir D, Tanksley SD (2005) The SOL genomics network: a comparative resource for solanaceae biology and beyond. Plant Physiol 138(3):1310–1317

Okabe Y, Asamizu E, Saito T, Matsukura C, Ariizumi T, Bres C, Rothan C, Mizoguchi T, Ezura H (2011) Tomato TILLING technology: Development of a reverse genetics tool for the efficient isolation of mutants from micro-tom mutant libraries. Plant Cell Physiol 52(11):1994–2005

Perez-Rodriguez P, Riano-Pachon DM, Correa LG, Rensing SA, Kersten B, Mueller-Roeber B (2010) PlnTFDB: Updated content and new features of the plant transcription factor database. Nucleic Acids Res 38(Database issue):D822–D827

Quackenbush J, Liang F, Holt I, Pertea G, Upton J (2000) The TIGR gene indices: Reconstruction and representation of expressed gene sequences. Nucleic Acids Res 28(1):141–145

Rodriguez GR, Munos S, Anderson C, Sim SC, Michel A, Causse M, Gardener BB, Francis D, van der Knaap E (2011) Distribution of SUN, OVATE, LC, and FAS in the tomato germplasm and the relationship to fruit shape diversity. Plant Physiol 156(1):275–285

Rouard M, Guignon V, Aluome C, Laporte MA, Droc G, Walde C, Zmasek CM, Perin C, Conte MG (2011) GreenPhylDB v2.0: comparative and functional genomics in plants. Nucleic Acids Res 39(Database issue): D1095–D1102

Saito T, Ariizumi T, Okabe Y, Asamizu E, Hiwasa-Tanase K, Fukuda N, Mizoguchi T, Yamazaki Y, Aoki K, Ezura H (2011) TOMATOMA: a novel tomato mutant database distributing micro-tom mutant collections. Plant Cell Physiol 52(2):283–296

Sakurai N, Ara T, Enomoto M, Motegi T, Morishita Y, Kurabayashi A, Iijima Y, Ogata Y, Nakajima D, Suzuki H, Shibata D (2014) Tools and databases of the KOMICS web portal for preprocessing, mining, and dissemination of metabolomics data. Biomed Res Int 2014:194812

Singh DK, Calvino M, Brauer EK, Fernandez-Pozo N, Strickler S, Yalamanchili R, Suzuki H, Aoki K, Shibata D, Stratmann JW, Popescu GV, Mueller LA, Popescu SC (2014) The tomato kinome and the TOKN ORFeome: resources for the study of kinases and signal transduction in tomato and solanaceae. Mol Plant Microbe Inter MPMI 27:7–17

Smith RN, Aleksic J, Butano D, Carr A, Contrino S, Hu F, Lyne M, Lyne R, Kalderimis A, Rutherford K, Stepan R, Sullivan J, Wakeling M, Watkins X, Micklem G (2012) InterMine: a flexible data warehouse system for the integration and analysis of heterogeneous biological data. Bioinformatics 28 (23):3163–3165

Stein LD (2010) The case for cloud computing in genome informatics. Genome Biol 11(5):207-2010-11-5-207. Epub 2010 May 5

Tecle IY, Menda N, Buels RM, van der Knaap E, Mueller LA (2010) solQTL: a tool for QTL analysis, visualization and linking to genomes at SGN database. BMC Bioinform 11:525

Thimm O, Blasing O, Gibon Y, Nagel A, Meyer S, Kruger P, Selbig J, Muller LA, Rhee SY, Stitt M (2004) MAPMAN: a user-driven tool to display genomics data sets onto diagrams of metabolic pathways and other biological processes. Plant J 37 (6):914–939

Tomato Genome Consortium (2012) The tomato genome sequence provides insights into fleshy fruit evolution. Nature 485(7400):635–641

UniProt Consortium (2010) The universal protein resource (UniProt) in 2010. Nucleic Acids Res 38 (Database issue):D142–D148

Winter D, Vinegar B, Nahal H, Ammar R, Wilson GV, Provart NJ (2007) An "electronic fluorescent pictograph" browser for exploring and analyzing large-scale biological data sets. PLoS ONE 2(8):e718

Zhong S, Fei Z, Chen YR, Zheng Y, Huang M, Vrebalov J, McQuinn R, Gapper N, Liu B, Xiang J, Shao Y, Giovannoni JJ (2013) Single-base resolution methylomes of tomato fruit development reveal epigenome modifications associated with ripening. Nat Biotechnol 31(2):154–159

Zimmermann P, Hennig L, Gruissem W (2005) Gene-expression analysis and network discovery using genevestigator. Trends Plant Sci 10(9):407–409

Prospects: The Tomato Genome as a Cornerstone for Gene Discovery

14

James J. Giovannoni

Abstract

The tomato genome was sequenced at a time when next generation sequencing technologies were replacing prior state-of-the-art methodologies for genome sequencing and assembly. The result was a strategy merging both old and new approaches. Because biologists guided this effort with an eye on maximal utility of the resulting product, one of the most complete and accurate plant genome sequences was developed for tomato as a model for plant biological inquiry. This volume details both the process and the outcome of the tomato genome sequencing effort as a cornerstone for discovery that will continue to be improved and that will serve researchers for years to come.

Keywords

Tomato · Genome sequence · Breeding

Very few scientific advancements result from insights and conclusions developed de novo. Like the architecture of ancient yet living cities, scientific discovery builds on the foundations left by those who explored before us. At the conclusion of this volume, it is as important to take a look back so as to put a forward view in proper perspective. As beautifully described in Chap. 2, tomato and its relatives capture a vast array of complexity and genetic diversity. Professor Charles Rick (1915–2002), while a professor at the University of California at Davis was instrumental in capturing this diversity and bringing it to the attention of researchers, geneticists and plant breeders over his 60 year career. Equally as important as his collections, descriptions and genetic characterization, was his openness and generosity. Prof. Rick knew the intrinsic value of sharing toward the synergistic

J.J. Giovannoni (✉)
United States Department of Agriculture – Agricultural Research Service, Robert W. Holley Center, Cornell University Campus, Tower Road, Ithaca, NY 14853, USA
e-mail: jjg33@cornell.edu

J.J. Giovannoni
Boyce Thompson Institute for Plant Research, Cornell University Campus, Tower Road, Ithaca, NY 14853, USA

M. Causse et al. (eds.), *The Tomato Genome*, Compendium of Plant Genomes,
DOI 10.1007/978-3-662-53389-5_14

advancement of science as exemplified by his development and support of what is now known as the C. M. Rick Tomato Genetic Resource Center. The Rick Center currently operates in this same generous spirit under the direction of Prof. Rick's successor, Prof. Roger Chetelat, and is a central and reliable source of tomato wild species germplasm, in addition to various true-breeding populations and monogenic mutant stocks for researchers the world-over. The same spirit of openness, generosity, and collaboration has been at the core of two influential tomato researchers who are also Prof. Rick's Ph.D. students, Profs. Steven Tanksley (Cornell University), and Dani Zamir (Hebrew University of Jerusalem) who have been leaders in the areas of plant molecular genetics, genomics, and breeding with much of their work focused on tomato. Both were instrumental in the early planning, organization, and implementation of the tomato genome sequencing effort and contributed through their careers toward both the development of maps and markers necessary for anchoring and orienting the genome sequence in addition to development of stable populations and gene mapping/isolation tools and methodologies comprising the fulcrum upon which much future exploitation of the genome sequence will be leveraged. Indeed these individuals and many others, including the host of authors of this volume and their colleagues, have contributed to not only the development of an important genome sequence relevant to a major economic and nutritional crop (Chap. 1) but also to the tomato experimental system as a model for plant biology that is worthy of prior and future investment and scientific endeavors, including those that helped justify and launch this effort (Chaps. 3, 4). Without question, prior seminal work on pathogen response, fruit development and ripening, leaf morphology, root physiology, hormone biology, and light perception, combined with the practical and genetic tractability of tomato presented its genome as an obvious target for sequencing. Finally, it is critical to note the important fact that the sequencing project was largely driven by biologists who ultimately endeavored to utilize the genome in future research, as opposed to simply scratching another genome off the "to do" list, and thus insured that high quality (Chaps. 5–10), open access (the first drafts of the genome were made public 3 years prior to publication) and ease of use were all requisite features of the genome and its enabling interface, SGN or the SOL Genomics Network (https://solgenomics.net/; Chap. 13). The genome sequence of tomato initially revealed biological insights into the evolution of genes with functions pertaining to fruit biology. Furthermore, comparison to the grape genome in particular, provided evidence of a genome triplication event specific to the tomato lineage (Chap. 11). Genome sequences of additional Solanaceae species have been developed both just before (potato) and shortly after (pepper, eggplant, tobaccos, and petunia) the publication of the tomato genome sequence in 2012, revealing the similarities among these genomes (Chap. 12) anticipated from prior comparative mapping studies.

The exponentially expanding impact of the tomato genome sequence

The tomato genome sequence, following its initial release in 2009, had rapid and profound effect on the ability of researchers to map single and quantitative traits, in large part due to the reality that DNA sequence necessary for the development of molecular markers was no longer a limiting factor in gene localization and discovery efforts. It is noteworthy that the number of tomato-related publications listed on the NCBI database (http://www.ncbi.nlm.nih.gov/pubmed/?term=tomato) has increased approximately 40 % on an annual basis since 2009 to present and those emphasizing genetics have doubled. Following on the heels of the cultivated tomato (*S. lycopersicum*) sequence and its wild ancestor (*S. pimpinellifolium*) (TGC 2012) was the de novo genome sequence of *S. pennellii*, the wild relative parent of one of the most widely exploited introgression populations deployed by numerous scientists and breeders for a range of objectives including cultivated tomato improvement, QTL mapping, and gene discovery (Alseekh et al. 2013). The reference genome has since provided the foundation for resequencing efforts that have been recently published (100 TGSC et al. 2014; Lin et al. 2014) in addition to ongoing efforts resulting in hundreds of additional genomes. These sequences reveal genetic polymorphisms that have facilitated a shift toward advanced molecular breeding by many tomato seed companies culminating in the rapid development of new varieties that are, and will continue to provide, a growing range of choices filling a broader spectrum of consumer preferences. These sequences have also shed light on the traits and associated loci instrumental in tomato domestication, and reveal the genetic architecture providing the core foundation of the cultivated tomato genome (Lin et al. 2014). It is certain that in the future additional *S. lycopersicum* genomes will be sequenced providing both a broader breeding resource and more powerful genomic infrastructure for genome association and genome selection activities (Pascual et al. 2016). The ongoing de novo sequencing of additional wild species, including many of the parents of widely used introgression, recombinant inbred and back-cross inbred lines, will further enable these germplasm resources and facilitate assessment of more recent tomato evolution. Such studies will further our understanding of important agricultural and biological traits including exploitable features of stress tolerance, climate adaptation, and plant development.

The tomato genome and evolving metagenome resulting from ongoing sequencing, refinement and resequencing activities also provides an important reference for ongoing and future sequence-enabled analyses including transcriptome profiling and characterizations of genome interactions with regulatory and structural components and their dynamics. Numerous studies have yielded tomato transcriptome data on hundreds of tissue/development/treatment/genotype combinations (see the TomExpress database; http://gbf.toulouse.inra.fr/tomexpress/www/welcomeTomExpress.php), profiles of small RNAs and epigenome dynamics including those influencing fruit development (Zhong et al. 2013). While the genome will continue to improve, at its core, the work described in this volume represents a cornerstone upon which future investigations in tomato and broader plant genome evolution and biological inquiry has and will continue to be built.

References

100 Tomato Genome Sequencing Consortium, Aflitos S, Schijlen E, de Jong H, de Ridder D, Smit S, Finkers R, Wang J, Zhang G, Li N, Mao L, Bakker F, Dirks R, Breit T, Gravendeel B, Huits H, Struss D, Swanson-Wagner R, van Leeuwen H, van Ham RC, Fito L, Guignier L, Sevilla M, Ellul P, Ganko E, Kapur A, Reclus E, de Geus B, van de Geest H, Te Lintel Hekkert B, van Haarst J, Smits L, Koops A, Sanchez-Perez G, van Heusden AW, Visser R, Quan Z, Min J, Liao L, Wang X, Wang G, Yue Z, Yang X, Xu N, Schranz E, Smets E, Vos R, Rauwerda J, Ursem R, Schuit C, Kerns M, van den Berg J, Vriezen W, Janssen A, Datema E, Jahrman T, Moquet F, Bonnet J, Peters S (2014) Exploring genetic variation in the tomato (Solanum section Lycopersicon) clade by whole-genome sequencing. Plant J 80:136–148

Alseekh S, Ofner I, Pleban T, Tripodi P, Di Dato F, Cammareri M, Mohammad A, Grandillo S, Fernie AR, Zamir D (2013) Resolution by recombination: breaking up Solanum pennellii introgressions and references therein. Trends Plant Sci 18:536–538

Lin T, Zhu GT, Zhang JH, Xu XY, Yu QH, Zheng Z, Zhang ZH, Lun YY, Li S, Wang XX, Huang ZJ, Li JM, Zhang CZ, Wang TT, Zhang YY, Wang AX, Zhang YC, Lin K, Li CY, Xiong GS, Xue YB, Mazzucato A, Causse M, Fei ZJ, Giovannoni JJ, Chetelat RT, Zamir D, Stadler T, Li JF, Ye ZB, Du YC, Huang SW (2014) Genomic analyses provide insights into the history of tomato breeding. Nat Genet 46:1220–1226

Pascual L, Albert E, Sauvage C, Duangjit J, Bouchet JP, Bitton F, Desplat N, Brunel D, Le Paslier MC, Ranc N, Bruguier L, Chauchard B, Verschave P, Causse M (2016) Dissecting quantitative trait variation in the resequencing era: complementarity of biparental, multi-parental and association panels. Plant Sci 242:120–130

Tomato Genomics Consortium (TGC) (2012) The tomato genome sequence provides insights into fleshy fruit evolution. Nature 485:635–641

Zhong S, Fei Z, Chen Y, Zheng Y, Huang M, Vrebalov J, McQuinn R, Gapper N, Lui B, Xiang J, Shao Y, Giovannoni J (2013) Single-base resolution methylomes of tomato fruit development reveal epigenome modifications associated with ripening. Nat Biotechnol 31:154–159

www.ingramcontent.com/pod-product-compliance
Lightning Source LLC
Chambersburg PA
CBHW080734150926
51501CB00005B/165
* 9 7 8 3 6 6 2 5 3 3 8 7 1 *